CREATION
REDISCOVERED

"And God said: Let the earth bring forth the living creature in its kind, cattle and creeping things, and beasts of the earth, according to their kinds. And it was so done."
—Genesis 1:24

CREATION
REDISCOVERED

EVOLUTION AND THE IMPORTANCE OF THE ORIGINS DEBATE

By

Gerard J. Keane

SECOND AND EXPANDED EDITION

"*In the beginning, God created heaven, and earth.*" —Genesis 1:1

TAN BOOKS AND PUBLISHERS, INC.
Rockford, Illinois 61105

Copyright © 1999 by Gerard J. Keane.

All rights reserved. No part of this book may be reproduced or transmitted in any form or by any means, electronic or mechanical, including photocopying, recording, or by any information storage or retrieval system, without permission in writing from the Publisher, except that brief selections may be quoted or copied without permission, provided full credit is given.

Library of Congress Catalog Card No.: 98-60324

ISBN 0-89555-607-3

The author may be contacted at:
P. O. Box 451, Doncaster, Victoria 3108, Australia.

Printed and bound in the United States of America.

TAN BOOKS AND PUBLISHERS, INC.
P.O. Box 424
Rockford, Illinois 61105
1999

ACKNOWLEDGMENTS

To our great Creator and Saviour, without whom nothing is possible.

To my dear wife and children, who had to endure all the inconvenience, but nevertheless helped me to get through the work.

To all those individuals who kindly provided so many incisive comments during the development of the manuscript. I am grateful for all the help given to me; the book could not have been brought together without the insights and wisdom provided by others.

Scripture quotations are taken from the *Douay-Rheims Bible,* published by TAN Books and Publishers, Inc., Rockford, Illinois.

Paragraph numbers of the English edition of the *Catechism of the Catholic Church* (1992) are given in parentheses as follows: (400).

CONTENTS

Foreword to the Revised Edition (by Professor Maciej Giertych) ix
Preface (by Rev. Fr. Peter Damian Fehlner) xiii
Introduction xxi

PART I — THE BASIC QUESTION
1. The Basic Question 3
2. Evolution Theories 12
3. Evolutionism 39
4. The Concept of Special Creation 45

PART II — THE DISCOVERIES OF SCIENCE
5. The Discoveries of Science 97
6. The Fossil Record 99
7. Genetics 118
8. Entropy 130
9. How Old Is the Universe? 136
10. Pointers to the Creator 166

PART III — CHRISTIAN INSIGHTS
11. Christian Insights 173
12. Problems in Theistic Evolution 175
13. The Position within Catholicism 185
14. The Inerrancy of Scripture 207
15. The Question of Age 241
16. Problems in Progressive Creation 270

PART IV — THE INFLUENCE OF EVOLUTION ON BELIEF SYSTEMS
17. The Influence of Evolution on Belief Systems 281
18. Nazism and Communism 283
19. Humanism 288
20. Christianity 296

PART V — THE SEARCH FOR MEANING IN LIFE
21. The Search for Meaning in Life 321
22. Religion and Meaning 323
23. The Problem of Evil 333

24. Existentialism 341
25. Creation Rediscovered 354

Appendix A: Pontifical Biblical Commission on *Genesis* 363
Appendix B: Modernism—A Snapshot 367
Bibliography 371
Index 381

FOREWORD
By Professor Maciej Giertych

Sometime in 1955, when I was taking Honor Moderations in Science (Botany, Chemistry and Geology) at Oxford University, the O. U. Biology Club announced a lecture against the theory of Evolution. The largest auditorium in the Biology Labs was filled to capacity. When the speaker was introduced (I regret I do not remember his name), it turned out he was an octogenarian with a Ph.D. in biology from Cambridge, obtained in the 19th century.

He spoke fervently against the theory of Evolution, defending what was for us an obviously indefensible position. He did not convince anybody with his antique arguments; he did not understand the questions that were fired at him; he rejected science as we knew it. We all had a good laugh hearing this dinosaur. He fought for his convictions against a sophisticated scientific environment, deaf to any opinions inspired by religious beliefs. Today his views are being vindicated by new evidence from natural sciences. May his soul rest in peace.

In 1955, like all in my generation, I was fully convinced that Evolution was an established biological fact. The evidence was primarily paleontological. We were taught how to identify geological strata with the help of fossils, specific for a given epoch. The rocks were dated by the fossils, the fossils by the strata. A lecturer in stratigraphy, when asked during a field trip how the strata were dated, explained that we know the rate of current sedimentation, the depths of strata and thus the age of rocks. In any case, there are new isotopic techniques that confirm all this. This sounded very scientific and convincing.

In my studies I went on to a B.A. and M.A. in forestry, a Ph.D. in plant physiology and finally a D.Sc. in genetics. For a long time I was not bothered by geology, Evolution or any suspicious thoughts. I had my own field of research in population genetics of forest trees, with no immediate relevance to the controversy over Evolution.

Gradually, as my children got to the stage of learning biology in school and discussing their problems with Dad, I realized that the evidence for Evolution had shifted from paleontology and embryology to population genetics. But population genetics is my subject! I knew it was used to explain how Evolution progressed, but I was not aware it is used to prove it. Without my noticing it, my special field had become the supplier of the most pertinent evidence supporting the theory.

If Evolution were proved in some field I was not familiar with, I understood the need to accommodate my field to this fact, to suggest explanations how it occurred in terms of genetics. But to claim that these attempted explanations are the primary evidence for the theory was quite unacceptable to me. I started reading the current literature on the topic of Evolution. Until then I was not aware how shaky the evidence for Evolution was, how much of what was "evidence" had to be discarded, how little new evidence had been accumulated over the years, and how very much ideas dominate facts. These ideas have become dogma, yet they have no footing in natural sciences. They stem from materialistic philosophies.

My primary objection as a geneticist was to the claim that the formation of races, or microevolution, as it is often referred to, is a small scale example of macroevolution—the origin of species. Race formation is, of course, very well documented. All it requires is isolation of a part of a population. After a few generations, due to natural selection and genetic drift, the isolated population will irreversibly lose some genes, and thus, as long as the isolation continues, in some features it will be different from the population it arose from. In fact, we do this ourselves all the time when breeding, substituting natural with artificial selection and creating artificial barriers to generative mixing outside the domesticated conditions.

The important thing to remember here is that a race is genetically impoverished relative to the whole population. It has fewer alleles (forms of genes). Some of them are arranged into special, interesting, rare combinations. This is particularly achieved by guided recombination of selected forms in breeding work. But these selected forms are less variable (less polymorphic). Thus what is referred to as micro-evolution represents natural or artificial reduction of the gene pool. You will not get Evolution that way. Evolution means construction of new genes. It means increase in the amount of genetic information, and not reduction of it.

The evolutionary value of new races or selected forms should be demonstrable by natural selection. However, if allowed to mix with the general breeding population, new races will disappear. The genes in select combinations will disperse again; the domesticated forms will go wild. Thus there is no evidence for Evolution here.

FOREWORD

Mutations figure prominently in the Evolution story. When in the early '60s I was starting breeding work on forest trees, everyone was very excited about the potential of artificial mutations. In many places around the world, special "cobalt bomb" centers were established to stimulate rates of mutations. What wonderful things were expected from increased variability by induced mutations. All of this work has long since been abandoned. It led nowhere. All that was obtained were deformed freaks, absolutely useless in forestry.

Maybe occasionally some oddity could be of ornamental value, but never able to live on its own in natural conditions. A glance through literature on mutations outside forestry quickly convinced me that the pattern is similar everywhere. Mutations are either neutral or detrimental. Positive ones, if they do occur, are too rare to be noticeable. Stability in nature is the rule. We have no proofs for Evolution from mutation research.

It is sometimes claimed that strains of diseases resistant to antibiotics, or weeds resistant to herbicides, are evidence for positive mutations. This is not so. Most of the time, the acquired resistance is due to genetic recombination and not due to mutations. Where mutations have been shown to be involved, their role depends on deforming part of the genetic code, which results in a deformed, usually less effective protein that is no longer suitable for attachment by the harmful chemical.

Herbicides are "custom made" for attachability to a vital protein specific for the weed species, and they kill the plant by depriving the protein of its function when attached to it. A mutation that cancels attachability to the herbicide and does not totally deprive the protein of its function is in this case beneficial, since it protects the functionality of the protein. However this is at a price, since in fact the change is somewhat detrimental to normal life processes. At best it is neutral. There are many ways in which living systems protect functionality. This is one of them. Others include healing or eliminating deformed parts or organisms. Natural selection belongs here. So does the immunological adaptation to an invader. Of course such protective adaptations do not create new species, new kinds, new organs or biological systems. They protect what already exists, usually at a cost. Defects accumulate along the way.

Within the genome of a species, that is, in the molecular structure of its DNA, we find many recurrent specific nucleotide sequences, known as "repeats." Different ones occur in different species. If this variation (neutral as far as we know) arose from random mutations, it should be random. How then did the "repeats" come to be? If mutations are the answer, they could not have been random. In this context "genetic drive" is postulated, as distinct from "genetic drift." But Who or what does the driving? The empirical science of genetics knows only random mutations.

Currently there are new suggestions that molecular genetics provides evidence for Evolution. Analyses of DNA sequences in various species should show similarities between related ones and big differences between systematically far-removed species. They do exactly that. Molecular genetics generally confirms the accuracy of taxonomy. But at the same time, it does not confirm postulated evolutionary sequences. There are no progressive changes, say from fishes to amphibians, to reptiles to mammals. Molecular genetics confirms systematics, not phylogeny; Linnaeus, not Darwin.

No. Genetics has no proofs for Evolution. It has trouble explaining it. The closer one looks at the evidence for Evolution, the less one finds of substance. In fact, the theory keeps on postulating evidence and failing to find it, and moves on to other postulates (fossil missing links, natural selection of improved forms, positive mutations, molecular phylogenetic sequences, etc.). This is not science.

A whole age of scientific endeavor was wasted searching for a phantom. It is time we stopped and looked at the facts! Natural sciences failed to supply any evidence for Evolution. Christian philosophy tried to accommodate this unproved postulate of materialist philosophies. Much time and intellectual effort went in vain, leading only to negative moral consequences. It is time those working in the humanities were told the truth.

Gerard J. Keane is doing exactly that. In clear and simple language, he reviews the present status of the Evolution-Creation controversy. I am very happy to be able to recommend this book. Indeed, *Creation Rediscovered* by science comes to the rescue of Christianity.

> Professor Maciej Giertych,
> B.A., M.A. Oxon, Ph.D. Toronto, D.Sc. Poznan
> Polish Academy of Sciences,
> Institute of Dendrology,
> 62-035 Kornik, Poland

PREFACE
By Rev. Fr. Peter Damian Fehlner, S.T.D.

Why a theological introduction to a book about Evolution and Creation? Most people would instinctively reply: not because Evolution is a theological question, but because it is assumed to be a scientific question posing a threat to traditional belief in Creation, in a particular way to the doctrine of the unique dignity of Adam and Eve and their descendants, based on creation of the soul and special divine formation of the bodies of Adam and Eve, and therefore to belief in the existence of God and the very possibility of the Incarnation and Salvation as the ultimate goal to which Creation is ordered. Thus there arises a problem of apologetics: are evolutionary hypotheses about the origin of the world, of the differentiation of the species and of man in particular a threat to the traditional dogmatic theism of Catholic theology?

In modern times two ways of approaching this problem have become usual: one is to deny any valid basis for evolutionary theories of origin. The other is to admit as plausible some theories of evolution, those precisely which are not incompatible with Theism. Whence the term *Theistic Evolution*.

In recent years this second approach has gained great popularity among Catholics, in particular among Catholic clergymen and religious. One can subscribe to all the articles of the Catholic creed, so the claim for Theistic Evolution runs, and not be pre-occupied with the final outcome of the scientific debate over the evolutionary hypothesis. For if one day "Evolution" should be proved factual, the only evolutionary thesis so to be demonstrated scientifically will be theistic rather than atheistic in thrust. One even hears the (very strange) assertion that God created the world by means of Evolution! Hence, *Atheistic* Evolution stands condemned by the Church.

But *Theistic* Evolution is not condemned, so it is further claimed, because the Church makes no judgment on the intrinsic merits of scientific hypotheses not contrary to faith and morals. And further, say its supporters, Evolution understood theistically uniquely underscores the prerequisite purpose and intelligence in the world which demonstrates the existence of God.

Now Mr. Gerard Keane's study: *Creation Rediscovered*, thoroughly revised and expanded, shows that no evolutionary hypothesis has been conclusively demonstrated as factual. Far from it: scientific theorizing about Origins tends more to favor the creationist version than the evolutionist one.

But there is one other, often overlooked point about such "scientific" theorizing about the origins of the world and of the species, very telling for the future direction the discussion of Origins will take. The point is this: These scientific theories of origins cannot be verified or falsified definitively on scientific grounds.

What is the significance of this point? An hypothesis incapable of scientific demonstration, of being verified as true or false, is not, strictly speaking, a scientific hypothesis. It may be true, but the truth or falsity of the theory must be decided on grounds and with methods of reflection proper to other branches of learning: those dealing with the theological, above all dogmatic theology, if the hypothesis is primarily theological. For the question of Origins—of the world and of man—is not a question of science, but of theology (including sound metaphysics).

Sound science recognizes its limits, even in regard to the sensible. Empirical science does not, because it cannot, tell us all that might be known about the material world. Wherever there is a question of the supernatural, of the miraculous, there it is beyond the limits of empirical science to tell us about material reality and what are the principles of its operation. For example: Creation as a distinctively divine mode of producing; the virginal Motherhood of Mary as a true, but higher mode of begetting; Transubstantiation of bread and wine into the Body and Blood of Christ; the glorified state of the risen human body.

In a word, empirical science has as its object the study of the natural operations of creatures, not the creative or miraculous operations of God, which these processes either presuppose for their existence and operation or which transcend these operations.

Dispassionately viewed, the current debate shows that neither the origin of the world in general nor of man in particular is primarily a question of empirical science. It is being decided, one way or the other, on theological-historical premises. Hence the prior truth of such premises is crucial to the entire debate. This is because both origins primarily involve creative and/or miraculous actions possible only to the Creator. It is not right—indeed, it is tragically wrong—to conceive of the origin of the world and of man as a scientific experiment and so something to be known *per se primo* "scientifically." Rather, the origin of the Universe, the origin of Adam and Eve, and the origin of every human person at conception is a wonderful, "miraculous," historic event, carefully planned and stupendously executed by the Creator (and in the case of Adam's children, with the procreator parents).

Now the term "Evolution" is commonly employed to designate certain explanations of the question of Origins on scientific grounds. Such an approach, because it attempts to explain scientifically the theological and miraculous, inevitably leads to conflict with traditional belief, leaving only the options of rejecting Evolution as false or of reinterpreting fundamental points of dogma so as to introduce a radically new system of belief.

More closely examined, the initial impression that Evolution Theory in some form might be supportive of traditional Christian teleology is revealed as misleading. For Evolution as the explanation of Origins prioritizes *change* as the basis of existence; whereas genuine teleology prioritizes the unchanging. Before any process can be posited, either as the principle or instrumental cause of *existence*, there stands the necessary being of the Creator, and those unique acts of production known as Creation and as miracle, which do not fall within the scope of science to explain.

This being so, it will be helpful, while pondering Mr. Keane's study, to keep in mind some basic truths of Catholic doctrine about Origins drawn from dogmatic theology and Christian metaphysics, prior to and transcending empirical science of any kind. Far from being an obstacle to "progress," these truths or dogmas will assist immeasurably to appreciate the real contribution of empirical science to understanding the truth about our origins.

Sound metaphysics, viz., Christian metaphysics, to employ the term of St. Bonaventure, tells us that something cannot come from nothing except by a creative act; and that the more perfect can only come from the less if the Creator acts miraculously to form the "higher" species as He formed the body of Adam from the slime of the earth. No natural process—read Evolution—can explain this because it cannot do what it necessarily presupposes to exist and act: Creation. That is why the origin of man is an historical event, not a term appearing at the end of an evolutionary process.

Traditional Catholic theology tells us that the Universe, visible and invisible, was created out of nothing by the triune God and subsequently structured and adorned in the work of six days, culminating in the formation of Adam's body directly from *inorganic* matter and the body of Eve directly from the unique body of Adam. All this: the creation of the world, the differentiation of the species and the ordering of the Universe within limits and for ends set by the Creator (not determined and progressively broadened by the operation of the creature) was principally the work of the Creator alone. Only after the Creator "rested" from this specific kind of action can the world be said to have begun to function "on its own," under the direction of men and angels, and so, in respect to its visible operations, to be the object of empirical science.

The great Fathers, East and West, the scholastics like St. Bonaventure and St. Thomas, are unanimous in their literal, not mythical, interpretation of the first chapters of *Genesis* on the origin of the world and of our first parents, in the sense just stated. For only thus can the uniqueness and dignity of human nature, in the body as well as in the soul, be securely demonstrated.

Some say the teaching of these Doctors in this regard has no more value than their teaching on questions scientific: that of an antiquated opinion. Such persons are mistaken. The question of Origins is not a scientific, but a *theological* question, uniquely so, for it involves a question of what God did freely and what only He could do when there were no witnesses. Hence, the importance of divine testimony in Revelation, attested by the Fathers, on this point.

There is only one reason for dissenting: the possibility that "science" might one day demonstrate an evolutionary theory of human origins to be factual in reference to the bodies (not souls) of Adam and Eve. But of this there is no reasonable expectation. Mr. Keane's study illustrates scientifically that reasonableness.

The same thing, however, can easily be done theologically, in a manner accessible to any well-instructed believer to whom it might seem the Creator could plausibly have formed Adam's body in any number of ways.

What should convince him that the narrative describing the actual formation of Adam's body should be taken "literally?" That not only were the souls of Adam and of Eve created, but that the body of Adam from the slime of the earth and the body of Eve from the side of Adam were formed miraculously by the Creator? That they were not the term of a natural, evolutionary process? Why is it that human nature is beyond the effective limits of merely material agents? It is this: *The transcendent character of the human body in respect to any other living body, even the most sophisticated! That body, informed as no other by a soul capable of knowing and loving the Creator, is animated spiritually.*

To be so animated requires a prior formation, something quite beyond the limits of any natural process—read Evolution—to produce. The transcendent character of the human body, the "image" of God as no other material being, is directly proportionate to its miraculous origin. It is not the term of a natural or evolutionary process, but of a miraculous action from on high, in which the Creator is the principal agent.

This is why God formed (not created out of nothing) the body of the first man from the virgin earth (as the Fathers unanimously understand "slime of the earth"), or inorganic matter, and why the Creator formed the body of the first woman miraculously from the body of the first man, *so that there might be no misunderstanding of the different causalities entailed in forming a species, above all the human species, and its subsequent operation within its natural limits.* No human body can exist except by way of descent (generation, procreation) from the first man through the first woman. Man alone procreates; animals only breed. Human intercourse is not merely "biological." It is primarily a moral action. That is why fidelity is the essential component of the marriage bond, and why every aspect of marriage is affected by the presence or absence of this virtue.

To this consideration a second of the Fathers of the Church must be added. The miraculous formation of the body of the first Adam from the "virgin earth" is a *type* of the even more miraculous formation of the body of the second Adam, Christ, from the Virgin Mother, viz., through a virginal conception and virginal birth. In a word: type and anti-type, figure and reality, prophecy and fulfillment are of the same order—historical and miraculous.

Denial by many scholars of the historicity of *Genesis* has ushered in a widespread form of "closed" Origins mindset, which is now largely self-perpetuating among Christians: question the truth of the *Genesis* account as the accurate description of a miracle, and one will be disposed (despite himself) to question the historicity of the miracle of the Virgin-birth, and with that the truth of the Incarnation as an historical rather than merely symbolic statement. Similarly, deny the historicity of the Virgin-birth and one will be predisposed (despite himself) to relegate the narrative of *Genesis* to the status of "myth" about Origins in justification.

The tendency of all scientific formulations of evolutionary theory for human origins to affirm some form of polygenism for the sake of "scientific" plausibility confirms this. So, too, in regard to the end of human life, evolutionary theory tends to affirm the mere "naturalness" of human death, thus fudging and indeed erasing the essential, unbridgeable difference between vestige and image of God, between mere animal and human person, between a duration that is mere succession of moments and a duration entailing eternity, between nature and grace and between human nature before and after the Fall.

These confusions and errors, in particular the denial of the numerical individuality of Adam and Eve, entails the denial of the universal need of redemption by a single Redeemer in a single Church, the new Eve, taken from the side of the New Adam in the sleep of death on the Cross. According to Pope Paul VI, a theory of Evolution is only plausible for a believer to the degree it does not contradict what his faith tells him is simply true, without qualification. Since the uniqueness and individuality of the first Adam are among such truths, and since the inner logic of evolutionary theorizing tends to contradict these, it is difficult to see how such speculation can be reconciled with faith.

With this we see that the question of Evolution is not merely, or primarily, of apologetic interest to believers. Evolution, as it is ordinarily taken to indicate a certain kind of scientific hypothesizing about Origins, is a doctrinal error parading in scientific guise. That is why, as Mr. Keane so ably shows, genuine science either tends to falsify theories of macroevolution, or simply declare that such theorizing is not properly the object of science.

Does the term "Theistic Evolution" have a legitimate place in Christian discourse, or might it designate some insight of Christian reflection, other than being a generic synonym for change or progress? Perhaps it might, but in that case it will be necessary to define the term carefully and explain why it does not entail the radical revisions of doctrine and revealed history which nigh universal convention about this word entails. It is difficult to see, however, how in practice the devilish Hegelian substitution of *becoming* for *being* thereby deifying *change* and directly contradicting the immutability of God and eternity of truths as taught by *James* 1:17—can be exorcised without abandoning the use of this phrase.

This means, therefore, that the phrase is misleading, possesses a built-in ambiguity and is "two-faced." *Theistic* suggests faith in God, the Creator; *Evolution* suggests just the opposite. Thus, the phrase is a *parte rei*, apart from the good intentions of its users, misleading. It points to an understanding of the world in terms of progress, an ever upward, spiral-like unfolding of the inner potentiality of matter until it reaches man, and in the version of Teilhard de Chardin, Christ Himself. What primarily and proximately energizes this process is from within the process itself, the existential—only incidentally supported and perfected by divine "intervention," an "intervention" defined and conditioned by the process, instead of the process being defined and limited by the prior act of creation and differentiation of essences. The classic, modern formulation of this view is the Hegelian.

How different this strange view is from the traditional vision of a created Universe hierarchically structured by the Creator from without, in terms of His own eternal counsels. Each order (grade of being) of that Universe is the direct work of the Creator and by His foreordination subordinated to and recapitulated by the higher orders, each of which is a grace in respect to the lower, the highest being the Incarnate Saviour and His Mother, the immaculately conceived Virgin. Another word for this "teleological" action is mediation, and it is the only basis for a true, and so humane vision of Origins and existence.

It is no accident that so many prominent promoters of the evolutionary perspective as the basis for a total reconstruction of Christian thought and life are Marian minimalists, indisposed to a hierarchical, mediational vision of the Universe, tending always to collapse the higher orders of grace into a single, naturalistic level of existence. Nor is it a coincidence that many promoting Theistic Evolution in the Church are radically opposed to a dogmatic definition of the universal mediation of the Virgin Mother centering on her role as co-redemptrix with her Redeemer Son on Calvary and during the celebration of the Eucharist.

That role, traditionally defined, excludes in any form an evolutionary vision of the world and confirms the ancient approach of Christian metaphysics in terms of hierarchical, graded levels of being, understood primarily as essence, rather than as existence. This recalls the doctrine of St. Anselm, where First Essence, greater than which none can be conceived, necessarily includes existence, and gives existence to finite or contingent essence by creating and ordering grades of being, the lower to the higher, as much or as little as He wills. Parallelwise, that Saint and Doctor speaks of the purity of the Virgin Mother as greater than which none can be conceived. Only the pure of heart can see God, and the only purity in fact adequate for this is the Marian. That is why, not Evolution, but the Virgin's mediation brings us to the Saviour and salvation.

In fact, historically, according to Bl. John Duns Scotus, He wills that His Son become Incarnate Saviour, and so the King and Master of all orders of being and their Redeemer by being born virginally of the Virgin, spouse of the Holy Spirit. Indeed, the Christ is the end of history, not via evolution, but via the grace of being predestined Incarnate Savior, born of the Virgin.

Deny this ancient Christian approach to finite being: its origin and structure, and the entire Universe tends more and more to be seen as the product of Evolution. Admit that metaphysics—supported by the first article of the Creed—and the illusion of Evolution disappears.

It is most important that Catholics have available to them studies of Origins such as Gerard Keane's *Creation Rediscovered*, which is free of errors in faith and morals and advances sound Origins arguments on the premises of Catholic theology.

Rev. Fr. Peter Damian Fehlner, S.T.D.[1]
Franciscan Friars of the Immaculate
Our Lady's Chapel
600 Pleasant St.
New Bedford, MA 02741-3003

---

1. Fr. Fehlner holds a doctorate in Sacred Theology from the Seraphicum in Rome (the Pontifical Theological Faculty of St. Bonaventure). He has taught dogmatic theology since 1959 and contributed to many journals in Europe and North America (*Miscellanea Francescana, Wissenschaft und Weisheit, Città di Vita, Miles Immaculatae, Christ To The World, Theological Studies, Homiletic and Pastoral Review, The Cord, Franciscan Educational Conference*) and was chief editor of *Miles Immaculatae* (1985-1989). His scholarly work on Origins, *"In The Beginning,"* was published in *Christ To The World*, Rome (1988).

INTRODUCTION

Since the first edition of *Creation Rediscovered* was published in 1991, an extensive number of fascinating developments have arisen—and so it is timely once again to survey the Origins[1] debate for these reasons:

- To present a Christian overview of matters relating to Origins—without attempting to address every aspect, and concentrating especially on the situation within the Catholic Church.
- To show that the concept of Special Creation is truly scientific and provides a better explanation of the data than evolution theories.
- To demonstrate how belief in Evolution has had a marked effect upon the consciousness of mankind and upon Christian beliefs.
- To draw attention to the need for clarification of various senses in *Genesis* by the teaching Magisterium of the Catholic Church.

While Catholicism is primarily addressed in this book, the book is not intended only for Catholics, for the Origins controversy is important to all those who consider themselves Christians. For Catholics generally, Origins has become something of a "forgotten" issue. Few see the need to study it fully, apparently because it is considered irrelevant. Not surprisingly, it seems clear that there is now much confusion concerning the doctrine of Original Sin. (Some information is presented here of what has been pronounced in papal encyclicals and in Catholic Tradition regarding Origins.)

The controversy surrounding the Origins debate still remains all about *beliefs* and only secondarily about empirical science. For those who are concerned about the widespread, on-going collapse of religious practice among Catholics, which erupted openly in the 1960s, it is hoped that this book will contribute towards genuine restoration

---

1. For the purposes of this book the capitalized word, "Origins," is used to designate "the origins of life as we now know it" for the sake of brevity and conciseness of meaning.

within the Catholic Church. (The writer is not a scientist or theologian, but a layman interested in conceptual problems affecting doctrinal beliefs.)

A strong case can be made that something has been amiss, affecting beliefs within the Catholic Church, since the early 19th century. The rise of pluralist democracies effectively brought an erosion of belief in Christianity. Many came to believe that promotion of Catholic doctrine in society, amidst a multitude of competing beliefs, was an undemocratic and unfair imposition of one's views upon others. But something also went astray within the Catholic Church concerning the comprehension and dissemination of matters relating to Origins.

A clear picture of the Origins controversy has taken about 200 years gradually to emerge. For example, the discovery of multitudes of fossils took many years to uncover, and information gained via molecular biology has only come to light in the last few decades. In many respects, therefore, the Origins debate is still very new, and its relevance to the crisis within Catholicism still is not widely understood.

The devastating collapse of faith within the Catholic Church since the 1960s was no doubt influenced by many factors, but in the opinion of the writer, the collapse is not fully explicable unless seen in the historical context of the last 500 years. Nor is the seeming enigma of many "conservatives" who, by not accepting that *Genesis* is primarily historical, may now be functioning as unwitting "carriers" of Modernism while yet being strongly opposed to its overall cancerous effects. (See Appendix B for a brief summary of Modernism.) The historical factors involved can be briefly described in various stages and aspects:

- The transformation within *Humanism* by the late 15th century, and its manifestation in the *Renaissance* era, helped establish the now common idea that man is no mere creature of God, but rather is an unfinished being capable of "creating" himself apart from God.
- The advent of heliocentrism in the 16th century set in train a certain revolution about man's place in the cosmos which helped give rise to philosophical *Idealism*: "I think, therefore I am" became a benchmark for separating subjective inner beliefs from objective reality outside of man's mind, following the speculations of René Descartes (1596-1650) and later of Immanuel Kant (1724-1804).
- The late 18th century humanist *Enlightenment* sought to place man before God in the conduct of society. Liberal democracies would be governed primarily by the collective wishes of the people (or, more realistically, by whichever group could wield domi-

nant power within the system at any point in time). The Kingship of Christ would not be deferred to when framing laws.
- The early 19th century assumptions of *uniformitarianism* brought doubts upon the idea of rapid catastrophism, thus making the *Genesis* account of Creation days and of the Flood of Noah seem unbelievable and in need of revision.
- The mid-19th century impact of rationalist *Higher Criticism*, emanating from liberal German Protestants, influenced by belief in uniformitarianism and Evolution, radically challenged the literal-as-given understanding of the *Genesis* account of Creation.
- The mid-19th century rise of *Darwinism* appeared to give a plausible evolutionary explanation of the descent of species by natural processes. Many assumed that it also provided an explanation for the *origin* of life, on the false ground that God was not required.
- The late 19th century rise of *Modernism*, influenced by all the above aspects, constituted a form of religion which by definition is completely opposed to, and tries to supplant, the Christian religion as taught by Christ and handed down for 2,000 years. Almost 100 years after Pope St. Pius X (1903-1914) moved to condemn Modernism, its revisionist impact has undergone a virile rebirth within the Catholic Church since Vatican Council II.
- *A revolution in consciousness* occurred in the Church as a result of these influencing factors, challenging the very meaning of long-held religious beliefs and practices. In particular, the overturning of the Latin rite of Mass in 1969 had the unfortunate effect of much crucial Catholic culture being lost to many individuals, resulting in alienation from authentic Catholicism.

But the relevance of Origins does not affect only Christian beliefs—the very cohesion of society is involved. If a higher, transcendent Authority is not recognized, society will continue to experience great strain:

> If members of a civilized society considered themselves as the product of blind random chance mutations, they could hardly be expected to believe they were the special creation of God. It follows, quite logically, that they could not reasonably feel themselves subject to the commands of their Creator, if that creator was time, coupled with chemicals and natural selection. If nobody owns them, they are free to make their own rules. Without any absolute authority to guide their moral decisions, they are only constrained by a rel-

ative authority, that of the State, whose rules they influence by their vote. Those rules would usually be the result of consensus, and would reflect the wish of the majority. Without any Christian ethic to influence the State . . . laws against divorce, pornography, homosexuality, abortion and suicide would be expected to be removed from the statute book. To reduce pressure upon the health services, a limited movement towards euthanasia would take place. In fact, all the social and moral phenomena we see in society today would be expected.[2]

The importance which individuals place upon Origins beliefs can thus impact greatly upon society. In many countries, abortion on demand is now the *de facto* reality, and the world is surely in desperate need of rediscovery of the authority of God. (One wonders how long the abortion holocaust can continue with impunity—4,000 surgical abortions each day in the USA alone!) But Protestant Christianity is poorly placed to satisfy decisively the latent yearning for true spiritual values or to counter the humanist lifestyle. Divided into thousands of splinter groups, many of which now accept abortion, and beset by liberal theology which denies the divinity of Christ, the quest within their ranks for social justice often seems to predominate over beliefs about doctrine handed down from Christ.

In the Catholic Church, there has been a marked loss of the sense of the sacred, and Mass attendance continues to fall, with little prospect of significant improvement. Modernist forces largely control the Church's institutions, and authentic doctrine tends to be poorly communicated. On the last page of *The Desolate City*, Anne Roche Muggeridge ends her disturbing book about the state of Catholicism with a plea:

> Catholicism is dying. If the Church of Christ is to survive as a visible light to the world, there must be, there will be, a Catholic counter-revolution. In God's good time. May it be soon.[3]

The fact that evolutionary philosophy had an extremely bad impact upon Catholicism has been recognized by Cardinal Ratzinger, Prefect of the Congregation for the Doctrine of the Faith. Addressing members of the European Doctrinal Commissions held near Vienna in May,

---

2. Peter Wilders, "Evolution: End of the Story?" *Christian Order,* April, 1990, p. 235.
3. Anne Roche Muggeridge, *The Desolate City: Revolution in the Catholic Church* (New York: Harper & Row, Publishers, Inc., 1990), p. 237.

1989, he asked where the difficulties lay which people have with the Faith today, and he went on to discuss the roots of the problems.

He spoke of the almost complete disappearance of the doctrine of Creation and its replacement by a secularized philosophy of Evolution. The resultant decline also meant that the figure of Jesus Christ was reduced to a purely historical person. The Cardinal stressed his concern that *a renewed Christianity could only be accomplished if the teaching on Creation is developed anew*—"such an undertaking ought to be regarded as one of the most pressing tasks of theology today."

But what actually is "Evolution"? Because of widespread confusion about its true meaning, a definition of terms is important:

- *Evolution is a molecules-to-man natural transformation in which new, "higher" genetic information is gained which was not possessed by one's ancestors.* However well or poorly grasped in detail, the idea of change to something vastly different (e.g., reptiles supposedly changing into birds) is the understanding now commonly held across society.
- *Natural Selection is not Evolution.* New, higher genetic information is not gained, but instead tends to be lost; at best, Natural Selection only conserves existing genetic information in life forms.
- *Variety within kind is not Evolution.* The wide variety found within each "kind" of creature or plant, due to reshuffling of genes (recombination), should not be confused with Evolution, because new, higher genetic information is not gained in the process giving rise to variety.
- *Change of an ecosystem is not Evolution.* Changes of faunistic and floristic composition which occur either progressively (in succession) or after a catastrophe (e.g., a forest fire) do not involve evolutionary change.
- *Growth to maturity is not Evolution.* The normal pattern of growth from conception to adult (e.g., seeds growing into mature plants or trees) involves an unfolding and change of shape and size, but new, higher genetic information is not gained in the process.
- *"Theistic Evolution" is not Evolution.* Ironically, this concept is forced to abandon natural Evolution and resort instead to innumerable divine interventions. (It necessarily rejects the global Flood of Noah and holds that violent bloodshed and death were always part of the "good" Creation, irrespective of the sin of Adam.)

Prior to the pro-Creation stance of the 1992 *Catechism of the Catholic Church* (in which the word "evolution" was not *specifically* mentioned even once), the last major pronouncement made by the teaching Magisterium of the Catholic Church, affecting Origins, was in the encyclical *Humani Generis*, issued by Pope Pius XII in 1950.

Since then, scientific research has gained many new insights as a result of an immense amount of new discoveries in many disciplines—including biochemistry, molecular biology, genetics, geology and astronomy. It is now known, with a high degree of certainty, that the Creator's design of DNA will not allow natural Evolution to occur.

The Catholic Church teaches that the rational souls of Adam and Eve were created by God in acts of special creation, but Pius XII (*Humani Generis*—1950) taught that Adam and Eve were real human beings, the first parents from whom all of mankind have descended; they are not symbolic representations of mankind. Most importantly, he did not *ex cathedra* declare Evolution as the *official* teaching of the Church. He did, however, allow discussion between specialists about the possible evolution of the body of Adam. The research has taken place, but full discussion within the Church has not yet occurred. What is there to fear from truth? It is time for views other than those of the *Pontifical Academy of Sciences* and the *Pontifical Biblical Commission* to be heard.

The Catholic Church can never teach that Eve's body evolved, nor tamper with the doctrine of Original Sin. And yet, despite the instructions of Pius XII to the contrary, Evolution is being presented, one-sided, virtually as fact in many Catholic academic institutions. This censorship ensures that the vital doctrine of Original Sin is not imparted in all its rigor.

Since most textbooks and TV documentaries take for granted that Evolution definitely occurred, it is hardly surprising that many individuals accept Evolution without question. In reality, not only are the required intermediate forms between the various species absent from the fossil record, but also many such supposed forms are conceptually untenable. Evolution Theory now stands exposed as both the worst mistake made in science and the most enduring *myth* of modern times. Though many still believe that Evolution has been proved, the arguments in support of it have been shown to be untenable. *Evolution is portrayed as a fact to be believed rather than as hypotheses to be tested, but its crucial mechanism continues to be ever-elusive. Ironically, if Evolution cannot occur, there is no mechanism to find!*

(The notion of ongoing Creation, where parents are seen as co-partners with God in the creation of new human beings, is not at issue here. Nor are we trying to determine what God could have done. Rather, we seek to understand what He actually chose to do in implementing Creation. *One profound reason why God would not use a method of naturalistic Evolution is that it could convey the mistaken idea that matter is eternal and thus there is no need for God.*)

Evolution beliefs may have had little impact on the doctrinal beliefs of many people, but for many others belief in Evolution has led directly to a loss of Christian faith. If natural Evolution is accepted as historically true, this belief can lead to confusion about the Fall of mankind. There is now a widespread impression that the concept of Original Sin is only religious "myth," devoid of genuine historical reality, which has been exposed by theologians. Without the Fall, the idea of redemption and a Saviour makes little sense, and one's faith is undermined.

Contrary to the views of most naturalistic evolutionists, it is indeed fully scientific to deduce the existence of a transcendent Creator. But faith in the Creator-God is itself a mysterious gift from God, and so disbelief in Evolution will not necessarily result in conversion to Christianity. Nevertheless, a widespread recognition that Evolution is *myth* is important to achieve throughout secular society.

In addition to this, however, is the fact that the secular humanist beliefs which dominate modern society cannot be effectively countered unless the basics of doctrine are once again proclaimed in schools and from pulpits. *A clear grasp of Origins is of crucial importance to both the recovery of nerve and the very teachings to be imparted.*

From a Christian perspective, should the Faith as handed down by the Apostles be retained, or should it be overturned to conform with the scientifically unsupportable evolutionary world-view? For the Catholic Church, there are two clearly incompatible alternatives at issue:

- Evolution really did take place; the first books of the Old Testament contain errors and are only religious "stories"; Adam and Eve are symbolic terms for the many early evolving human beings; and Scripture is now open to radical revision, despite 2,000 years of consistently held beliefs handed down from Christ.
- Evolution did not take place and was not part of the method chosen by God during His creation; the first books of the Old Testament contain a blending of both natural (i.e., true) history and religious truth, with no errors whatsoever; Adam and Eve were the first two human beings created by God, and interpretation of Scripture can never be open to radical revision.

Terms such as "evolutionist" and "creationist" are, of course, very simplified labels and their use can give rise to confusion. Like political labels, they are used out of convenience to categorize a range of personal views and general concepts broadly representative of a movement or a coalition of interests. As with political parties, individuals on all sides may differ substantially on a number of specific points while nevertheless sharing a broad overall position.

Evolutionists disagree substantially about the elusive yet-to-be-discovered mechanism of Evolution, but almost all of them agree on an age of billions of years for the Universe. On the other hand, those who believe in Special Creation agree that Evolution cannot occur, but they tend to disagree substantially about the age of the Universe. In the writer's opinion, the question of the age of the Universe cannot be left aside as though it is irrelevant.

- The assumption of long ages is crucial to the insupportable idea of Evolution, and the growth in popularity of presumed eons of time has helped validate Evolution in the thinking of those who have not made a careful study of this theory.
- This assumption of vast amounts of time involves concepts which ultimately challenge the teachings of the Catholic Church on death and secondary causes.
- These inconceivable time-frames have to be *read into* Scripture against the majority opinion in Tradition from Church Fathers that the sacred writer(s) of *Genesis* (including God as the principal author) intended to assert a literal-as-given meaning for *yom* creation days of 24 hours.
- Belief in an age of billions of years is, more than any other reason, *the* major factor preventing the truth of Origins from being taught rigorously in many Catholic educational institutions. This ensures that *Genesis* 1-11 remains widely regarded as virtually unbelievable mythology, explicable only by revisionist exegesis.

The question of "Age" should not be regarded as unimportant, nor should support for a "young" Universe be regarded as divisive. On the contrary, since Pope Leo XIII formally directed that the literal and obvious view must hold pride of place until rigorously disproved, *those who support an age of billions of years have the onus of proof upon them to prove their case.* Discussion of information on the question of "Age" is warranted and desirable within Catholic institutions. (Evolution has long been presented in the public arena as "fact"—even though the crucial mechanism of Evolution is missing—and so has the "fact" of a billions of years age for the Universe been presented as though beyond any credible doubt. Unchallenged acceptance of such "facts" has enabled some aspects of revisionist theology to appear credible to many in the Church.)

The Origins debate has often been portrayed wrongly as one between Christian "fundamentalism" and science à la the Scopes "Monkey Trial." Some perhaps do so in an attempt to control the debate agenda.[4] While atheists say that creationists' arguments are based on superstition, liberal Christians say they are based upon a simplistic, overly literal view of Scripture. Nevertheless, many highly qualified scientists, Christians and non-Christians, have pointed out fundamental flaws in Evolution Theory.

The concept of Special Creation holds that the elements and all living things were made by a Creator, who also revealed in *Genesis* a partial account—*not a detailed scientific textbook*—of the events of Creation. In addition to faith in God, scientists in their respective disciplines can investigate the empirical data and deduce that an intelligent Designer *must* have created the Universe.

Belief in Special Creation does not mean that Scriptural passages must be understood *only* in the literal-as-given meaning. Overall, however, the controversy over *Genesis* ought to be about which passages are *not* described by the sacred writer(s) in this literal sense.

There is much talk today about "myths" and "errors" in Scripture and much hostility to the idea that true history is described in *Genesis*. Many scholars, whether conservative or liberal,[5] tend to regard it almost exclusively in terms of supposed "salvation history" alone, with little or no place admitted for true history, and this attitude can easily result in acceptance of the idea that errors exist in Scripture. (The supposedly differing Creation accounts in *Genesis* 1 and 2 are not contradictory and indicative of errors, but are in fact complementary accounts.)

---

4. For example, those who believe in naturalistic philosophy often try to frame the terms of debate by claiming in effect that only their views are scientific. Other views which recognize the existence of God, or simply deduce the existence of an unseen designer, are dismissed by them as religious in nature and thus unscientific by definition.
5. It is acknowledged by the writer that terms such as conservative and liberal are very imprecise and can be misleading and unfair, since a wide range of views tend to be grouped together under one label. However, in the interest of brevity the following definitions are used in this book: Conservatives are deemed to be those who hold that the meaning of Scripture cannot be radically revised, and Liberals are those who regard it as being open to radical revision.

If even one aspect of liberal theology is accepted (e.g., that there are errors in the Bible), on what grounds then are liberal theologians to be rejected when they attempt to demolish such beliefs as miracles, the divinity of Christ and the Resurrection? Where is the consistency in that reasoning which accepts one revisionist aspect but rejects others? Why insist, for example, upon belief that angels rolled the rock away from Christ's tomb, and yet deny the historical reality of the Flood?

The idea that errors exist in Scripture, which has arisen from revisionist theories of Higher Criticism, did enormous harm to doctrinal beliefs. Revisionism is itself erroneous, by definition, because God is the Principal Author of Scripture. God is by nature Truth Itself and is incompatible with error and chaos.

Also, how can one justify the idea that God intended *Genesis* to be understood only in terms of supposed "religious mythology?" Though the laws of nature were not revealed and had to be discovered by human endeavor, information about the Creation events had to be revealed by God. There were no *human* witnesses to the Creation events, except to some extent Adam and Eve, and thus only the partial revelation by God in Scripture could provide man with some idea of what took place. *Let us not forget that the divine Creator-Redeemer is an absolutely reliable eye-witness, incapable of deception!*

Pope Pius XII was quite firm in his teaching that true history is described in *Genesis,* though not recorded in the way of modern historians, and Catholic Tradition right from the time of Christ has always upheld the historicity of *Genesis*. But many "conservative" Catholics tend to disregard Tradition and may be compromised with elements of Modernism because they are content to consign *Genesis* to the status of mythology, rather than defend its true historicity and foundational importance to the Church founded by Jesus Christ.

This attitude—however unintended in its effect on beliefs—is thought appropriate to ensure that *Genesis* cannot conflict with discoveries of modern science. (The cry, "Remember Galileo!" echoes on and on.) In reality, this attitude only ensures that Modernists go mostly unchallenged in their suppression of crucial Origins doctrine in schools and institutions of higher education.

The confusion over Origins and the foundational importance of *Genesis* lies close to the heart of the many problems in the Catholic Church today, and hinders a complete diagnosis of what has been amiss for many years. Until such matters are addressed fully, the harm coming from Modernist theology seems likely to continue unabated, and appeals for adherence to Church authority will be ignored.

The concept of Special Creation has not been tried and found wanting within the Catholic Church. It has been misjudged as little more than a simplistic answer to complex problems, and thus thought irrelevant and not considered seriously. Nevertheless, we live in an era when the very distinctiveness of Catholic beliefs in the modern world has been profoundly eroded, and doctrinal unity within the Church is now in a lamentable state. By "rediscovering" Creation doctrine in all its many features, there is nothing to lose and much to gain, because truth has a liberating and enlightening effect upon the human mind.

Let us hope and pray that the Magisterium (following the pro-Creation stance of the 1992 *Catechism*) will see fit soon to re-examine comprehensively all aspects relating to Origins, and that an encyclical will be issued, further clarifying relevant doctrinal beliefs.

After all, the Church founded by Christ is commissioned to work for the salvation of souls, and to promote truth irrespective of popularity. Any attempt to bring God the Creator back to center stage and facilitate moral renewal within this troubled materialistic world can only have good fruits. In contrast to the culture of death and violence which pervades the modern world, the rediscovery of the true story of Creation offers a beneficial impact upon both Church and society.

CREATION
REDISCOVERED

> *"And the Lord God formed man of the slime of the earth: and breathed into his face the breath of life, and man became a living soul."* —Genesis 2:7

—PART I—

THE BASIC QUESTION

—Chapter 1—

THE BASIC QUESTION

While many theories are held about the Origins of life as we now know it, in general they can be reduced to *three basic beliefs: Atheistic Evolution, Theistic Evolution and Special Creation.* Only one of these can be the truth, for the three beliefs are mutually incompatible.

Although he originally was a Christian, Charles Darwin came to embrace positivism[1] (the belief that only knowledge gained through empirical science is valid and that other forms of knowledge which admit the existence of the supernatural are not legitimate). He thus sought to define causation only in terms of naturalistic philosophy. But the range of aspects in Origins unavoidably involves philosophy and theology—and so the debate is all about beliefs in general, and not simply about empirical science.

Science can be defined as knowledge, the study of reality. It is the study of what "is," including things which can be perceived *beyond* this world. Despite those who would impose the view that nothing exists other than the material Universe, science *per se* cannot be defined only as empirical science; questions of philosophy and theology are also proper subjects for investigation.

Msgr. John F. McCarthy, O.S. (editor of *Living Tradition*, Rome) explains the crucial importance of *reality* in comprehending the nature of "science":

> Science is the knowledge of the meaning of reality, and it may be divided into the knowledge of the various kinds and meaning of reality. Intelligence can distinguish between the identity of a sensible object and its form. It can, for instance, distinguish between the cow and its whiteness and blackness. Again, it can distinguish the sensible form of the cow and what it has narrowed down to be the "essence" of cows as such. . . .

---

1. The terms *positivism, naturalism* and *scientism* are so similar in meaning as to be virtually interchangeable within the Origins debate.

> *Reality* is first and foremost a concept in the mind identified with that portion of mental objects which cannot be recognized to be illusory, deceptive, or fantastic, and referring to their sources as known by intellectual inference to exist extramentally. *Being* is the general term which includes both the conscious and the extra-conscious modes of existence, including illusions, deceptions, and fantasies. Thus, Gibraltar has real existence, while Oz has only fantastic existence. . . . The number five does not have extramental substantial existence, but it has a real place within the human intellect . . . . [It] fits within the concept of reality as a real conscious feature of the concretely existing intellect, standing outside of consciousness and constituting a part of the substance, man. Reality, then, is not identified with physical reality; it is rather a genus whose meaning becomes clear as it is divided into the two species of physical reality and intellectual reality. . . .
>
> Reality does not, then, mean merely verification in sense experience; it means also verification in intellectual experience. . . . Reality is the experience of the intelligibility of things. It imposes itself upon the mind, not only as the existence of sensory objects, but also as the meaning which lies behind them.[2]

Arguing that the power and success of modern empirical science has been itself "phenomenal" to the point of establishing well-entrenched disbelief in the validity and usefulness of any other type of science, Msgr. McCarthy shows to the contrary that

> The fact that reality is an intellectual *object* allows the intelligence to study it *as an object*, and not as identified with one's own subjectivity. . . .[3]
>
> Science is composed of *insight* on the part of the knowing subject, *meaning* on the part of the real objects that he knows, and *understanding* on the part of the intellect which provides his medium of thought. It is not a mere collection of unrelated facts verified by experience. It is structured knowledge, and the structure arises from the natural development of the mind itself. Material science is the collection of facts; formal science is the understanding of the facts in the intellect of the knower. . . .
>
> The recognition of the difference between *what* the intellect knows and *how* it knows what it knows divides the field of science into material and formal knowledge of reality. It also divides the field into

---

2. Msgr. John F. McCarthy, O.S., *The Science of Historical Theology: Elements of a Definition* (TAN Books & Publishers, Inc., 1991), pp. 37-39.
3. *Ibid.*, p. 41.

the lower level of knowledge of the facts (*scientia*) and the higher level of understanding of the facts (*intellectus*). It is understanding that advances science towards ever greater intelligibility and protects its conclusions from those forms of unscientific understanding called pseudo-science.[4]

The procedure used in empirical science is that an hypothesis is proposed as a tentative explanation for certain phenomena, and then, attempts are made to disprove it. A scientific theory may result, involving a number of observations which in some way can be tested by experiments.[5] The theory may come to be regarded as a law of nature, but later be superseded when better insights are achieved.

While empirical science seeks to develop mankind's understanding of the behavior of matter, it is only because matter is law-abiding that it is possible to pursue such investigation. Empirical science is thus a growing collection of knowledge about the laws of nature and hence is a derived knowledge. Science is the discovery and description of those rules by which the physical Universe operates. These laws of nature must be acknowledged as having objective existence, and are in no way the product of man's scientific endeavors.

Empirical science cannot proclaim *absolute* certainty, and it is misleading to refer to scientific "facts"; rather, empirical science is about "degrees of near certainty." It cannot explain the existence of matter or of laws of nature; nor can it explain why *particular* laws of nature are in operation and not *some other* set of laws. It cannot say anything about such abstract things as good, evil, truth, justice, beauty and love, for these exist *beyond* its scope of examination. Empirical science thus cannot be expected to explain fully the question of Origins, but yet it can discern the existence of truth *beyond* empirical science.

Empirical science can, by deduction, shed light on the existence of an intelligent Force at work in the Universe. The existence of coded information impressed upon matter provides a clue to the presence of an intelligent Designer. For example, the fantastic complexity and orderliness of the DNA code—condensed into an incredibly tiny size—suggests the work of a brilliant Intellect, rather than random chance processes. The sheer density of information packed into tiny cells sug-

---

4. *Ibid.,* pp. 48, 49.
5. If it is not *falsifiable*—open to attempts being made to disprove it—then it is not the subject of empirical science as such. Beliefs about the origins of space, matter and time cannot be tested and can only be accepted in faith, and thus are unfalsifiable. But theories about the behavior of matter and possible mechanisms for evolution can and have been subjected to testing and have been found wanting.

gests that powerful thought has gone into their design—just as human beings use sophisticated intelligence to design and construct "jumbo jets," spacecraft and other intricate equipment.

No one believes for a moment that the wonders of computer technology are the result of random chance processes, but rather, such wonders are clearly the design of extremely intelligent human designers. It is hardly unreasonable or unscientific, therefore, and it is consistent with common sense to deduce the existence of a super-brilliant Intelligence at work *beyond* the material Universe.

Another aspect of empirical science also warrants consideration: An established mode of procedure in scientific research is that of the investigation of effects, even though the intrinsic nature of the cause behind the effect may remain a puzzle. For example, gravity is a phenomenon which can be investigated mainly by means of its actions and effects on other bodies. While its effects are known, its intrinsic nature is still only partly understood by scientists.

If radio signals were to be received from outer space, they would be regarded by scientists as evidence of an intelligent source. It should not be regarded as unscientific to postulate an unseen force behind, say, the genes and their coded order and behind the existence of laws of nature. Like investigation by a detective, it is indeed scientific to deduce an intelligent Cause behind the bewildering complexity of life forms (e.g., the message sequence on the DNA molecule alone should be regarded as *prima facie* evidence of unseen Intelligence).

Walter ReMine argues powerfully that nature was intentionally constructed to look like it is the product of a single-source Designer:

> An artist uses brush strokes, composition, style, and coloring that are often unique to that artist. The chance combination of these features by any other painter would be most unlikely. . . . This same reasoning applies to life. Diverse life forms display strikingly similar characteristics. For example, there is the nearly universal use of: DNA as the carrier of inheritance; the expression of that information as proteins via an RNA intermediate; the genetic code; the use of left-handed amino acids in proteins; and the bi-layered phosphatide construction of cell membranes. The biochemical similarities extend to proteins and to the cellular metabolism of the most diverse living beings. Adenosine triphosphate (ATP), biotin, riboflavin, hemes, pyridoxin, vitamins B_{12} and K, and folic acid are used in metabolic processes everywhere. Furthermore, amino acid sequences of common proteins are similar among different organisms. For example, the protein cytochrome-*c* contains 104 amino acids, yet 64 of these

are identical between yeast and horses. Even more impressive is a protein, appropriately called *ubiquitin*, present in all organisms, tissues, and cells studied so far—and it has an absolutely identical amino acid sequence in each case. . . .

The unity of life could not possibly result from chance, nor from multiple sources, nor from multiple designers acting independently. Life must have come from some single common source. Evolutionists say "common descent." Creationists say "common designer." . . . This is not happenstance. It is premeditated design.[6]

ReMine has also shown that Creation Theory truly qualifies as scientific and that naturalistic Evolution has been shown to be false according to rigorous scientific methodology. But *illusion* of proof is facilitated by the way some evolutionists constantly shift ground between various conflicting, supposed mechanisms for Evolution.

It is sound practice for scientists to recognize the validity of other scientific disciplines, and so the argument from design cannot be dismissed as irrelevant or invalid—belief in an unseen Creator is not a blind "leap of faith," but rather *faith based on reason*. However, an important question must be addressed: In what way does theology, once hailed as the "Queen of Sciences," lay claim to being scientific? Arguing that the science of historical theology is equipped to delve into the meaning of supernatural events, Msgr. McCarthy points out,

> In the most fundamental sense, God is either known infinitely or He is not known at all, *as He is in Himself*. Traditional theology admits this fact, yet it finds a middle ground. It is possible for God to reveal something about Himself as He is in Himself that is above the natural comprehension of a created intellect and yet does not require infinite intelligence to comprehend. This is the revelation of God, and, *as revealed*, God *is an object* of man's knowledge. Theological science searches for this object within its own specific realm of objectivity.[7]

To turn towards the objectivity of God entails a recognition of the importance of God in the world we face. God is the efficient, final, and exemplary cause of that world. We may find traces of God in nature and the image of God in man. But, above all, we find God revealed in the word of Sacred Scripture, presented to us by the Church. We achieve the fulfillment of our existence by searching for

---

6. Walter James ReMine, *The Biotic Message* (St. Paul, MN: St. Paul Science, 1993), p. 18.
7. McCarthy, *op. cit.,* p. 129.

God in the word of His revelation and by finding Him in the objective meaning of His existence, as it is hidden within the word of divine revelation. This is the task of theology.[8]

Since the Origins debate is all about beliefs, it may be worthwhile to recall here that even those who hold that nothing exists apart from the physical Universe have also to wrestle with *faith*. Are we, for example, to believe that matter has always existed? How have the laws of nature come into operation? Did the bewildering complexity of DNA arise somehow by itself? Questions such as these ultimately require answers which unavoidably involve faith in something. Consider the lengths to which some theoretical physicists are driven to speculate (reminiscent of the theories of Stephen Hawking) in the futile hope of explaining reality without a transcendent Creator-God:

> The heady debate shifts, unresolved—one of many that sporadically erupt among the theoretical physicists gathered at the Aspen Center for Physics in the Colorado Rockies. A sense of barely suppressed excitement fills the air. The Theory of Everything, or TOE, the theorists believe, is hovering right around the corner. When finally grasped—the fantasy goes—the TOE will be simple enough to write down as a single equation and to solve. The solution will describe a universe that is unmistakably ours: with three spatial dimensions and one time dimension; with quarks, electrons and the other particles that make up chairs, magpies and stars; with even the big bang from which everything began. . . . Grand promises were also heard a decade ago, when "string theory" gained favor as a TOE. . . .[9]
>
> The search for a genuinely unique Theory of Everything that would eliminate all contingency and demonstrate that the physical world must necessarily be as it is, seems to be doomed to failure on grounds of logical consistency.[10]

Regarding another aspect of the Origins debate, Christians know that the Bible cannot clash with science. Some hold that empirical science has contradicted the Bible, but this conclusion is wrong. It *must* be wrong, by definition, for God—who is the principal Author of the Bible—is omniscient, Truth Itself, and free from all error. Since

---

8. *Ibid.*, p. 148.
9. Madhusree Mukerjee, "Explaining Everything," *Scientific American* (Scientific American, Inc., January, 1996).
10. Paul Davies, *The Mind of God* (London: Penguin Books Ltd., 1992). p. 167.

God is both the Creator of the Universe (including space, time and matter*) and the principal Author of Scripture, the Bible cannot contradict science.

In what way is the inerrant Bible relevant to science, both empirical and Scriptural forms? Firstly, it concerns *faith in the trustworthiness of God as a reliable eye-witness to Creation.* Such faith can enable Christians to begin a particular hypothesis about ancient Origins events with something more than a hunch. Secondly, it also informs us unerringly about the human and divine natures of Jesus Christ.

Regardless of the particular style employed to record the historical events, the biblical testimony on Creation and historical geology cannot be dismissed as irrelevant to modern man. (Archaeological discoveries in the Middle East have established that the Old Testament is surprisingly accurate in its documentation of people, places and events.) If a researcher begins with the belief that the Bible contains a description of true history, he cannot dismiss what it says about Creation events—an entirely different basis than if one started from Evolution assumptions, or from the simplistic idea that *Genesis* only informs us of the *who and why* and not the *when and how* of Creation.

As with much of everyday life in which we have to trust news reports from places inaccessible to us personally, Origins beliefs necessarily involve *faith* decisions. The researcher was not there in the ancient past to see if the processes operating today have always held or whether other process rates were in operation from time to time. However, Christians believe also that faith itself, or the ability to believe, is in some mysterious way also a gift from God:

> The power of faith is not another faculty added to the intellect of man. It is rather a new ability to see, instilled by an act of God into the intellect of man. The existence of revealed truth cannot be shown, except by indirect arguments, to a person who does not have the power of faith. Since faith is a power of intellectual sight different from the power of "natural reason" but, like natural reason, analogous

---

* It is a general way of speaking to say that God created space, time and matter. In scholastic philosophy, *space* and *time* are considered to be mental constructs, having no reality in and of themselves, *time* being "the measure of motion according to before and after," and *space* being simply "the distance between physical bodies." *Matter* (improperly understood to mean "material" by those not philosophically trained) has existence, but only as a principle of being, as in "prime matter and substantial form"; matter does not exist alone by itself, but always with a "form," or essence, which gives matter definition as part of a particular being.—*Editor, 1999.*

to the power of physical sight, the existence of the object of faith is evident only to those who have the power of faith, just as the existence of natural meaning is evident only to those who have natural intelligence and the existence of color is evident only to those who have the power of physical sight. . . .

Since the attainment of the goal of man [the vision of God in Heaven] requires the use of intellect and will by each individual to be saved, that personal theology which consists in the comprehension by the individual of the existence and meaning of his goal beyond the confines of his natural existence is also absolutely necessary.[11]

This Christian belief about the nature of faith does not sit easily within pluralist democratic societies, in which various belief systems offer vastly conflicting explanations of reality (all requiring faith from the believer), in an era when there is much confusion about Origins.

Put simply, the central claim of each basic belief—*Atheistic Evolution, Theistic Evolution* and *Special Creation*—is ultimately that of providing an explanation of the origin of the Universe and of the destiny of human beings. A person can believe either that matter was created by a transcendent God or that matter has always been in existence. Belief in either Creation or Evolution has to be made on the basis of faith, either faith in God or faith in random chance.

The two-model approach advocated by opponents of Evolution (i.e., the use of Creation/Evolution conceptual models as a comparative basis for prediction and falsification) has been criticized by some evolutionists as unacceptable, on the ground that disbelief in Evolution is not proof of Creation. Well, disbelief in a Designer is not proof of Evolution, either. Evolutionists are themselves open to the charge of double standards, because their arguments are not so much about the validity of Evolution, but rather are arguments against a Designer. By supposedly proving Evolution while firmly denying the existence of an unseen Designer, they can then charge that Evolution *must* have occurred, as there is no other option.

The scientific arguments of Origins ought to be imparted in colleges (e.g., typology versus transformism comparison). Why should a relatively few humanists be allowed to impose an *Evolution-only* syllabus by claiming that only Naturalism qualifies as science, or why should some Christian opponents of Special Creation do likewise by claiming that such Creation beliefs are unscientific? There is no more reason for

---

11. McCarthy, *The Science of Historical Theology*, p. 90.

excluding evidence for Creation beliefs on the ground that it may produce conservative religious believers, than there is for excluding evidence for Evolution beliefs (if any existed) on the ground that it may produce atheists.

— Chapter 2 —

EVOLUTION THEORIES

To most people, the idea of Natural Evolution essentially means that all present species upon Earth have emerged from very simple forms of life. It should be understood, however, that there is not just *one* theory of Evolution, but rather there are many conflicting versions. As older theories run into difficult conceptual problems, newer versions emerge, their proponents hoping to find the elusive mechanism for Evolution. This chapter looks at some commonly held ideas about Evolution and discusses how such beliefs have been either overturned or face great strain.

While many Christians believe in Evolution, it is essentially a set of ideas promoted by atheistic or agnostic scientists. Their views vary considerably, but all hold that the Universe has evolved by chance processes, without a transcendent Creator. (On the other hand, some Christians still argue that Evolution Theory has been challenged and is not yet proved, and that reasoning and scientific examination have long been superseded. As will be shown, the modern arguments against Evolution are compelling.)

The idea of *spontaneous generation*—the hypothesis that life arose from non-living matter—was widely believed throughout society as being true, until experiments conducted in 1668 by Francesco Redi and later, in the 1860's, by Louis Pasteur, showed beyond doubt that only life begets life. Nevertheless, some scientists still believe that matter has always existed and has been going through an endless process of change.

The Universe is thought by some to have "begun" with a gigantic explosion (the Big Bang Theory) somewhere between 10 and 20 billion (i.e., 20,000 million) years ago. At the beginning, according to this theory, in some form of molecular "soup," chemicals somehow came together by chance, and the Earth began to form about 4.5 billion years ago.

Evolutionists believe that living cells eventually emerged, and then life began slowly to diversify. Not only did chemical evolution give rise to biological evolution, but also random disorder gave way to order of ever-higher complexity. *Time-Life's* textbook on Evolution defines the emergence of life as

> ... the concept that there is a kinship among all forms of life because all evolved in an amplitude of time from one common ancestry ... There are differences between them because they have diverged from that ancestry in taking over the Earth, its air and its waters.[1]

Similarly, evolutionists believe that human beings and other primates evolved from a common progenitor:

> Over 10 million years ago a versatile monkey of the Proconsul type sired two distinct lines: the forest apes and pre-humans, such as *Australopithecus,* who prowled the prairies and lived in caves. One of many branches of *Australopithecines* survived to become true men, like Peking Man, a probable precursor of modern Orientals.[2]

The general belief among evolutionists is that biological evolution is still occurring, but since the process is extremely slow, it is impossible to observe any significant changes now taking place.

Evolution has long been linked to *uniformitarianism*, the idea that various features of the Earth's strata, including the fossil record, only formed slowly over millions of years by the same processes that can now be observed at work upon the Earth. This concept is known also as "the present is the key to the past." Whether it is valid or not to project from current observable rates back some billions of years is one of the major controversies in the Origins debate, having regard to the fact that vast ages are indispensable to the evolutionary mindset.

There is now widespread disagreement among evolutionists about the supposed mechanism responsible for Evolution. If the elusive mechanism of upwards Evolution still defies explanation, it is hard to see how Naturalistic Evolution can ever explain the *origin* of life.

The Argument from Design

A profound problem for Naturalistic Evolution Theory can be observed in the question of Design. Order of the type found in biology clearly does not reside within matter, but rather is imposed upon it, and

---

1. Ruth Moore and the Editors of *Life, Evolution* (Nederland: Time-Life International, 1964), p. 10.
2. *Ibid.*, p. 159.

the very existence of laws of nature, by which all matter behaves, suggests strongly that an intelligent Agency has designed the Universe and caused these laws to function.

Although naturalistic evolutionists are unwilling to believe in a transcendent Creator, this has not hindered the emergence of the so-called *Anthropic Principle*. This "principle" assumes that there can be many universes in existence, and ours is one in which the physical laws and conditions were suitable for life to evolve. It is considered to be strongly biocentric (i.e., it is amazingly "just right" for mankind):

> Through my scientific work I have come to believe more and more strongly that the physical universe is put together with an ingenuity so astonishing that I cannot accept it merely as a brute fact. There must, it seems to me, be a deeper level of explanation. Whether one wishes to call that deeper level "God" is a matter of taste and definition. Furthermore, I have come to the point of view that mind—i.e., conscious awareness of the world—is not a meaningless and incidental quirk of nature, but an absolutely fundamental facet of reality. That is not to say that *we* are the purpose for which the universe exists. Far from it. I do, however, believe that we human beings are built into the scheme of things in a very basic way.[3]

But the argument from Design goes much further than simply having the appropriate physical laws and environment. Design is of central importance in the Origins debate. In 1802, William Paley advanced the idea[4] that if one came upon a watch lying on the ground, it could be fairly concluded that there must have been a designer who comprehended its construction and made it. He compared the eye with telescope design, and argued for the existence of an intelligent Creator.

Both Paul Davies and Richard Dawkins reject Paley's conclusion, but they fail to explain the *origin* of organs (such as the eye), each of which is, in reality, an extremely complex system of interdependent systems. Dawkins' core argument of selfish "gene selection" is simplistic and unable to account for complex genetic interactions. Both scientists appeal simply to blind forces at work in nature:

> Alas, we all know about the speedy demise of this argument [from Design]. Darwin's theory of evolution demonstrated decisively that complex organization efficiently adapted to the environment could

---

3. Davies, *The Mind of God*, p. 16.
4. William Paley, *Natural Theology: Or Evidences of the Existence and Attributes of the Deity Collected from the Appearances of Nature* (1802).

arise as a result of random mutations and natural selection. No designer is needed to produce an eye or a wing. Such organs appear as a result of perfectly ordinary natural processes.... In its new form the argument [from Design] is directed not to the material objects of the universe as such, but to the underlying laws, where it is immune from Darwinian attack.[5]

Paley's argument is made with passionate sincerity and is informed by the best biological scholarship of the day, but it is wrong, gloriously and utterly wrong. The analogy between telescope and eye, between watch and living organism, is false. All appearances to the contrary, the only watchmaker in nature is the blind forces of physics, albeit deployed in a very special way. A true watchmaker has foresight: he designs his cogs and springs and plans their interconnections with a future purpose in his mind's eye. Natural selection, the blind, unconscious, automatic process which Darwin discovered, and which we now know is the explanation for the existence and apparently purposeful form of all life, has no purpose in mind. It has no mind and no mind's eye. It does not plan for the future. It has no vision, no foresight, no sight at all. If it can be said to play the role of watchmaker in nature, it is the blind watchmaker.[6]

Such naturalistic arguments have been refuted by modern discoveries in biochemistry, as Michael Behe explains in *Darwin's Black Box*:

The relevant steps in biological processes occur ultimately at the molecular level, so a satisfactory explanation of a biological phenomenon—such as sight, digestion, or immunity—must include its molecular explanation.... it is no longer enough for an evolutionary explanation of that power to consider only the *anatomical* structures of whole eyes, as Darwin did in the nineteenth century (and as popularizers of evolution continue to do today). Each of the anatomical steps and structures that Darwin thought were so simple actually involves staggeringly complicated biochemical processes that cannot be papered over with rhetoric.... Thus biochemistry offers a Lilliputian challenge to Darwin.[7]

---

5. Davies, *The Mind of God*, p. 203.
6. Richard Dawkins, *The Blind Watchmaker* (Essex, England: Longman Scientific & Technical, Longman Group UK Limited, Longman House, 1986), p. 5. Emphasis added.
7. Michael J. Behe, *Darwin's Black Box: The Biochemical Challenge to Evolution* (New York: The Free Press—A Division of Simon & Schuster Inc., 1996), p. 22.

The reality of *irreducibly complex* systems at the molecular level cannot be ignored; absence of any one interdependent part causes the system to cease functioning. (Behe illustrates the point with comparison to a simple mouse trap—if any one part is missing, the trap cannot work.) Evolution theorists are now faced with profound conceptual problems arising from biochemistry:

> In many biological structures, proteins are simply components of larger molecular machines . . . many proteins are part of structures that only function when virtually all the components have been assembled. A good example of this is a cilium. Cilia are hairlike organelles on the surfaces of many animal and lower plant cells that serve to move fluid over the cell's surface or to "row" single cells through a fluid. In humans, for example, epithelial cells lining the respiratory tract each have about 200 cilia that beat in synchrony to sweep mucus towards the throat for elimination. . . .
>
> Cilia are composed of at least a half dozen proteins: alpha-tubulin, beta-tubulin, dynein, nexin, spoke protein, and a central bridge protein. These combine to perform one task, ciliary motion, and all of these proteins must be present for the cilium to function. If the tubulins are absent, then there are no filaments to slide; if the dynein is missing, then the tubulin remains rigid and motionless; if nexin or the other connecting proteins are missing, then the axoneme falls apart when the filaments slide. *What we see in the cilium, then, is not just profound complexity, but also irreducible complexity on the molecular scale* . . .
>
> And just as scientists, when they began to learn the complexities of the cell, realized how silly it was to think that life arose spontaneously in a single step or a few steps from ocean mud, so too we now realize that the complex cilium cannot be reached in a single step or a few steps. . . . Since the irreducibly complex cilium cannot have functional precursors, it cannot be produced by natural selection, which requires a continuum of function to work. We can go further and say that, *if the cilium cannot be produced by natural selection, then the cilium was designed.*
>
> Other examples of irreducible complexity abound, including aspects of protein transport, blood clotting, closed circular DNA, electron transport, the bacterial flagellum, telomeres, photosynthesis, transcription regulation, and much more. Examples of irreducible complexity can be found on virtually every page of a biochemistry textbook.[8]

---

8. Michael J. Behe, "Molecular Machines: Experimental Support for the Design Interface," *Watchmaker* (Shade Gap, PA: Morning Star Newsletter, Catholic Origins Society, Jan.-Feb. 1996). Emphasis added.

Behe also points out that, of 886 papers published in the *Journal of Molecular Evolution* in the past ten years, there were *zero* papers discussing detailed models for intermediates in the development of complex biomolecular structures; nor are such papers found in *Proceedings of the National Academy of Science, Nature* or *Science*.[9]

Walter ReMine, also in support of Design, argues that complex life forms were devised by an unseen Designer to look deliberately *unlike* Evolution and thus thwart naturalistic explanations. *Life was designed to resist all naturalistic interpretations, not just Darwin's:*

> Message theory is explicit. It says life was fashioned to look unlike evolution. To successfully accomplish this, the designer had to avoid creating life forms too densely packed within morphology space. A densely packed system of life—without gaps—would look like evolution. Message theory says the gaps in life are intentional, because they look *unlike* evolution. If life forms could be significantly spanned by linking them together with experimental demonstrations, then message theory would be wrong. If gradual intergradations of small steps linked life on a large scale, then message theory would be refuted. Message theory is testable science. The biotic message cannot peacefully coexist with evolution. They both cannot be true. It is one or the other, and clear-cut evidence of large-scale evolution would be clear-cut evidence against the biotic message.[10]

The argument from Intelligent Design can be taken further. Textbooks often carry sketches of embryos, limbs and skulls of various creatures, and evolutionists claim that apparent similarities are proof of descent. But assertions about the "fact" of Evolution are not enough; the difficult task still awaiting Evolution theorists is to provide specific details of how the mechanism of Evolution actually occurred.

So how did these similarities come about? By chance? Or from creative design by a masterful Designer who also created the awesome, mind-boggling wonders of the Universe? (Naturalism can only take for granted the existence and nature of physical laws within the Universe, as well as matter and energy. It is silent about the very laws of physics and chemistry to which it appeals for coherence.)[11]

Evolutionists still have to show a clear pattern of descent with mod-

---

9. *Ibid.* See also *Darwin's Black Box*, pp. 173-179.
10. ReMine, *The Biotic Message*, p. 259.
11. This aspect is discussed at length by Prof. E. H. Andrews in *God, Science & Evolution*, Anzea Books, Anzea Publishers, 3-5 Richmond Road, Homebush West NSW 2140 Australia (1981 Edition).

ification, both conceptually and from evidence in the fossils. If evolution did occur, it should be possible to show lineages and to establish phylogeny (i.e., closely identified "tree" structure ancestry), but this has not been done. This is hardly surprising:

> Life was designed to resist all naturalistic interpretations. Therefore, the biomessage sender had to defeat the appearance of lineage. This was done with diversity [which] is the antithesis of lineage. Diversity destroys the semblance of lineage. . . . Diversity thwarts phylogeny.[12]
>
> The pattern of life at the molecular level of proteins and genes follows message theory precisely. It could hardly be more potent evidence. Life's many molecular phenograms and cladograms form a smooth, distinct pattern that refutes transposition and unmasking processes. That pattern allows the absences of gradual intergradation and phylogeny to take on real force as evidence against evolution. That pattern also unifies life and reveals an incredible degree of planning and design.[13]

In contrast to evolutionary depictions of "tree-like" or "bush-like" ancestry, the actual fossil evidence of creatures and plants found in the rocks is one of "parallel" vertical ancestry. A vast amount of fossils has now been uncovered, and a clear pattern has been established. Most of the phyla appear, all of a sudden, in the Cambrian Period—the field evidence is that of well-preserved unique types of fossils and uniform absence of intermediate forms.

The Geological Column—Fact or Theory?

The theory of stratigraphical geology assumes that strata on the Earth were gradually formed by layers being deposited above each other intermittently, with the lowest levels being laid down much earlier than upper levels. Various ages were assigned to the strata, and the fossils found in each stratum were dated accordingly.

This concept became crucial to 19th century Evolution Theory, and in turn helped popularize the idea that the Universe *must* be billions of years old because Evolution requires many millions of years for it to occur. Other dating methods have since been developed, to try to verify the ages assigned to the geological column. Francis Hitching (non-cre-

---

12. ReMine, *The Biotic Message,* p. 259.
13. *Ibid.,* p. 466.

ationist) noted the very close relationship which arose between uniformitarianism and Evolution Theory:

> Uniformitarianism had its birth with the publication of Charles Lyell's *Principles of Geology* in 1830. A lawyer, he had studied for a time under the creationist geologist William Buckland, subsequently Dean of Westminster, who taught that "Geology is the efficient auxiliary and handmaid of religion," and who saw widespread evidence of "direct intervention by a divine creator." Lyell sought an alternative that avoided the miraculous, and said that geological history could perfectly well be understood by observing processes still going on—the drip of rain, the wind-blown grains of sand, the pulverizing descent of glaciers. "The present is the key to the past" became the geologist's catchphrase, and in due course this innocuous statement of method duly took its place in post-creationist science. But Lyell went on to infer something much more arguable: that the processes on Earth had *always* been much the same as now, and that the rates of geological change had been uniform through time. It was a concept eagerly adopted by Darwin, who needed slow and steady environmental changes for natural selection to work in finely graduated steps; and with the triumph of Darwin's book, this aspect of uniformitarianism became accepted, too. Ever since, it has dominated the study of geology to what can be seen as a pernicious degree.[14]

The concept of the geological column rapidly became dominated by a mindset which had little place for *catastrophism*, and it gained general acceptance long before sufficient research had been conducted in the field. As to whether the theoretical model is accurate, it is not a matter of denying the reality of strata upon Earth, but rather of how formation of the actual strata is best explained. Why should it be assumed as proved that strata formed over vast ages of time?

Criticism of the original assumptions upon which the geological column theory was first formulated is not confined to creationist writers, as shown in comments made by evolutionist Edmund Spieker:

> I wonder how many of us realize that the time scale was frozen in essentially its present form by 1840? . . . How much world geology was known in 1840? A bit of Western Europe, none too well, and a lesser fringe of North America. All of Asia, Africa, South America, and most of North America were virtually unknown. How dared the

---

14. Francis Hitching, *The Neck of the Giraffe: Or Where Darwin Went Wrong* (London: Pan Books Ltd., 1982), p. 156. Emphasis in original.

pioneers [of this theory] assume that their scale would fit the rocks in these vast areas, by far most of the world? Only in dogmatic assumption—a mere extension of the kind of reasoning developed by Werner from the facts in his little district of Saxony. And in many parts of the world, notably India and South America, it does not fit. But even there it is applied! The followers of the founding fathers went forth across the Earth and in Procrustean fashion made it fit the sections they found, even in places where the actual evidence literally proclaimed denial. So flexible and accommodating are the "facts" of geology.[15]

Experiments on sedimentology have been carried out at the State University of Colorado, under the direction of French scientist Guy Berthault.[16] They indicate that much of the layers on Earth are very likely to have initially formed in sideways sloping strata series, under the action of rushing water, and were not simply deposited one above the other.

In the process, the larger particles collected horizontally at the bottom of each new surge of deposition and then graded upwards according to size of particles in each layer. Therefore, the assumed principle of superposition in fact does not apply to all strata. The strata thus cannot be used to date rocks and fossils because various layers and fossils may have been deposited simultaneously in these sloping layers.

If the strata cannot be identified with vast periods of time, then neither can the fossils contained in them. The credibility of uniformitarianism is directly challenged, and so are the related presuppositions assumed as true by specialists involved in other fields. Not only does the premise used in originally assuming that the Earth *must* be billions of years old have no validity, but the supposed evolutionary tree of evolving creatures (based on assumed ascending order of fossils found in the strata) also has no basis in reality.

The experiments filmed at Colorado were presented for scientific peer review at the 4th French Congress of Sedimentology (November 19, 1993) and the 14th International Congress of Sedimentology

---

15. Edmund M. Spieker, "Mountain-Building and Nature of Geologic Time-Scale," *Bulletin of the American Association of Petroleum Geologists*, Vol. 40 (August 1956), p. 1803.
16. See the video, *Evolution: Fact or Belief?* produced for Cercle Scientifique et Historique, distributed in the U.S. by Keep the Faith, Inc., Fairfield, N.J. See also the video, *Drama in the Rocks*, Sarong Ltd. (Jersey) 42 Bd. d'Italie MC 98000 Monaco.

EVOLUTION THEORIES

(August 27, 1994) at Recife, Brazil, but new discoveries which are very damaging to Evolution theory are seldom given media attention. Nevertheless, this sedimentology research is most important and adds another dimension to the growing volume of information being assembled by Origins researchers about the early Earth.

The Missing Mechanism

According to Darwin's theory, Evolution takes place naturally, when those organisms which are best adapted to the environment reproduce and pass on their advantageous characteristics to their offspring. Evolution unavoidably means that *higher*, more genetically complex forms of life emerged from simpler forms of life. But how?

The environment was held by Darwin to have been an important part of the process, because it was by this means that he thought new species could emerge in the struggle for survival:

> As more individuals are produced than can possibly survive, there must in every case be a struggle for existence, either one individual with another of the same species, or with the individuals of distinct species, or with the physical conditions of life. It is the doctrine of Malthus applied with manifold force to the whole animal and vegetable kingdoms; for in this case there can be no artificial increase of food, and no prudential restraint from marriage. Although some species may be now increasing, more or less rapidly, in numbers, all cannot do so, for the world would not hold them.[17]
>
> It may be said that natural selection is daily and hourly scrutinizing throughout the world every variation, even the slightest; rejecting that which is bad, preserving and adding up all that is good; silently and insensibly working, whenever and wherever opportunity offers, at the improvement of each organic being in relation to its organic and inorganic conditions of life. We see nothing of these slow changes in progress, until the hand of time has marked the long lapses of ages, and then so imperfect is our view into long past geological ages, that we only see that the forms of life are now different from what they formerly were.[18]

It is fair to say that the theory of biological Evolution is itself still evolving, as theorists strive to understand the ever-elusive mechanism

---

17. Charles Darwin, *The Origin of Species* (Middlesex, England: Penguin Books Limited, 1979), p. 117.
18. *Ibid.*, p. 133.

giving rise to Evolution. As problems are encountered which tend to negate previous theories about Evolution, new ideas are devised to re-establish its feasibility.

It seems that most scenarios have now been considered, but the crucial mechanism is still missing. With the development of the electron microscope and the growth in understanding of the science of genetics, Darwin's views of natural selection, combined with Herbert Spencer's idea of *survival of the fittest*, have been superseded. Darwin did not live long enough to see the modern development of the science of genetics and was unaware of the experiments on inheritance carried out by Fr. Gregor Mendel, the "father" of the science of genetics:

> It is unfortunate for Darwin that he did not learn about Mendel's work sooner. And it is especially ironic, since he had a copy of Mendel's papers in his library, which he never read. (Recent examination showed that its joined pages were left uncut.)[19]

While Darwin was popularizing the idea of constant change giving rise to *higher* life forms, the unheralded Mendel was discovering and demonstrating the remarkable stability which prevents the emergence of such life forms and precludes incredibly complex life forms evolving in the first place from single-celled organisms.

The concept of "survival of the fittest" has been much debated. A species does not know in advance which traits would make it more fit for survival if it did not yet have them in a useful form. It does not, for example, know whether it should grow wings or stronger legs. Only over an incredibly long period could the evolution process show how each combination of traits would work out, and all this has to happen by chance processes. Some critics suggest that—supposing Evolution to be true—those species which have survived might be only those who just happen to have survived—in other words, it may be only a matter of the *survival of the luckiest*. The ones who perished would have been in the wrong place at the wrong time.

Most evolutionists now recognize that the feasibility of any supposed evolution process would depend not so much upon the struggle for existence but, rather, would require the transmission of genetic material that would be capable of imparting new and improved characteristics to offspring. These offspring would rely upon beneficial genetic changes

---

19. Percival Davis, *et al.*, *Of Pandas and People: The Central Question of Biological Origins* (Dallas: Haughton Publishing Company, 1989), p. 60.

being possible through mutations, so that creatures could adapt constantly to changing conditions.

Natural Selection

As pointed out by creationist Edward Blythe long before Darwin, *natural selection* is known to occur, but it does not result in Evolution. Darwin later mistakenly concluded from the variety which had arisen on the Galapagos Islands that all species, from amoeba to man, had evolved from common ancestry. In fact, he actually observed *adaptation through reduction of genetic variation* (now often commonly referred to simply as *variety within kind,* or *genetic variation*).

Instead of adding new information to the gene pool, natural selection can result in loss of information and at best acts only to *conserve* the existing types, rather than facilitate the emergence of "higher" creatures. In what is known as the *founder effect*, small numbers of isolated creatures can quickly become specialized as genetic information is lost, and they may not thereafter be able to interbreed with other descendants of the original species. A similar result, known as a *bottleneck effect*, occurs when a disaster wipes out most of a population. The resultant subspecies do not gain genetic information in the process, and so the processes of natural selection and genetic variation provide no mechanism for *vertical* evolution.

Despite this, evolutionists often avoid making clear the crucial distinction between Evolution and genetic variation, and simply claim all change as "Evolution." This blatant misuse of terminology contributes greatly to confusion in the public perception of the Origins debate, as most people generally understand Evolution as meaning somehow a radical change of one creature into another of *higher* type.

Thus, in contrast to the evolutionary view, which suggests that the environment forces each species to adapt and somehow acquire new and *higher* information to its gene pool in order to survive, natural selection only works within the existing gene pool of each species and has a stabilizing effect, since novelties tend to be eliminated.

The natural environment is a most important factor. The peppered moth of England was mistakenly thought to become darker as trees became blackened by soot and as atmospheric pollutants inhibited growth of light-colored lichens during the Industrial Revolution, and this was cited as Evolution in progress. In fact, lighter members of the species became more vulnerable to predators and were reduced in numbers, while those already possessing genes which result in darker shades were better suited to survive and began to increase in numbers.

Climate and isolation are also important factors. The squirrels living on the north and south sides of the Grand Canyon are thought likely to have descended from one original population, but are now effectively separated and can thus become quite different. As Gary Parker notes, the crucial point about natural selection is that, whereas Evolution requires expansion of the gene pool, natural selection results in a more specialized, reduced gene pool (i.e., with a narrower range of genetic information than the parent population):

> Differences from average percentages can come to expression quickly in small populations (a process called "genetic drift"). Take the Pennsylvania Amish, for example. Because they are descendants of only about 200 settlers who tended to marry among themselves, they have a greater percentage than the American average of genes for short fingers, short stature, a sixth finger, and a certain blood disease. For similar reasons, plants and animals on opposite sides of mountains, rivers, or canyons often have variations in size, color, ear-shape, or some such feature that makes them recognizable as variations of a given type.
>
> All the different varieties of human beings can, of course, marry one another and have children. Many varieties of plants and animals also retain the ability to reproduce and trade genes—despite differences in appearance as great as those between collies and cocker spaniels, or wolves and wolfhounds. But varieties of one type may also lose the ability to interbreed with others of their type. For example, fruit flies multiplying through Central and South America have split up in many subgroups. And since these subgroups no longer interbreed, each can be called a separate species.[20]

Among human beings, those who tended to be fair-skinned perhaps gravitated toward cooler climates, whereas those who tended to be dark-skinned perhaps were more tolerant of hotter climates. Similarly, language barriers may have been a factor in certain traits becoming more dominant in particular tribes. Thus, in various ways, different genes may have become more common in certain tribes, especially if isolated from other tribes, but the gene pool spread over all mankind remained in existence. From the original first parents, Adam and Eve, a combination of genetic variation and natural selection has resulted in

---

20. Henry M. Morris & Gary E. Parker, *What Is Creation Science?* (San Diego: Creation-Life Publishers, 1983), p. 82.

the great diversity observable within mankind today.

As variation occurs within each kind, it tends to result in a somewhat more restricted gene pool than the parent variety and quite possibly a restricted ability to cope with changes in the environment. Extinction of some specialized varieties would be much more likely to occur than upwards evolutionary progress.

Genetic Variation Rather than Evolution

Evolution is often described as "macroevolution" and genetic variation as "microevolution," but these terms are rather ambiguous and perhaps should not be used. After all, macroevolution *is* Evolution, but microevolution *is not* Evolution. Since everyone agrees that genetic variation does occur, the point at issue is whether Evolution is possible. Unfortunately, for as long as some individuals continue to describe Evolution simplistically as "change," confusion will prevail and a distinction will have to be drawn between macro- and microevolution.

As indicated already, the concept of Evolution does not refer to variety within each basic kind, nor to the normal growth to maturity of creatures or plants. (The likelihood that some Asian populations are increasing in height due to better nutrition does not mean that human beings eventually will be eight feet tall.)

There seems little doubt that some characteristics in the gene pool can come to dominance over time and this can result in varieties of creatures which can no longer interbreed. Non-evolutionary scientists attribute the cause to natural selection—these "new" subspecies are really only specialized forms of the original kind.

An example of this can be observed in dogs. Whereas evolution from amoeba to man requires the ongoing addition of new hereditary information not previously possessed, the original dog type is now diversified into many varieties, and this diversity happens without any addition of new genetic information. Alsatians and Chihuahuas do not generally interbreed because they have become too specialized; different parts of the original genetic information have become dominant in each variety and, in the process, too much other genetic information has been bred out of existence.

The concept of genetic variation shows the immense variety which exists in the genes of each kind or type. It recognizes, however, that there are limits or boundaries (e.g., cats will always be cats, dogs will always be dogs) which prevent change into a "higher" entity with a radically new genetic structure. Each kind or type has its own specific

DNA code structure, which effectively precludes the possibility of Evolution. If this were not true, there would now be great difficulty in identifying separate species.

Since the science of genetics was first discovered by Gregor Mendel in the mid-19th century, many experiments have been conducted and it is now known how mixture of the genes of a particular species can result in, say, differences in feathers, beaks or skin. However, the basic kind does not really change, because no new genes arise, and so the accumulation of many small changes does not give rise to Evolution. Neither do *recombination* nor *mutations* result in Evolution:

> The genes can be combined and recombined in a vast number of different ways. . . . Recombination may produce interesting and useful varieties, but it takes its toll upon the animal in terms of strength and survivability. Highly bred animals are weaker, more prone to disease, and less fertile. In fact, if bred too far, the animal becomes infertile and dies out. There is a limit to how far animals can be bred from their original form. The changes do not accumulate indefinitely.
>
> Moreover, if an organism is first highly bred, and then allowed to breed freely, offspring quickly revert to the original form. The natural tendency in living things is to stay within definite limits. Although recombination is often cited as a source of new traits for evolution to work on, it does not produce the endless, vertical change necessary for evolution. Recombination is merely reshuffling of the existing genes.
>
> The only known means of introducing genuinely new genetic material into the gene pool is by mutation, a change in the DNA structure. Gene [or "point"] mutations occur when individual genes are damaged from exposure to heat, chemicals, or radiation. Chromosome mutations occur when sections of the DNA are duplicated, inverted, lost, or moved to another place in the DNA molecule. . . . Mutations do not create new structures. They merely alter existing ones.[21]

Neo-Darwinism—Valid or Invalid?

When it became clear that all organisms inherit their characteristics from the coded information in the genes possessed by each of their parents, Darwin's theory received a serious setback. Some have since

---

21. Davis *et al., Of Pandas and People*, p. 11.

claimed that, in combination with natural selection, species developed new traits through mutations of the genes and these became the building blocks of Evolution. *In the modern concept of neo-Darwinism, new traits are deemed to come about by chance and not by use and disuse.*

Neo-Darwinism also rejects the idea that acquired habits can determine evolutionary progress.[22] But neo-Darwinism is greatly dependent on the idea that mutations can provide the required genetic changes for such progress. This belief, however, remains in the realm of conjecture and runs against the observed data; mutations are nearly universally known to be bad features. The concept of Evolution requires that a large number of changes must be brought about by beneficial *"uphill"* mutations, which by definition would be both accidental and exceedingly rare (if indeed they exist at all). The odds involved are immense and impossibly high:

> Most mutations are harmful, and scientists estimate that only one in 1,000 is not. The probability of two non-harmful mutations occurring is one in 1,000,000. The odds of five non-harmful mutations occurring is one in one thousand million million. For all practical purposes, there is no chance that all five mutations will occur within the life cycle of a single organism.[23]

It seems unreasonable, therefore, to propose that such multiple, concurrent beneficial mutations, resulting each time in a truly new species, would occur in the frequency necessary to sustain the process of Evolution, even with the "guiding hand" of subsequent selection. The immensity of problems involved for Evolution to occur, with respect to the amount of time available, has been pointed out by Origins researchers:

> If we suggest, for example, that birds' feathers arose by evolution from reptiles' scales, we must postulate not one, but a very large number of mutations, involving not only the physical form, but the

---

22. In the early 19th century, before the time of Charles Darwin, the French naturalist Jean Baptiste Lamarck proposed the idea that changes in physique and habits acquired by the individual could be passed on to the next generation. The neck of the giraffe was thus thought to have grown longer as it reached higher for food. This concept is now discredited among evolutionists and creationists alike. After all, because of its long legs, the giraffe needs a long neck for reaching down to drink. The neck is specially designed to withstand great pressure.
23. Davis *et al., Of Pandas and People,* p. 13.

controlling muscles, the oil-secreting glands and so on. Yet the advantage of feathers over scales, that alone would enable natural selection to operate, could not emerge until the feather had progressed to the stage of having a different function from that of the scale. The likelihood of the many reinforcing mutations necessary to carry forward the transformation all occurring before the pressure of selection could operate is remote . . . The long neck of the giraffe could not have evolved without corresponding (and in evolutionary terms, quite independent) changes in the vascular system. This is because the difference in blood pressure between the "head up" and "head down" position is so great that the brain could not tolerate it without the very intricate system of valves which prevents this being a problem. It is no use just evolving a long neck. At the same time you have to evolve the appropriate anatomy and physiology to enable that long neck to give advantage to the animal. The chances of this happening by the coincidence of random mutations (in the various genes responsible for these different features of the animal) are incredibly small.[24]

Another formidable argument against Evolution can be deduced from the *foramen ovale* passageway and the *ductus arteriosus* vessel, which exist in or near the heart of the human fetus.[25] The following outline concerning these two "ducts" is necessarily brief and does not do justice to the amazingly complex functioning of the heart and circulatory system affecting the brain and other parts of the body.

The human heart functions as a dual pump, serving the pulmonary circulation and the systemic circulation. The first involves the motion of blood through the lungs, where it expels carbon dioxide and is oxygenated; the latter concerns the care and feeding of all cells in the body. Thus, though regarded as one organ, the heart is actually two pumps that function in extremely close harmony. The harmony is so precise that only 1% or 2% out of sequence means death in minutes.

While the baby is in the mother's womb, the fetal lungs are not functionally active, and oxygen is supplied to the fetal blood in the placenta. Bearing in mind that the right heart of the baby cannot lie dormant during this stage—for it must be fully developed to be able to assume a rigorous pumping activity upon birth—a problem exists which affects the activity carried on while it is not needed to pump blood through the pulmonary circulation.

It just so happens that, while in the womb, the unborn baby's heart

---

24. E. H. Andrews, *God, Science & Evolution*, p. 16.
25. See Marlyn E. Clark, M.S., *Our Amazing Circulatory System . . . By Chance or Creation?* (San Diego: Creation-Life Publishers, 1976).

has a temporary passageway (*ductus arteriosus*) between the pulmonary artery and the aorta and a temporary vessel (*foramen ovale*) connecting the left atrium and the right atrium. With these devices the heart is able to develop and function correctly while in the womb.

For several days after birth, the valve flap of the *foramen ovale* fuses with the wall of the atrium, and the *ductus arteriosus* begins to constrict, due to a complex interaction between the high oxygen blood and the walls of the ductus; both temporary devices thus cease operation fairly soon after birth.

The odds against all of this happening by chance and beneficial mutations are much too high. Clearly, these temporary devices associated with the fetal heart are quite crucial to the very existence of mankind. One must have them before birth, but not have them soon after birth, or the heart would very likely malfunction. This situation cannot reasonably be held to be the product of accidental events. There would be no time available to correct defects in the system; the human species would have perished before it could have even gotten started.

If it is argued that the evolutionary process just happened to get the human heart design exactly right the first time around, one would have to accept incredibly high odds. A far more reasonable deduction is: An intelligent Designer was involved beforehand in the design of the heart.

Despite the strength of such arguments to refute Evolution Theory, one is constantly confronted with some new theory to explain the supposed process of Evolution. Copious, highly technical arguments, which purport to convince the reader that biological evolution is historically true, are continually being published.

Often the evolutionary scientist involved will assert that, even though some parts of his or her proposal are still a puzzle—yet to be fully understood and sorted out—nevertheless, the genetic "jump" just might have happened, probably did in fact happen, and Evolution is thus supposed to be proved beyond reasonable doubt.

However, at the October, 1980 meeting in Chicago of some 150 evolutionary scientists, it became apparent that many reject the idea that the processes of microevolution gave rise over time to macroevolution. Paleontologists at that meeting were able to outline strongly how the principal feature of individual species in the fossil record is *stasis* and not *change* (i.e., there is no evidence in the fossil record of slow, barely perceptible changes).[26] Thus small, observed changes do not accumu-

---

26. See Roger Lewin, "Evolutionary Theory Under Fire," *Science*, Vol. 210 (Nov. 21, 1980), pp. 883-887.

late into big ones. For evolution to occur, it had to happen in big jumps by unobserved, unknown mechanisms.

The fossil record is now extensive, enough to validate the reality of *stasis* in "parallel," vertical ancestry. *In other words, species reproduce themselves only within kind.* This is powerful evidence against the idea of "tree" structured phylogeny descent. Accordingly, some evolutionists now advocate a "bush" structure of highly branching descent.

One theory now finding favor among these scientists is that of "punctuated equilibrium," a model which holds that a particular species would remain fairly constant for many thousands of years, but would occasionally be punctuated by bursts of evolution—and thus a new species with a "higher" genetic structure would appear in the fossil record, pointing to a series of progressive changes. On the one hand, this idea has the advantage for Evolution Theory of explaining the *absence* of transitional or intermediate forms. But on the other hand, there is no observable evidence for the theory.

The concept of punctuated equilibrium stakes everything on beneficial mutations and chance events occurring at rapid, unknown rates, and so is really a baseless hypothesis.

The Problem of Immense Odds

When Evolution Theory is closely analyzed, it soon becomes clear that the concept is totally dependent upon acceptance of incredibly immense odds in many unrelated situations. Consider the data relating to amino acids: depending on how an amino acid is put together, it can be either "left-handed" or "right-handed." The amino acids produced by various gas and spark experiments, conducted by scientists trying to recreate new life, always include equal numbers of left-handed and right-handed types, whereas living organisms (in another example of how life forms created by the unseen Designer thwart evolutionary explanation) utilize only left-handed amino acids:

> Living things use only left-handed amino acids. Right-handed ones don't "fit" the metabolism of the cell any more than a right-handed glove would fit onto your left hand. If just one right-handed amino acid finds its way into a protein, the protein's ability to function is reduced, often completely. (Oddly enough, by contrast, only right-handed sugars are used in DNA and RNA.) Researchers have found no way to produce only correct three-dimensional structure in simulation experiments. . . .

If amino acids formed on the early Earth . . . they would combine with other compounds in all sorts of cross-reactions, tied up and unavailable for any biologically useful function. This explains why, in actual experiments, the predominant outcome is large yields of non-biological goo.[27]

If a typical protein has 400 amino acids, the odds that all of them will be left-handed by chance (as required by Evolution) would be comparable to the odds against flipping a coin and getting 400 heads in a row. The chance of this is vanishingly small, only one over a figure of one followed by over 100 zeros. Even if a random protein of 400 left-handed amino acids were to coalesce spontaneously, it would have only the slightest chance of being formed of the proper left-handed amino acids (there are over 20 kinds) and in the proper order.

This great difficulty of odds facing Evolution cannot be over-emphasized. One chemist, John Grebe, highlighted the problem thus:

> The DNA assembly uses only 20 out of 64 possible sub-assemblies. The basic units are called nucleotides. They are arranged in a spiral rope ladder-like structure, made of a purine, pyrimidine, sugar, and phosphate unit, each group of four forming one of many rung sections. The 15,000 or more atoms of the individual sub-assemblies, if left to chance as required by the evolutionary theory, would go together in any of 10^{87} different ways. (This is ten billion with 77 more ciphers behind it.) It is like throwing 15,000 dice at one time to determine what specific molecule to make; and then to test each one for the survival of the fittest until the one out of 10^{87} different possibilities is proven by the survival of the fittest to be the right one.[28]

The astronomer Sir Fred Hoyle addressed the problem of odds in relation to the emergence of life and concluded that there has not been enough time for the degree of specificity to occur by random processes:

> The likelihood of the formation of life from inanimate matter is one to a number with 40,000 noughts after it . . . It is big enough to bury Darwin and the whole theory of evolution. There was no primeval soup, neither on this planet nor on any other, and if the

---

27. Davis *et al.*, *Of Pandas and People*, p. 5.
28. John J. Grebe, "DNA Studies in Relation to Creation Concepts," *Why Not Creation?* (Grand Rapids, MI: Baker Book House, 1970), p. 316.

beginnings of life were not random, they must therefore have been the product of purposeful intelligence.[29]

Numerous aspects can readily be cited for which Evolution Theory requires immense odds: The generation of living cells from non-living matter, the survival of these cells in an unsuitable chemical environment, the origin of photosynthesis, the dependence on extremely rare beneficial mutations, the need for evolution of interdependent traits in unrelated species, the need for concurrent evolution in both male and female forms at each new level of evolution, the many-times repeated evolution of eyes in unrelated species, the repeated evolution of flight, the existence of ducts associated with the fetal human heart—all these aspects and many more are dependent upon one chance in many, many millions being successful—not just once, but time and time again.

The problem of odds highlights a crucial conceptual dilemma. If the evolution process failed at any one of these many stages along the way, it could proceed no further. The conclusion therefore seems overwhelming that Evolution definitely did not take place. Against all this, however, an evolutionist can simply say, "So what! Surely, given enough time and enough throws of the dice, these things could have happened, and indeed must have happened; otherwise, we wouldn't be here!" But this reaction only begs the question and takes for granted the very point which has to be proved.

Such a reaction cannot be justified. A precedent exists, indisputably established for many decades now in law courts around the world, which rejects the practical possibility of immense odds. It is established in the science of fingerprints and is unquestioned by scientists around the world. But on what basis?

The Fingerprints Precedent

One tends to take for granted the common belief that each human being has a unique set of fingerprints. To understand the reasoning behind fingerprinting theory and its impact upon the credibility of Evolution Theory, it is necessary to restate here the basis involved in the science of fingerprinting. This science acknowledges three premises: fingerprints are formed well before birth, the patterns remain the same throughout life, and every human being is a unique creature.

---

29. "Hoyle on Evolution," *Nature,* 294, No. 5837 (Nov. 12, 1981), p. 148.

In an authoritative textbook on the subject of dermatoglyphics, *Finger Prints, Palms and Soles*, Harold Cummins and Charles Midlo illustrate a typical fingerprint, showing 25 labeled minutiae (i.e., specific characteristics) and state that identical fingerprints are a practical impossibility. They outline how the presence and locations of forks, swirls, bifurcations, endings and other ridge details may be considered from the same mathematical approach as that for the chance of throwing coins. They state that the chance of throwing 25 coins and having them each land head up on their previously designated square on a floor is 1/50 raised to the 25th power. Incredible odds are thus involved for an identical fingerprint to occur.

Using this calculation based on the tossing of 25 coins, the odds involved are far in excess of 10^{40}. The figure actually given by Cummins and Midlo is one chance in:
2,980,232,238,769,531,250,000,000,000,000,000,000,000,000.

This probability is based on only 25 minutiae, but Cummins and Midlo point out that if the sweat pores were also imprinted, the number of points of comparison would be "tremendously increased." And the odds shown above would also be tremendously increased!

Further, it should be borne in mind that the odds involved in the science of genetics, as highlighted in the emerging use of DNA "fingerprinting" in criminal science, are even much more immense than those given above. Since mathematicians seem to agree that, statistically, any odds beyond 1 in 10^{50} have a zero probability of ever happening,[30] it seems certain that the science of fingerprinting is soundly based and will never be overturned.

Nor does this data from Cummins and Midlo consider the nine other fingers of each human being. Interestingly, while the odds against identical sets of fingerprints are impossibly high, this very fact also gives some idea of just how unique is each human being. (Even "identical" twins from one split egg have different fingerprints.)

The facts suggest Special Creation rather than Evolution. What must the odds be against trillions-celled human beings evolving from "simple" life forms by undirected processes? Surely a brilliant Designer must have been involved, rather than chance beneficial mutations, to fit an unlimited number of combinations into the small area at

---

30. See I. L. Cohen, *Darwin Was Wrong—A Study in Probabilities* (Greenvale, NY: New Research Publications, Inc., 1984), p. 205. See also Emil Borel, *Elements of the Theory of Probability* (New Jersey: Prentice-Hall, 1965), p. 57.

the end of each finger! Further, from an Evolution point of view, unique fingerprints do not make sense, since they do not convey any survival value.

The fingerprint question is especially appropriate to the Origins debate, for in fact the basis of the fingerprint science is technically unfalsifiable, although it must rank at a very high degree of near certainty. (It is obviously impossible at any given point in time to sample the fingerprints of people who died in past centuries, let alone those who are yet to be born in the future.) Although it can neither be proved nor disproved, strictly speaking, there is universal agreement in this branch of science that the immense odds involved mean zero probability.

> Under the circumstances it is impossible to offer decisive *proof* that no two fingers bear identical patterns, but the facts in hand demonstrate the soundness of the working principle that *prints from two different fingers never are identical.*[31]

Scientists around the world do not question this belief that no two human beings will have identical fingerprints, because the odds against two people having the same combinations of loops, swirls, bifurcations, endings, lakes and the like are impossibly high. Each human being is incredibly unique. The theory is unanimously accepted, although it was never intended to be considered in terms of a precedent, as suggested here, against Evolution Theory.

Few theories are so strongly established as the science of fingerprinting, even though its basis is technically unfalsifiable. Further, it is common knowledge that the practical application of the science of fingerprinting is accepted internationally in courts of law. Accordingly, if a person can be convicted and sent to prison as a result, then its basis can be considered as having been very firmly established.

In analyzing the relevance of fingerprint science to the Origins debate, an objection can be raised that an identical print may occur perhaps millions of years after an earlier occurrence, even though Cummins and Midlo state that it will never happen. However, even if such a freak event did take place, it would not provide much comfort for the evolutionist position.

---

31. Harold Cummins & Charles Midlo, *Finger Prints, Palms and Soles: An Introduction to Dermatoglyphics* (Philadelphia: The Blakiston Company, 1943), p. 154. Emphasis in original.

Firstly, it is important to realize that there could never be enough time available to go on tossing all the coins. Even if billions of years were available, during which freak amino acids could form and life somehow get started, that vast amount of time would not be enough for the countless events (all involving astronomical odds) necessary for evolution to occur in the time available on Earth.[32]

Secondly, the amount of time available is misleading—for that is not the central issue; rather, it is the question of odds which is most pertinent. The sheer immensity of odds is the determining factor.

Since there exists such a powerful precedent unanimously established in one branch of science, namely forensic science, the following proposition deserves to be adhered to whenever the question of immense odds is considered in any other branch of science: *When the odds against an event happening reach an immense number, that event can be regarded with certainty as having zero probability.*

That this precedent should be drawn from a branch of empirical science intimately connected with the field of Law is especially fitting, for the profession of Law places high authoritative importance on the relevance of precedents. Thus, by deferring to this precedent established beyond reproof in Law, a consistent scientist cannot resort to the idea that given enough time and given enough throws of the dice some events (let alone the number of events required for Evolution) just might have occurred against the incredible odds.

This challenge to Evolution Theory arising from the science of fingerprints is therefore quite profound. It is inconsistent, on the one hand, to agree fully with the idea of condemning a person to long imprisonment (or even to capital punishment) on the basis of astronomical improbability of odds and yet, on the other hand, also not to reject Evolution Theory on the basis of astronomical improbability of odds.

The Illusion of Evolution

Further specific weaknesses in Evolution Theory will be discussed in Part II, and it will be argued that Evolution Theory is unlikely ever to be anything other than science fiction. Despite the inherent problems, however, Evolution beliefs still tend to be accepted without question—

---

32. The findings of the evolutionary geneticist J. B. S. Haldane cannot be ignored here. His calculations, known as "Haldane's Dilemma," show that not enough time has elapsed to allow an abundance of traits to evolve in various higher vertebrate species.

a 19th century hypothesis became a 20th century *myth*. Writing from a non-creationist position, Michael Denton points out that

> The raising of the status of Darwinian theory to a self-evident axiom has had the consequence that the very real problems and objections with which Darwin so painfully laboured in the *Origin* have become entirely invisible. Crucial problems such as the absence of connecting links or the difficulty of envisaging intermediate forms are virtually never discussed, and the creation of even the most complex of adaptations is put down to natural selection without a ripple of doubt.
>
> The overriding supremacy of the myth has created a widespread illusion that the theory of evolution was all but proved one hundred years ago and that all subsequent biological research—paleontological, zoological and in the newer branches of genetics and molecular biology—has provided ever-increasing evidence for Darwinian ideas. Nothing could be further from the truth. The fact is that the evidence was so patchy one hundred years ago that even Darwin himself had increasing doubts as to the validity of his views, and the only aspect of his theory which has received any support over the past century is where it applies to micro-evolutionary phenomena. His general theory, that all life on earth had originated and evolved by a gradual successive accumulation of fortuitous mutations, is still, as it was in Darwin's time, a highly speculative hypothesis entirely without direct factual support and very far from the self-evident axiom some of its more aggressive advocates would have us believe.[33]

But *the illusion of Evolution* is nevertheless most enchanting. Evolutionary beliefs are extremely elastic, able to be stretched to accommodate almost any data in order to validate Evolution:

> Evolutionary illusions are so thorough that evolutionists themselves are unaware. . . . The central illusion of evolution lies in making a wide array of contradictory mechanisms look like a seamless whole. There is no single evolutionary mechanism—there are countless. Evolutionary theory is a smorgasbord: a vast buffet of disjointed and conflicting mechanisms waiting to be chosen by the theorist. For any given question, the theorist invokes only those mechanisms that look most satisfying. Yet, the next question elicits a different response, with other mechanisms invoked and neglected.

---

33. Michael Denton, *Evolution: A Theory in Crisis* (London: Burnett Books Limited, produced and distributed by The Hutchinson Publishing Group, 1985, republished in the U.S. by Adler & Adler Publishers, Inc., Bethesda, MD, 1986), p. 77.

Evolutionary theory has no coherent structure. It is amorphous . . . accommodates data like fog accommodates landscape. [It] fails to clearly predict anything about life that is actually true. . . . *Evolution is not science.*[34]

Illusion actually depends upon a confusing misuse of terminology and concepts to distract attention from reality. Consider other examples cited by Walter ReMine in his incisive book:[35]

- Creating the impression that phylogeny exists, clear and identified, when its absence is a major unexplained problem for evolution.
- Misdirection, by shifting attention to use embryology as evidence against a designer instead of addressing the severe difficulties posed for evolution by von Baer's laws of embryology.
- Misleading illustrations, using diagrams about the fossil sequence, as if no fossils are out of sequence with evolutionary expectations.
- The use of equivocation, by shifting the meaning of words back and forth, as in efforts to show that natural selection favors Evolution or that small biologic change can be extrapolated to the macro scale.
- Crucial evolutionary aspects are ever-elastic:

    Illusion is achieved by shifting between Tautology, Special Definition, Metaphysics and Lame formulations. In this way natural selection can appear to have all the good qualities one could want in science: empirical, measurable, explanatory, general, testable, non-tautologous, and true. This shift can happen rapidly during a book or lecture. Once we understand the principle, watching natural selection in action is like watching the three-shell game at the carnival. One never knows which of the walnut shells the pea will be under next.[36]

    Illusion is achieved by concealing life's diversity within the compact names of supraspecific groups (e.g., birds, mammals, and reptiles are higher taxa and not specific species) . . . Evolutionists list these as ancestors and descendants in lineages. It then sounds as if lineages have been specifically identified, when they have not. . . . Illusion is created by misusing the key words of the origins debate: ancestral, primitive, advanced, derived, intermediate, transitional, lineage, and phylogeny. Evolutionists have redefined all these terms so

---

34. ReMine, *The Biotic Message,* p. 24. Emphasis in original.
35. *Ibid.,* pp. 460, 461.
36. *Ibid.,* p. 107.

that no ancestors ever need be identified. These words are used to convey the sound and imagery of direct ancestry, without supplying the evidence. The evolutionists' peculiar definitions of terminology also served a strategic purpose. The definitions made it awkward for an anti-evolutionist to communicate. By taking away all the key words, evolutionists effectively silenced opponents.[37]

Theistic Evolution

Theistic Evolution is a general term used to describe theories which hold that God created the first matter and then directed Evolution to take place. Evolution is thus considered to be part of the means of Creation used by God. To avoid the problem of death revealed in *Romans* 5:12 ("Wherefore as by one man sin entered into this world, and by sin death; and so death passed upon all men, in whom all have sinned."), theistic evolutionists are forced ultimately to abandon the idea of Natural Evolution and appeal implicitly to intermittent divine intervention, on innumerable occasions, over billions of years.

The concept tends to be uncritically accepted by many Catholics, on the assumption that it is compatible with both empirical science and Tradition and that it demonstrates the grandeur of God even more than Creation itself. However, when carefully examined, Theistic Evolution can be shown to have inherent conceptual problems.

Theistic Evolution is in reality a compromise position which seeks to accommodate evolutionary beliefs, but it is more accurately defined as *Theistic Intervention*, because the process of Natural Evolution has to be abandoned. (See Chapter 12.)

---

37. *Ibid.*, pp. 300, 301.

—Chapter 3—
EVOLUTIONISM

Evolutionism can be defined as the philosophy of Evolution. If one is convinced that "Evolution" is a fact to be accounted for, rather than a set of hypotheses to be tested, then we are dealing with beliefs and not empirical science. Evolutionism is nothing less than unwavering faith in macroevolution, and faith by definition tends often to be rather impervious to arguments disagreeable to one's beliefs. The influence of the philosophy of Evolutionism may extend to politics, sociology, anthropology, religion and any other field which touches upon questions of existence.

The idea that scientific analysis of natural phenomena is the only valid path to knowledge is central to the "closed" mindset of Naturalistic Philosophy (known alternatively as *Naturalism* or *Scientism*). In this mindset, belief in the existence of a transcendent creative Force *beyond* this world is deemed invalid because it is regarded as unscientific. Naturalism sees no need for a transcendent God, as the *Encyclopaedia Brittanica's* commentary on Evolution notes:

> Darwin did two things: he showed that evolution was a fact contradicting scriptural legends of creation and that its cause, natural selection, was automatic with no room for divine guidance or design. Furthermore, if there had been design, it must have been very maleficent to cause all the suffering and pain that befall animals and men.[1]

Charles Darwin was not the founder of Evolution Theory, but he seemed to provide a plausible Origins concept for which a transcendent God was not required. Darwin's ideas and those of his colleagues became popular among atheists, agnostics and liberal theologians who did not believe in various central Christian teachings.

To Naturalists, Darwin undoubtedly was largely successful in his

---

1. *Encyclopaedia Brittanica, Macropaedia,* Vol. 7 (1979), p. 23.

quest to do away with the need for a Creator. In reality, however, he failed in his attempt to prove scientifically that there had been a purely naturalistic "creation." He wrote extensively about variety, but never did explain the *origin* of species. Even in his own lifetime, Darwin's macroevolutionary ideas were strongly rejected by renowned scientists (such as typologists), who have since been proved correct.

If the science of genetics had been known in the mid-nineteenth century, Darwin's ideas (and macroevolution *per se*) would have been seen as obviously mistaken. The theme of descent with modification has in fact failed Darwin's own test of credibility, and naturalistic evolutionists must live in hope that the elusive mechanism will eventually be found. Meanwhile, in some camps the *illusion* of Evolution is very appealing and enchanting and helps keep at remote distance questions of belief in a Creator.

In view of developments across a range of disciplines which are disagreeable to Evolution Theory, Evolutionism should now be redefined as faith in the *myth* of Evolution. Despite the crucial mechanism remaining ever elusive, Evolution dogma is nevertheless insisted upon by many Naturalists, who refuse to admit the possibility of intelligent design by a transcendent Creator. Phillip Johnson comments:

> The fossil record on the whole testifies that whatever "evolution" might have been, it was not the process of gradual change in continuous lineages that Darwinism implies. As an explanation for modifications in populations, Darwinism is an empirical doctrine. As an explanation for how complex organisms came into existence in the first place, *it is pure philosophy. If empiricism were the primary value at stake, Darwinism would long ago have been limited to microevolution*, where it would have no important theological or philosophical implications.
>
> . . . [But] *empiricism is not the primary value at stake*. The more important priority is to maintain the naturalistic world view and with it the prestige of "science" as the source of all important knowledge. Without Darwinism, scientific naturalism would have no creation story. A retreat on a matter of this importance would be catastrophic for the Darwinist establishment. . . .
>
> To prevent such a catastrophe, defenders of naturalism must enforce rules of procedure for science that preclude opposing points of view. With that accomplished, the next critical step is to treat "science" as equivalent to truth and non-science as equivalent to fantasy. The conclusions of science can then be misleadingly portrayed as refuting arguments that were in fact disqualified from consideration

at the outset. As long as scientific naturalists make the rules, critics who demand positive evidence for Darwinism need not be taken seriously. They do not understand "how science works."[2]

... To question whether naturalistic evolution itself is "true," on the other hand, is to talk nonsense. Naturalistic evolution is the only conceivable explanation for life, and so the fact that life exists proves it to be true. ... The important question, however, is whether this philosophical viewpoint is merely an understandable professional prejudice or whether it is *the* objectively valid way of understanding the world. That is the real issue behind the push to make naturalistic evolution a fundamental tenet of society, to which everyone must be converted.[3]

Many thinkers have taken for granted the "fact" of Evolution, and this helped facilitate all sorts of abstract ideas to develop about the human psyche and unconscious instincts. The rejection of belief in a transcendent Creator opened up renewed pantheistic speculation (or is it wishful thinking?) about the existence of an immanent consciousness throughout the Universe and of the eternal existence of matter.

Paul Davies (physicist), writing in *The Mind of God*, accepts without question that Evolution has occurred and that our mental processes have evolved to reflect the nature of the physical world where *homo sapiens* is located. In particular, he argues in support of the philosophical belief, popular among many naturalistic scientists, that the Universe is capable of creating itself; the laws of physics have an existence "out there" transcending the Universe, and thus there is no need for a transcendent Creator God.[4]

Such speculative theories aside, Evolutionism becomes very real when it is argued that, since human beings are the highest level of intelligence in existence and there is no omnipotent God, it is vital to take control of mankind's evolutionary progress. This can easily translate into the death culture, including Malthusian policies of population control, abortion on demand, infanticide and financial aid to poor countries being conditional upon implementation of these policies. The logic of Evolutionism can thus lead to drastic consequences for human beings. For example, an environmental scientist may consider that the safeguarding of the human species lies in genetic engineering:

---

2. Phillip E. Johnson, *Darwin on Trial* (Washington D.C.: Regnery Gateway, 1991), p. 115. Emphasis added.
3. *Ibid.*, p. 121.
4. Davies, *The Mind of God*, p. 73.

> It is becoming more and more evident that man can no longer rely solely on Darwinian natural selection and other "natural" processes to ensure his fitness to cope with the hazards of the environments in which he lives. Here mankind faces possibly the most portentous challenge of its whole history as a biological species. The time is not far off when man will have to regulate his numbers, and control his genetic patrimony in order to sustain his bodily and mental vigor.[5]

It is one thing to tamper with the genes of plants or animals, but quite another to experiment with human life. A Pandora's Box is opened up, and all sorts of difficult questions soon arise: If evolutionary scientists believe that mankind can be "retooled," which implies an improvement in the person, who is to determine the specifications for the improved model? Will a new elite arise in society with *de facto* control over the life and death of innocent human life? What of all the civil rights questions involved?

Quite apart from the impact of naturalistic philosophy in pluralist democracies, faith in the "fact" of Evolution is a common denominator of many of the belief systems which have been bad news for human beings this century. The especially pernicious effect of Evolutionism can be seen in the use made of Social Darwinism by totalitarian tyrants. To assert this is not conveniently to blame Evolution for every evil which exists, for there have always existed numerous aspects of evil in the world (e.g., various authoritarian regimes often inflict savage terror for reasons in no way related to Evolution beliefs) but to give due recognition—nothing more and nothing less.

E. F. Schumacher wrote against the influence of Evolutionism upon society. He evidently believed that biological evolution was historically true, but he could not accept the atheistic belief that Evolution "explains" everything and that there is no God:

> The Doctrine of Evolutionism is generally presented in a manner that betrays, and offends against, all principles of scientific probity. It starts with the explanation of changes in living beings; but without warning, as it were, it suddenly purports to explain not only the development of consciousness, self-awareness, language and social institutions but also the origin of life itself . . . .
>
> Evolutionism, purporting to explain all and everything solely and exclusively by natural selection for adaptation and survival, is the most extreme product of the materialistic utilitarianism of the nine-

---

5. Theodosius Dobzhansky, *Evolution*, p. 7.

teenth century. *The inability of 20th-century thought to rid itself of this imposture is a failure that may well cause the collapse of Western civilization.* For it is impossible for any civilization to survive without a faith in meanings and values transcending the utilitarianism of comfort and survival—in other words, without a religious faith.[6]

C. S. Lewis perceptively noted a conceptual weakness in Evolutionism: If Evolution operates by random chance processes, then man's mind itself is thus the result of accidental processes. How then can one be sure that the mind can be trusted, especially when we know we all make mistakes? How can one be sure about belief in Evolution Theory?

> If my own mind is a product of the irrational—if what seem my clearest reasonings are only the way in which a creature conditioned as I am is bound to feel—how shall I trust my mind when it tells me about Evolution? They say in effect, "I will prove that what you call a proof is only the result of mental habits which result from heredity which results from biochemistry which results from physics." But this is the same as saying: "I will prove that proofs are irrational"; more succinctly, "I will prove that there are no proofs."[7]

Various Origins researchers have also noted this point. Either mind is a consequence of the electrical impulses and organization of an anatomical organ, or else it is a self-existent phenomenon which "rides upon" brain function without deriving from it.[8]

Evolutionism has contributed significantly to the blurring of perception of objective truth. The notion that everything in existence is constantly changing tends to facilitate attitudes of moral relativism (the idea that values are not absolute, but can vary according to the circumstances) and to undermine the notion of absolute principles.

In the context of pluralistic democratic societies over the last two centuries (with various incompatible belief systems competing in uneasy co-existence), the undermining of absolute principles has itself been a crucial factor in the gradual collapse of Christian beliefs and practices and has helped facilitate the widespread public acceptance of humanist policies such as abortion on demand. (In only several gener-

---

6. E. F. Schumacher, *A Guide for the Perplexed* (London: Sphere Books Limited, 1978 Abacus Edition), pp. 130, 133. Emphasis added.
7. C. S. Lewis, *Christian Reflections* (Glasgow: Collins Fount Paperbacks, William Collins Sons & Co. Ltd., 1985), p. 118.
8. See E. H. Andrews, *God, Science & Evolution*, p. 38.

ations, modern societies have emerged from the tragedy of many people dying at a young age to the grim situation where millions of innocent unborn human beings are now put to death.)

Since the 19th century, increasingly dominant naturalistic forces in pluralistic societies have sought to overwhelm belief in Christian teachings and to establish a common mindset which regards belief in Evolution as scientific fact, and belief in an intelligent Creator as religious non-science. Not surprisingly, Phillip Johnson shows that evidence of this can be seen both outside and within the Church:

> The combination of absolutism in evolutionary science and relativism (or selective relativism) in morals perfectly reflects the established religious philosophy of late 20th century America. *Naturalism in science provides the foundation for liberal rationalism in morals, by keeping the possibility of divine authority effectively out of the picture.* Belief in Naturalistic Evolution is foundational to a great deal else, and so it can hardly be presented as open to doubt. The schools accordingly teach that humans *discover* the profound truth of Evolution, but they *invent* moral standards and can change them as human needs change. . . . Today we take for granted that secular universities, both public and private, are pervasively dominated by naturalistic thinking and pervasively unfriendly to theism. This is true, or even especially true, in universities that have religious studies departments. "Religious belief" is an acceptable subject of study, but religion is almost always studied from a naturalistic and relativistic perspective.[9]

In view of the ever-developing and widespread emphasis upon *information technology* in the modern world (which includes the unquestioned acceptance of the concept of intelligent design in such things as computer code and satellite technology), to deny an unseen, transcendent, Creative Intelligence at work "beyond" the bewilderingly complex Universe is most unscientific. The inherent contradictions which come with belief in Evolution Theory seem likely to become increasingly evident to all as the truth about Origins becomes known.

---

9. Phillip E. Johnson, *Reason in the Balance: The Case Against Naturalism in Science, Law & Education* (Downers Grove, IL: InterVarsity Press, 1995), p. 166. Emphasis added.

—Chapter 4—
THE CONCEPT OF SPECIAL CREATION

The Creation of the Universe

A concept of Creation will now be outlined which is harmonious with the official teachings of the Catholic Church, but which also makes note of the extensive data compiled by scientists.

Christians believe there is only one God—the Divine Trinity—the infinite First Cause who created all that exists, including space, time and matter. The Gospel of St. John describes Jesus Christ, the Second Person of the Trinity and Creator/Redeemer, as the Word Incarnate:

> In the beginning was the Word, and the Word was with God, and the Word was God. The same was in the beginning with God. All things were made by him: and without him was made nothing that was made. In him was life, and the life was the light of men. And the light shineth in darkness, and the darkness did not comprehend it. There was a man sent from God, whose name was John. This man came for a witness, to give testimony of the light, that all men might believe through him. He was not the light, but was to give testimony of the light. That was the true light, which enlighteneth every man that cometh into this world. He was in the world, and the world was made by him, and the world knew him not. He came unto his own, and his own received him not. But as many as received him, he gave them power to be made the sons of God, to them that believe in his name. Who are born, not of blood, nor of the will of the flesh, nor of the will of man, but of God. And the Word was made flesh, and dwelt among us, (and we saw his glory, the glory as it were of the only begotten of the Father,) full of grace and truth. (*John* 1:1-14).

St. Paul wrote of the Kingship of Christ: "In him were all things created, in heaven and on earth, visible and invisible, whether thrones or dominations or principalities or powers—all things were created by him in him. He is before all, and by him all things consist." (*Col.* 1:16). *Hebrews* 1:1-3 also refers to the role of the Second Person in the work of Creation. And *The Catechism of the Council of Trent* declared:

> The Second Person of the Blessed Trinity, equal in all things to the Father and the Holy Ghost; for in the Divine Persons nothing unequal or unlike should exist, or even be imagined to exist, since we acknowledge the essence, will and power of all to be one.[1]
>
> The Apostle [St. Paul] sometimes calls Jesus Christ the second Adam, and compares Him to the first Adam; for as in the first all men die, so in the second all are made alive: and as in the natural order Adam was the father of the human race, so in the supernatural order Christ is the author of grace and of glory.[2]

Vatican Council I taught that God primarily instituted the Creation for the revelation of His perfection, and secondarily so that angelic and human creatures could share permanently in the bliss of the Beatific Vision of God.[3] But rebellion through pride occurred among some of the Angels, and they were banished to Hell.

God's Creation of the Universe is recorded in the first book of Moses, commonly called *Genesis*.[4] Chapter 1 provides a panoramic view of the unfolding of Creation; Chapter 2 focuses primarily on the creation of Adam and Eve; and Chapter 3 describes the tragedy of Original Sin. *Genesis* 1 to 2:3 records the order of Creation as follows:

Day 1	(1:3-5)	Light
Day 2	(1:6-8)	Water and sky
Day 3	(1:9-13)	Land and vegetation
Day 4	(1:14-19)	Sun, moon and stars
Day 5	(1:20-23)	Fish and birds
Day 6	(1:24-31)	Beasts and man
Day 7	(2:1-3)	The day God "rested"

As well as *Genesis* and *John,* there are other statements indicating that God created everything. The mother of seven martyred sons told

---

1. *The Catechism of the Council of Trent* (1923 edition; Rockford, IL: TAN Books & Publishers, Inc., 1982), p. 37.
2. *Ibid.*, p. 46.
3. Dr. Ludwig Ott, *Fundamentals of Catholic Dogma* (Cork: The Mercier Press, Ltd., 1955; Rockford, IL: TAN Books & Publishers, Inc., 1974), p. 82.
4. All references are taken from the *Douay-Rheims Bible* (1899 edition; Rockford, IL: TAN Books & Publishers, Inc., 1971).

her youngest boy, "I beseech thee, my son, look upon heaven and earth, and all that is in them: and consider that God made them out of nothing, and mankind also." (*2 Maccabees* 7:28).

Isaias also provides vital clues about the events at Creation, which also affect the age of the Universe: "*I am the Lord, and there is none else: besides me.*" (*Is.* 45:5). "*I made the earth, and I created man upon it: my hand stretched forth the heavens, and I commanded all their host.*" (*Is.* 45:12). "*For thus saith the Lord that created the heavens, God himself that formed the earth, and made it, the very maker thereof: he did not create it in vain: he formed it to be inhabited. I am the Lord, and there is no other.*" (*Is.* 45:18).

Exodus 20:11 is quite specific about the time chosen by God in which to unfold His Creation: ". . . in six days the Lord made heaven and earth, and the sea, and all things that are in them, and rested on the seventh day."

The Creation *events* were a series of efficient Divine creative acts which may have made use of rapid creative processes to bring about complementary features. These *events* should not be portrayed as divine *interventions,* for such terminology tends to assume implicitly that eons of years separated these Creation events. Nor should the unfolding of the Creation events be misrepresented to harmonize with the "process theology" favored by some Naturalistic scientists who favor the *Anthropic Principle*:

> Process thought is an attempt to view the world not as a collection of objects, or even as a set of events, but as a *process* with a definite directionality. *The flux of time thus plays a key role in process philosophy, which asserts the primacy of "becoming" over "being"*. . . Central to [Alfred North] Whitehead's philosophy is that God is responsible for ordering the world, not through direct action, but by providing the various potentialities which the physical universe is then free to actualize. . . . Traces of this subtle and indirect influence may be discerned in the progressive nature of biological evolution, for example, and the tendency for the universe to self-organize into a richer variety of ever more complex forms. *Whitehead thus replaces the monarchical image of God as omnipotent creator and ruler to that of a participator in the creative process. He is no longer self-sufficient and unchanging, but influences, and is influenced by, the unfolding reality of the physical universe.*[5]

---

5. Paul Davies, *The Mind of God*, pp. 181, 183. Emphasis added.

As well as creating many levels of angels, God created galaxies and stars, the solar system, all types of plants, animals, birds and other life forms—and human beings complete with rational souls. Since Earth was to be home for the crowning achievement of Creation (i.e., mankind), this would explain why Earth was given pride of place in the Creation account. The Universe, of course, would have been created in perfect condition, as a backdrop to life on Earth and to enable human beings to ponder the awesomeness of the Creator.

Though what we know about God is very incomplete, we know that He exists in a state of timelessness—"*I am who am.*" (*Ex.* 3:14), "*Before Abraham was, I am.*" (*John* 8:58). Put simply, we believe that God created space, matter and time from nothing (*ex nihilo*). Not only was matter created, but also *organization* of matter was created. Laws of nature were placed into operation by which matter and energy, space and time both exist and interact.

A fascinating aspect worth pondering is this: What actually constitutes living matter? Since all the particles within each cell are separate units, each not alive as such—but yet working harmoniously together—what is it that animates the total entity? In contrast to the idea of purely *mechanistic* functioning of matter, or to *vitalistic* notions which hold that nature has its own inherent life-force, Christianity can provide a coherent explanation—that the Creator has impressed complex *information* onto cells, which can reproduce and pass on that information to the next generation via *secondary causes*.

Rather than continually intervene thereafter with innumerable creative acts, God chose to employ *secondary causes* through which the Universe would function and life forms would propagate each new generation. Thus, laws of nature govern the way that matter performs, and autonomy was granted to life forms so they could effectively act as partners with God in the creation of new offspring. However, although He has put secondary causes into effect, God can nevertheless intervene at any time in the operation of the Universe to bring about a desired effect (e.g., as when Christ raised the dead Lazarus back to life, or when He brought on the global Flood).

In 1909 the *Pontifical Biblical Commission* (then an organ of the teaching Magisterium) declared that, although *Genesis* 1-3 contains a narrative of things which truly happened, the sacred writer did not intend to record the innermost nature of things. God is the principal Author of Scripture, but the intention of the human writer(s) of *Genesis* was to give a popular account, adapted to the senses, of Creation events. (For example, using the language of appearance, the moon is depicted as a great light, when in fact it really only reflects sunlight.)

Most importantly, although there is no need to look for scientific exactitude, the *Genesis* texts are primarily historical and are foundational to Christian doctrine—they are not mythology or true only in a vague "religious" sense.

Complete details of the Origins events are, of course, not fully known; various aspects remain unclear and may not be fully known this side of eternity. God did not reveal in scientific terms how He created the Universe or the body of man, nor did He reveal fundamental laws of nature. He left knowledge about laws of nature to be sought and discovered by human endeavor, painstakingly, involving great effort by many individuals.

While such things as atomic structure and the study of genetics—crucial aspects little understood 100 years ago—are now known in detail to a very high degree of certainty, much remains to be deduced by scientists about the early Earth and the still-mysterious Universe. Even the timing and placement of the Flood in relation to the actual geological strata/fossil record is still much debated among creationist geologists.[6]

This is hardly surprising. After all, gravity itself is still not fully understood by physicists. But the overall picture is now understood in greater focus than ever before, and as each year passes, more details are emerging—rather like a jigsaw puzzle gradually coming together.

Attempts made prior to recent decades to find harmony between eons of years and the *Genesis* passages, mainly to justify evolutionary beliefs, were hampered by lack of precise information and failed even to please scholars who regard *Genesis* as mythology. Nevertheless, *there must exist an authentic concordism between true empirical science and the historical events recorded in Genesis,* and Origins researchers endeavor to discern the objective truth about these events.

Designed to Look Unlike Evolution

As already discussed, Walter ReMine has made a compelling case that the unseen Designer made life forms specifically *unlike* Evolution, in order to thwart evolutionary explanations. But what of the Universe *per se*?

---

6. See Roy D. Holt, "Evidence for a Late Cainozoic Flood/post-Flood Boundary," *Creation Ex Nihilo Technical Journal* (P. O. Box 6302, Acacia Ridge D.C., Qld 4110 Australia, 1996), Vol. 10, Part 1.

A very strong argument can also be made that the unseen Designer created the solar system specifically *unlike* planetary "evolution." Because the solar system is so vast, compared to the wonders of the natural world which we can more easily see and marvel at, there is a tendency not to truly appreciate the awesomeness of it. And if the solar system is far from being fully understood, how can we make assumptions about the origins of the mysterious and immense Universe (so big that it is almost beyond human comprehension) which dwarfs it by comparison?

To appreciate fully the idea of Special Creation in its proper context, it is worthwhile first to stand back and reflect upon some of the extraordinary facts now known about the solar system, including astonishing aspects which have only been discovered in recent years following interplanetary flights such as the *Voyager* spacecraft. The facts show that the concept of *uniformitarianism* is invalid, and the complexity within the Universe could not possibly have arisen from a supposed gargantuan explosion simplistically labeled "The Big Bang."

The following details are taken from *Planets—A Smithsonian Guide*,[7] a book packed with information and excellent photographs of moons and their satellites. The planetary facts are well presented, but no consideration is given to the possibility of an unseen Designer; evolutionary origins are simply assumed. On the contrary, bias is evident in the discussion of the 1976 *Viking* soil samples on Mars, which revealed no organic compounds that might indicate biological origin. To assert that "no conclusive evidence of life was found" (pp. 109, 124) is misleading, since the results showed there is no evidence whatsoever of life on Mars.

(The Mars-rock found in Antarctica in 1984, and supposed to have been flung to Earth after Mars was hit by an asteroid, gained major worldwide media publicity in mid-1996, but the rock is sterile and only vaguely suggestive of microbial life. Such sensational reporting of highly fanciful "evidence" for Evolution and an age of billions of years is now commonplace almost weekly in major newspapers and magazines and on television and radio. Such constant reporting must impact on beliefs in the general population. Seldom, however, do creationist researchers receive commensurate attention in the media.)[8]

---

7. Thomas R. Watters, *Planets—A Smithsonian Guide: The Story of Our Solar System—from Earth to the Farthest Planet and Beyond* (New York: Macmillan USA, a Simon & Schuster Macmillan Company, a Latimer Book, 1995).
8. See two articles by Jonathan Sarfati, entitled "Life From Mars?" *Creation Ex Nihilo Technical Journal*, Vol. 10, Part 3, 1996 and Vol. 19, Nos. 1 & 2, Dec., 1996.

THE CONCEPT OF SPECIAL CREATION   51

Whereas evolutionary explanations require order and complexity to have arisen randomly from the supposed Big Bang explosion, the solar system instead shows signs of deliberate design, rather than random planetary "evolution."

- Venus rotates backwards compared to all other planets.
- Uranus and Pluto lie on their side compared to other planets. (Earth's axis inclination is 23.5°, Uranus 97.9°, Pluto 122.5°).
- Mercury rotates slowly on its axis—exactly three times while it orbits the Sun twice. A ratio of 3:2. (Each year is 88 Earth days.)
- The orbital plane of Pluto lies in a markedly different plane from all other planets, but yet the orbital periods of Pluto and Neptune are almost exactly in the ratio of 3:2, Pluto making two orbits while Neptune makes three. (Each Pluto year is 248 Earth years.)
- *Pluto is a pair.* Pluto's satellite, Charon (which, like Pluto, is thought also to have an atmosphere) "is so large in relation to Pluto that many scientists consider the two bodies to be a double planet. . . . The masses of the two bodies are so closely matched that the center of gravity is outside Pluto. Thus, *the two spiral around a common center of gravity located in the space between them."* (p. 181).
- Within the asteroid belt between Mars and Jupiter, there are 2,000 asteroids with elliptical orbits which can affect Earth. One, called "243 Ida," has a moon of its own. (p. 28).
- All four gas giants—Jupiter, Saturn, Uranus and Neptune—have thin "rings" of particles, and at least some of these rings have companion pairs of satellites. "Shepherd" satellites constantly interact with the rings and perform their function through gravitational attraction. For each ring, an outer satellite, moving more slowly because of its greater distance from the planet, attracts particles which then lose energy and slow down. Thus, an outer limit is set for that ring. But an inner satellite moves faster than the ring and gives the particles it attracts a boost in energy, causing them to move to slightly higher orbits. The rings are thus contained. (p. 161).
- Neptune's satellite, Triton, is steeply inclined to the plane of that planet's equator, its rings and other satellites. Further, Triton orbits Neptune clockwise, opposite to the planet's rotation and opposite to the direction of the other satellites.
- Of Jupiter's 16 satellites, the four outer ones orbit the planet in the opposite direction to that of its other satellites, and opposite to the

planet's rotation.
- Saturn's farthest satellite, Phoebe, also orbits the planet in the opposite direction to that of its other satellites and opposite to the planet's rotation. Another satellite, Titan, has its own atmosphere.
- Of Saturn's other satellites, two perform a most amazing feat: "The jagged and irregular *Janus* and *Epimetheus* are nearly the same size, at 118 miles and 75 miles [diameter] respectively. They are called *co-orbitals* because they share very nearly the same orbit. Their orbits are separated by only 30 miles. One makes the trip in 16.66 hours, while the other takes 16.67 hours. This slight difference creates an interesting relationship. *The faster satellite catches up with the slower one every four years. Then, instead of colliding, they exchange orbits. The outer, slower, one slips into the inner orbit and speeds up. The inner, faster, one moves into the outer orbit and slows down. Four years later they meet and switch orbits again.*" (p. 160).
- Does planet X exist? Some astronomers now theorize that another planet (perhaps several) exists far away, beyond Pluto. "Indeed, six objects were discovered orbiting the Sun beyond Pluto between 1992 and 1993." (p. 190).
- Much reference is made in the book to astonishing complexity and variety in the solar system. Uranus's satellite, Miranda, "has a surface like no other in the solar system." (p. 168). Of Jupiter's satellites, Io is the most volcanically active body in the solar system; Europa shows little signs of impact craters, but is covered by an ice crust probably about 47 miles to 62 miles thick; Ganymede is an ice giant—half ice, half rock; and Callisto is the most heavily cratered body yet observed in the solar system. And lightning 10,000 times more powerful than upon Earth has been detected in Jupiter's clouds. (pp. 134-140).
- With rare exceptions, all the hard planets and the hard satellites, and even the asteroids themselves (which are mostly covered in soil), all have impact craters showing bombardment on a mind-boggling scale. (The Moon has 30,000 craters and Earth must have sustained about five times that amount of bombardment.) The Manicougan impact site in Quebec is 40 miles across—the impact force was so great it melted the rock in the central plateau region. What is the source of this mysterious bombardment? No one knows for sure, but clearly an incredibly large planetary object must have exploded.

Many, if not all of the above-listed solar system phenomena defy explanation as resulting naturally from a Big Bang explosion. *What best explains the reality of the solar system—planetary "evolution," or intelligent Design? And what explains the reality of life on Earth—Naturalistic Evolution or intelligent Design?*

Unique Kinds in Complementary Creation

A central feature of Special Creation is that of separate distinct "kinds." *Genesis* refers to each being made according to its "kind"—human beings, plants, sea creatures, animals, birds, creeping things, which often interact in complementary interdependence. The very uniqueness of life forms points to their direct creation, as the following example of plants illustrates:

> Flowering plants are profoundly different from non-flowering plants, the former having seeds inside their fruit and very different reproductive systems, and yet no clues to their origin have come from the fossil record. There are no transitional forms between the flowering plants and plants without flowers. It is as if the flowering plants were abruptly introduced and immediately spread to all habitats on the Earth.[9]

St. Paul, in teaching about the future resurrection of the body, refers to different "kinds" in his first letter to the Corinthians, and uses an analogy from nature: "That which thou sowest is not quickened, except it die first. And that which thou sowest, thou sowest not the body that shall be; but bare grain, as of wheat, or of some of the rest. But God giveth it a body as he will: and to every seed its proper body. All flesh *is* not the same flesh: but one *is the flesh* of men, another of beasts, another of birds, another of fishes. And *there are* bodies celestial, and bodies terrestrial: but, one *is the* glory of the celestial, and another of the terrestrial. One *is the* glory of the sun, another the glory of the moon, and another the glory of the stars. For star differeth from star in glory." (*1 Cor.* 15:36-41).

A "kind" includes creatures or plants that are derived from each of the originally created life forms. But precise definition has proved most difficult; problems arise in classifying groups where differences are subtle:

---

9. Davis, *et al.*, *Of Pandas and People*, p. 106.

The division into kinds is easier the more the divergence observed. It is obvious, for example, that among invertebrates the protozoa, sponges, jellyfish, worms, snails, trilobites, lobsters, and bees are all different kinds. Among the vertebrates, the fishes, amphibians, reptiles, birds, and mammals are obviously different basic kinds. Among the reptiles, the turtles, crocodiles, dinosaurs, pterosaurs (flying reptiles), and ichthyosaurs (aquatic reptiles) would be placed in different kinds. Each one of these major groups of reptiles could be further subdivided into the basic kinds within each. Within the mammalian class, duckbilled platypuses, opossums, bats, hedge hogs, rats, rabbits, dogs, cats, lemurs, monkeys, apes and men are easily assignable to different basic kinds. Among the apes, the gibbons, orangutans, chimpanzees, and gorillas would each be included in a different basic kind. When we attempt to make fine divisions within groups of plants and animals where distinguishing features are subtle, there is a possibility of error. Many taxonomic distinctions established by man are uncertain and must remain tentative.[10]

There may now be various species within a particular "kind," and variety definitely does occur—but not *beyond* kinds. For example, cats and dogs may vary in size, shape and markings, but they always remain either cats or dogs. (The word "mankind" is an apt description for human beings, who all belong to one species.) Since it is very difficult to categorize succinctly the various types of plants and creatures, and this problem tends to be a source of confusion, the word "baramin" was coined to facilitate definition of terms:

> From the Hebrew roots—*"bara"* (created) and *"min"* (kind) . . . The word "baramin" has been suggested as a modern name for the basic created types of living things. Creationists believe that a definite boundary was established between the various baramins, and that this boundary has made possible the classification of living things into distinct groups . . . "baramin" is not exactly equivalent to any single category in modern (evolutionistic) taxonomy, such as the terms: "species," "genus," "family," or "order". . . new species can arise, but new baramin cannot.[11]

---

10. Duane T. Gish, *Evolution, The Fossils Say No!* (San Diego, CA: Creation-Life Publishers, 1980), p. 37.
11. See Paul S. Taylor, *The Illustrated Origins Answer Book* (Films for Christ Association Inc., 2628 W. Birchwood Circle, Mesa, Arizona 85202, 1989), p. 28.

THE CONCEPT OF SPECIAL CREATION

Evolutionists often point to the fact that similar traits appear in unrelated species and claim that *convergence* therefore occurs in the process of evolutionary descent. But what creatures have ever truly converged? Another explanation for the widespread existence of similar traits is possible: even though all organisms are comprised of similar "building blocks"—elementary particles, cells and so on—the Designer has arranged things so that each kind is *unique*.

The platypus is a classic example of uniqueness which defies location in any supposed evolutionary phylogeny "tree" or "bush." It is a huge problem for such explanation because it is so *unlike* a product of Evolution. It is warm-blooded, suckles its young like mammals, lays eggs, has a single ventral opening (for elimination, mating, and birth), and a shoulder girdle like most reptiles. The platypus can detect electrical currents like some fish and has a bill like a duck, although the bill is rubbery, not hard. It also has webbed forefeet like an otter, a flat tail like a beaver, and males can inject poisonous venom like a viper.

Like the platypus, many life forms each contain similar features, which are combined in unique ways which defy evolutionary descent—but which are readily explicable as the direct product of Special Creation. For example, where precisely in the evolutionary tree does one place the extinct, six-foot high "wombats" which once lived in Australia (a skeleton of one is on display in the British Museum of Natural History, London); or the giant earth worms, up to 16 feet long, which still live today in the south Gippsland region of the State of Victoria, Australia?

Similarity is a fascinating aspect which produces headaches for taxonomy. On what basis should life forms truly be classified—similarity in function or similarity in structure?

> The fact that we can classify living things at all means that we perceive degrees of similarity among them. A dog is more like a wolf than it is like a fox; as a result, the dog and the wolf are classified in the same genus (*Canis*) and the fox is classified in a different genus. Yet a dog is more like a fox than it is like a cat; so they are classified in the same family (*Canidae*) and the cat is classified in a different family. But a dog is more like a cat than it is like a horse; they are placed in the same order (*Carnivore*), and the horse is placed in a different order. Still, a dog is more like a horse than it is like a fish; therefore, they share the same class (Mammal) and the fish is in a different class. But a dog is more like a fish than it is like a worm; both dog and fish belong to a single phylum (vertebrates) and the worm belongs to a different phylum. The dog has more in common with a

worm, however, than it has with an oak tree; therefore, they are in the same kingdom (animals) and the tree is in a different kingdom (plants).[12]

Consider the contrast between Australian marsupial mammals, who are very similar in skeletal structure to many North American placental mammals—but with a major difference, nevertheless. In contrast to placentals, the marsupials nurture their offspring in a pouch on the mother's belly and possess a pair of bones which support the pouch. The North American wolf (*placental*) and the extinct Tasmanian tiger (a wolf which had stripes on its back and also a pouch like a kangaroo) possess strikingly similar skeletal features. If compared on this basis alone, they could be classified as one species.

The two types of pandas provide another classic example of uniqueness. Both share various physical traits unique only to pandas, especially the skull and jaws, and both have a unique type of "thumb," which is really an enlarged bone admirably suited to help strip bamboo. But the giant panda is now regarded as a bear (although it bleats rather like a sheep) while the lesser, red panda is classified as a racoon. The giant panda has 42 chromosomes (compared to most bears' count of 74) and the red panda has 36 chromosomes.[13] Classification problems!

The wide prevalence across many life forms, of obvious similarities utilized in a patchwork variety of combinations, presents difficulties for Naturalistic comparison and classification:

> You can see why design proponents assert that [homology] is too confusing to settle questions of structural classification. Structures are classified as *analogous* [similarity in function] or *homologous* [similarity in structure] with no objective criteria for the choice. Homologous structures can only be identified with confidence if evolution is presupposed. But if one classifies organisms on the basis of assumed evolutionary relationships, how can that classification be used as evidence that evolutionary relationships are real? The reasoning is circular, the confidence misplaced. This means that neither the concept of homologous structures nor any concept derived from it can be used as evidence for evolution.[14]

---

12. Davis, *et al., Of Pandas and People*, p. 27.
13. *Ibid.*, pp. 31, 118.
14. *Ibid.*, p. 127.

THE CONCEPT OF SPECIAL CREATION

There seems little doubt that each original kind was created to be truly unique—each its own special self, and created with a unique combination of traits. *The idea of uniqueness presents profound conceptual problems which defy credible explanation by Evolution.* For example, consider just some of the curiosities known about flying creatures:

- Flight is thought to have evolved many times across many types of creatures. As if this were not enough, did the creatures learn to fly by jumping into the air or by leaping down off heights?
- The wings of puffins (colorful coastal birds) function in exactly the same way when flying through air as when swimming under water.
- Weaver birds actually thread knots when forming hanging nests.
- The bill of some hummingbirds is curved in shape to match the design of plants from which they feed.
- Young mutton birds leave southern Australia headed for Siberia—two weeks after the adults have left! How do they know where to go? (Migratory birds cannot "evolve" the stamina to fly across vast oceans. Either they have the ability to fly the enormous distance at the first attempt—or they could not survive the first such flight!)
- The amazing dragonfly has four beautifully designed lightweight wings which function exactly in harmony.
- Butterflies and moths (tens of thousands of types) are among the most amazing of all flying creatures. The metamorphosis of the caterpillar/butterfly takes place when genetic material transforms each time to a radically different formation—Evolution is not involved!

The eminent entomologist and taxonomist, Bernard D'Abrera, publisher of many beautifully illustrated volumes such as the *Butterflies of the World* series, argues strongly in favor of a Creator to explain the bewildering array of patterns, as well as mimicry and metamorphosis:

> The sheer volume of genetic programming and engineering involved in such spectacular deception and convergence of pattern cannot, except in the wildest imaginations, be the result of "blind chance." Even if we were to accept simplistic theories about mimicry being the sole reason for such a pageant, we are still faced with the inextricable and incontrovertible fact of an Intelligence at work. Since the butterflies (by universal agreement) do not consciously program themselves thus, then painfully obviously there must be an

Intelligence *external* to them, intimately involved in every stage of the mechanics and mechanisms of that pageant....

Putting it simply, the burden of proof that a Creator and Creation does not exist remains with those who dismiss both without compunction and without good reason—for in order to *finally prove* that Nature is an uncreated effect, they must use nature itself as incontrovertible evidence for, and of the absolute lack of, a creative First Cause.[15]

In this Computer Age—the consequences and implications of a program are self-evident in the axiom that a program needs a programmer. For metamorphosis is a *predictable* and logical sequence of steps of the transformation of a living organism, the existence of that logic more than suggesting the presence of an Intelligence behind it. A *riodinid* pupa could never have been programmed to produce a *lycaenid*, much less an ithomiid—but in this case the genetic patterns of ten riodinid *Stalachtis* pupae have been set to produce eight ithomiid and two acraeid look-alikes.

The alternating instar/ecdysis sequence is itself a program, but the astounding thing is that within that one program, we apparently have a further metamorphosis or a kind of sleight of hand behind a pupal screen, an illogical interruption to the syntax—an apparent biological solecism. The imago is surely genetically a riodinid—we saw it pass through riodinid early stages, perhaps even attended by ants—but wait, what emerges is an entomological trompe d'oeil, and as any evolutionist will tell us, ultimately the whole show had its origins in "pure chance"! A profoundly intelligent and complex miracle, reduced coldly and yes, maliciously, to the banal probabilities of the toss of a pair of dice.

What has evolutionary science told us about *Stalachtis*? Absolutely nothing—but what it has told us about itself is that it is entirely inadequate, incompetent and ignorant of the great truth of Cause and Effect.... *Stalachtis* (and all of the created Universe) not only belie the mumbo-jumbo of evolutionary science, but also cry out to the humbly intelligent mind that, as St. Thomas Aquinas has demonstrated, we may know God by His great act of love—His act of creation of *everything,* ex nihilo![16]

The Creation of Adam and Eve

Scripture records that human beings are all created "in God's image

---

15. Bernard D'Abrera, *Butterflies of the Neotropical Region, Part VI: Riodinidae* (Black Rock, Victoria, Australia: Hill House, 1994), p. 1038. Emphasis in original.
16. *Ibid.,* p. 1015. Emphasis in original.

and likeness." Of all the living creatures upon Earth, man alone has the capacity to "know" God.

It is believed that God specially created Adam as a fully developed adult male human being, and his rational soul was created at the same time. (As will be shown later, a strong case can be made that his body must have been brought into existence in adult form.) Also, it is thought that Eve was created as a fully developed adult female human being, also with a rational soul. Her creation is given in *Genesis* in a mysterious manner, described as being derived from a "rib" of Adam.

God's intention for creating Eve is given very clearly in *Genesis*—as a helpmate for Adam, who gave her the name of Eve because "she was the mother of all the living." (*Gen.* 3:20). Later, Christ referred to the first human parents in his teaching on marriage. (*Mark* 10:6-8).

How is *Genesis* to be understood where it declares that the body of Adam was made of dust (or "slime") of the ground (*Gen.* 2:7), and Eve made from a rib of Adam? Was it actually material taken from the ground, or is there some deeper meaning? Perhaps the creation of Adam has a simple explanation. Since matter is composed entirely of elementary particles, God could have made him by drawing sufficient inorganic elementary particles from matter in the ground and rapidly reassembling them into the shape of an adult male human being.

(In keeping with *Genesis* 3:19, our bodies return to the same "dust" after death when each one decomposes to elementary particles.)

Since God did not choose to reveal the laws of nature to man, but left these to be discerned by human deduction, then *Genesis* could hardly describe in detail such minute things as elementary particles. Nevertheless, to early mankind, the idea of "dust" may have seemed synonymous with the smallest things imaginable. This explanation does not suggest that Adam was made from nothing, nor from previously living matter. He was not a completely new creation, *ex nihilo*, brought into the already created Universe, for God used inorganic pre-existent matter when He created Adam.

Regarding the "rib" account, the real meaning is much more debatable. Was it actually a bone taken from Adam, or is a figurative sense used, perhaps also denoting the complementary nature of male and female human beings?

Some scholars have speculated that Adam was the exemplary cause of Eve. Being created first, he could in a sense be regarded as an example or pattern for the female partner already envisaged by God, but not yet created. It seems that the Hebrew word *tsela* (i.e., rib) is more properly translated as "side." This has led some scholars to propose that the primary significance of the "rib" account is one of complementarity. While having quite different bodies, the first male and female human beings were each one of a pair with an inherent liking for, and need for, the other sex. They were obviously designed so that procreation could occur and new human life could thus be transmitted via secondary causes.

God certainly could have created Eve directly from a bone taken from Adam, but is there a deeper explanation, involving DNA and the use of X and Y chromosomes? Perhaps God took a small portion of tissue/cell from the side of the sleeping Adam, removed the Y chromosome and duplicated the X chromosome so that the cell was now female, and then caused rapid growth to produce the body of a perfect woman, genetically compatible with Adam.

This concept is not inconsistent with the Church's thinking down the years that the body of Eve was definitely derived somehow from the body of Adam. It is not a Fundamentalist explanation; rather, this concept allows for a figurative sense used by the sacred writer. After all, the very design of Adam's body suggests the need for a female companion. His body *must* have been designed with a future female partner in mind. Since God is omniscient (all-knowing), when He created Adam He would have known that Eve was soon to be created, and thus both bodies would have been designed beforehand.

Whatever the actual details involved, it seems clear that some part of Adam's body was definitely involved when Eve was created. We are told in Scripture that Adam was aware of this aspect: "This now is bone of my bones and flesh of my flesh; she shall be called woman, because she was taken out of Man." (*Gen.* 2:23). Otherwise, why not create both Adam and Eve simultaneously?

The amazing complexity and interdependence within the human body, with markedly different but complementary male and female forms, raises a serious conceptual problem for Evolution Theory. If the sexual organs were once only at a rudimentary stage of evolutionary development, how could procreation have even been possible?

If the evolution of both sexes (involving markedly different bodies) was not exactly harmonious at each new level of evolution and close enough in time and geographical location (all of which is supposed to have happened on innumerable occasions by chance processes), the various species could not have advanced to a higher level of evolution. Organs can only be of use when adequately developed.

Concurrent evolution of both sexes would thus be essential. The animal parents of evolving creatures would have to repeat a series of extremely rare but complementary beneficial mutations in both male and female offspring, all affecting the different reproductive organs of both sexes. (Or various sets of parents would have to do the same thing, each for a male and a female, and born close enough in time and location to meet each other.) The odds against such complex interdependent evolution are surely much too high for credibility.

From an evolutionary point of view, why does "sex" even exist? The wide prevalence of *genetic mixing from two parents* (described in Origins terms as "sex") poses no problem for Special Creation, but it constitutes a virtually inexplicable mystery for Evolution Theory:

> Sex occurs in all major groups of life. It is the leading mode of reproduction in groups as different as arthropods, echinoderms, molluscs, and vertebrates. Yet the sexual process is highly similar throughout nature. Meiosis, with its intricate movement of chromosomes, is often almost identical in these diverse groups. If sex was a rare phenomenon, then evolutionists might interpret it as neutral or non-adaptive. . . . When a sexual organism forms gametes (sperm or egg cells) there is a meiotic division, in which half the genes are removed . . . only half a parent's genes are sent to each of its progeny. This is called the cost of meiosis . . . an asexual parent sends *all* its genes to each progeny. In the Darwinian struggle to pass on more of one's genes to future generations, asexuality is twice as efficient as sexuality. It is therefore difficult to explain how sexual reproduction might have become prevalent.[17]

Genetic mixing from two parents looks to be deliberately intended by design, for it apparently helps to control the spread of harmful mutations. Unlike asexual reproduction, there is a chance of progeny arising with fewer harmful mutations than in either parent. How could random processes "know" this?

The Reality of Original Sin

After creating the first human pair, God, who is infinitely perfect and who, of His nature, needs nothing, gave Adam and Eve the freedom to choose Him, even though this meant they could abuse their freedom of will and reject Him. It seems clear that, because of His commitment to justice and free will, God requires the beings He has created for an eternal destiny to demonstrate freely their love for Him in some way.

---

17. ReMine, *The Biotic Message,* p. 196.

Genesis records that Adam and Eve were given a Paradise to live in (the Garden of Eden) and were given dominion over all the fishes, birds, animals and creeping things of the Earth, and Adam was given the task of naming non-human creatures. *Genesis* 1:29-30 also records that all creatures above land were provided with a vegetarian diet. That God primarily intended for above-ground creatures to live as vegetarians is suggested in *Genesis* 9:3, which records that permission was extended *after* the Flood for human beings to eat non-human creatures—every moving thing, including fish.

If this idea of a vegetarian diet seems far-fetched, it should not be overlooked that Noah was instructed to stock up with food to feed all creatures on the Ark (birds, animals, every creeping thing of the ground—all according to their kinds) during the twelve months voyage: "take unto thee of all food that may be eaten, and thou shalt lay it up with thee: and it shall be food for thee and them." (*Gen.* 6:21). Since it would not have been practicable to stock recently killed creatures and keep them in edible condition for twelve months (other than possibly as dried meat) to feed animals such as lions and tigers, the food taken aboard the Ark seems more likely to have been vegetarian matter. If all the predatory animals aboard the Ark did in fact survive for twelve months on a vegetarian diet, why not indefinitely?

There is no clue given in *Genesis* that God *primarily* intended animals to kill each other for food. The fact that creatures have always possessed the ability to attack and devour other creatures does not necessarily mean that God preferred them to behave in this manner. Nevertheless, the question of the death of non-human life—if Adam had remained obedient—is extremely difficult to ponder fully because of such things as insectivorous plants and microscopic life forms at work devouring other such life forms within the bodies of creatures.

Such aspects notwithstanding, it will be suggested later in this chapter that the *behavior* of creatures was adversely affected by the Fall, and that *"kill or be killed"* was not the preference of God, but instead was brought on only by the disobedience of Adam. This is *not* to suggest that non-human creatures would not have died if Adam had remained obedient. However, the mode of their deaths in a state of Paradise on Earth is mysterious to us. (See also Chapters 12 and 16).

God instructed Adam to avoid eating "of the tree of the knowledge of good and evil" and warned him that "in what day soever thou shalt eat of it, thou shalt die the death." (*Gen.* 2:17). Was this a tree as understood today? The possibility cannot be discounted. It is conceivable that it could have been, say, a magnificent fruit tree which was set by God as a test of obedience for Adam, and whose forbidden mystique became a great fascination for him—and he determined finally to taste the fruit of this tree. On the other hand, both the "tree of life" and the "tree of the knowledge of good and evil" (*Gen.* 2:9) may have much deeper meaning.

Unfortunately for themselves and for all mankind, Eve was seduced by Satan, and through Eve, Adam was also seduced into disobeying God's Will. With this Original (i.e., first) Sin, not only death entered into the world, but also disease, violence and cruelty. Mankind since then has had to live with the effects of Original Sin—an earthly environment that can be extremely harsh, an imperfect human nature that can easily incline away from the pursuit of good and which needs God's grace constantly, an intellect that can make mistakes, and subjection to the temptations of Satan and other fallen angels. (Even the innate sense of modesty within human beings traces back to Original Sin; only after the Fall did they feel the need to wear clothes.)

The first human parents sinned through pride, and this tragedy has been inherited by all mankind, somewhat as the family of a rich person may be impoverished if he squanders wealth. The power previously possessed by Adam and Eve to control easily the many drives within the body was thrown away by sin; they could not pass on such power to their descendants.

To Adam God said, "Cursed is the earth in thy work; with labour and toil shalt thou eat thereof all the days of thy life. Thorns and thistles shall it bring forth to thee; and thou shalt eat the herbs of the earth. In the sweat of thy face shalt thou eat bread till thou return to the earth, out of which thou wast taken: for dust thou art, and into dust thou shalt return." (*Gen.* 3:17-19). Adam and Eve were subsequently driven from the Garden of Eden. Mankind was now alienated from God, and no amount of human effort could effect a reconciliation with Him.

We know, however, that this situation was not to prevail indefinitely. Mankind would be led to knowledge of salvation, and the Second Person of the Divine Trinity would personally pay the ransom for man's alienation. How amazing to comprehend and how consoling to know that each human being is so dearly loved by the awesome Creator of the Universe! In an act of great magnanimity, God eventually sent His only Son, Jesus Christ, to suffer and die on behalf of fallen man in order to redeem mankind and make it possible for human beings, body and soul, to enter Heaven.

Clearly, the Fall was no trivial event, nor was the price of Redemption which was paid by Jesus Christ. The voluntary atonement carried out by Jesus Christ on behalf of fallen man shows that God regarded the sin of Adam as a tragedy of catastrophic proportions, far beyond adequate human description. Perhaps the effects of Original Sin upon the Earth—and the whole Universe—were and are much greater than have been commonly supposed. St. Paul hints at this when he says that "every creature groaneth and travaileth in pain, even till now." (*Rom.* 8:22).

The Flood of Noah

The extent of the Great Flood is still very controversial. Despite the overwhelming field evidence in favor of a global Flood—a true global cataclysm—some scholars maintain that, even if the Flood account was a real fact of history, it was only a local event. On the other hand, since the weight of detailed arguments against Evolution Theory is very strong, it makes little sense to maintain that the Universe is billions of years old. And if the Universe is only about 6,000 years old, as per the Bible account, then the Flood account is not only a real event in history, but *must* have been of global, cataclysmic dimensions, if we are to believe the teachings of Jesus Christ about an enormous flood which, only some thousands of years ago, destroyed all except those in the Ark. (*Matt.* 24:37-39; *Luke* 17:26-27).

The charge often thrown at creationists—*anthropomorphism* (resulting in reducing the greatness of God down to limited human ideas)—can also be applied to those who reject out of hand the possibility of a "young" Universe and a global flood.

It is certainly within God's power to have created the Universe only some 6,000 years ago and to have caused a global flood to occur. What does the field evidence found upon Earth suggest? As will be outlined in Chapter 6, it suggests a record of violent upheaval and of death and destruction, rather than evidence of evolutionary progress. The evidence is of catastrophic proportions, as indicated in *Genesis*.

We are informed in *Genesis* that following the banishment of Adam and Eve from the Garden of Eden, mankind became more sinful. In view of this, the Lord "repented him that he had made man on the earth. And being touched inwardly with sorrow of heart" (*Gen.* 6:6), God decided to destroy all human beings, save for Noah and his family, and all the creatures that lived upon land, save for a few of each—seven pairs of all clean beasts and two pairs of each kind of beasts that are not clean, and seven pairs of birds, to keep their kinds alive. (Cf. *Gen.* 7:2-3). (The "clean" animals are thought to be those used for sacrifice.)

It is now thought to be most likely that when the Flood occurred (*Gen.* 7:6), vast quantities of underground water burst upwards. Scripture records that the whole Earth was covered by water. Many of the Earth's trees and much other vegetation were uprooted by a combination of volcanic activity and the action of Flood waters, and deposited at other parts of the Earth. They were then subjected to further rapid burial by sediment arising from rapid erosion of the Earth's outer crust by immense forces of water bursting up from underground. In the process, huge numbers of fish, birds, animals and other creatures were engulfed. Fossil beds were laid down and coal and oil deposits formed.

With respect to the action of God recorded in *Psalms* 104:5-8, it has been suggested by scientists that during the year-long events of the Flood, the pressures unleashed beneath the Earth's outer crust would have resulted in high mountains being pushed up and a corresponding sinking of what are now the ocean basins.

The dimensions of the Flood are graphically recalled in the account provided to the sacred writer by God, who was the only global Eyewitness and who is the principal Author of Scripture.

> "And the waters prevailed beyond measure upon the earth: and all the high mountains under the whole heaven were covered. The water was fifteen cubits higher than the mountains which it covered. And all flesh was destroyed that moved upon the earth, both of fowl, and of cattle, and of beasts, and of all creeping things that creep upon the earth: and all men. And all things wherein there is the breath of life on the earth, died . . ." (*Gen.* 7:19-23).

The memory of a huge flood lingers on in the folklore of many cultures around the Earth, and this points heavily against the idea of the Flood being either a local event or simply a mythical story:

> The most comprehensive compilation of global flood stories is that by Sir James George Frazer in his three-volume *Folklore in the Old Testament*. He cites evidence of Flood accounts from the Indians in North America to the Indians of India. Hawaii, Alaska, Indonesia, Europe, Asia, Australia, and Mesopotamia all have Flood accounts. Needless to say, as the tribes migrated farther and farther from Ararat, the stories became more and more distorted. In fact, this point has been carefully documented by John Warwick Montgomery. This impressive evidence would seem to substantiate the view that these stories do have a common origin and are not exaggerated tales of

local catastrophes, as some have maintained.[18]

Anthropologists have collected at least 59 Flood legends from the aborigines of North America, 46 from Central and South America, 31 from Europe, 17 from the Middle East, 23 from Asia, and 37 from South Sea Islands and Australia. All accounts hold three features in common—a worldwide flood destroyed both man and animals, a vessel of safety was provided, and only a small number of people survived.[19]

The catastrophe of a worldwide flood would explain why immense beds of fossils are found in all sorts of otherwise inexplicable places—they all date from the same time and would have been caught up in the enormous deposition turmoil. Animal fossils and large coal deposits exist in Antarctica, and this is consistent with lush vegetation growing in all parts of Earth in a pleasant setting before the Flood.

A catastrophic global cataclysm explains such huge phenomena as the Grand Canyon, with its uniform layers of sedimentary rock strata and great canyons. According to Creation geologists,

> The strata simply could *not* have remained so nearly uniform and horizontal over such great areas and great periods of time while undergoing such repeated epeirogenic[20] movements. By far the most reasonable way of accounting for them is in terms of relatively rapid deposition out of the sediment-laden water of the Flood. Following the Flood, while the rocks were still comparatively soft and unconsolidated, the great canyons were rapidly scoured out as the waters rushed down from the newly uplifted peneplains[21] to the newly enlarged ocean basins.[22]

(As an aside, these geologists do not believe that the Colorado River slowly eroded the Grand Canyon; rather, they suggest that the Canyon gave rise naturally to the river.)

---

18. Joseph C. Dillow, *The Waters Above: Earth's Pre-Flood Vapor Canopy* (Chicago: The Moody Bible Institute of Chicago, 1981), p. 114.
19. See James A. Strickling, "A Statistical Analysis of Flood Legends" (Terre Haute, IN: Creation Research Society Quarterly), Vol. 9, No. 3, Dec., 1972, pp. 152-155.
20. Concerning formation of continents, oceans, etc.
21. Land reduced to plain level by erosion.
22. John C. Whitcomb Jr., *The World That Perished* (London: Evangelical Press, 1974), p. 74.

When the Mount St. Helens (USA) volcano erupted in 1980, it exploded with a force of about 400 million tons of TNT in one day. In the aftermath, 600 feet of strata sequences formed in just a matter of months. Later, on March 19, 1982, a giant mudflow broke through a blocked canyon on the north fork of the Toutle River and in one afternoon formed a new canyon system over 100 feet deep, with features similar to those of the Grand Canyon, but about 1/40th scale size.[23] One has only to reflect upon the great damage that can be caused by tidal waves to see that a flood of worldwide proportions could result in formations such as the Grand Canyon.

In reference to the Flood, *Genesis* records that "all the fountains of the great deep were broken up, and the flood gates of heaven were opened: And the rain fell upon the earth forty days and forty nights." (*Gen.* 7:11-12). For rain to fall continuously upon the Earth for such a long time, it would be necessary for a vast quantity of underground water to be thrown up into the atmosphere. Similarly, the idea that it did not rain before the Flood also has a reasonable basis in *Genesis*: "for the Lord God had not rained upon the earth . . . but a spring rose out of the earth, watering all the surface of the earth." (*Gen.* 2:5-6).

As will be discussed, the model of Special Creation also provides a possible explanation of the so-called Ice Age, and of how animals could have travelled from distant parts of Earth to reach the Ark before the Flood. If the oceans contained less water before the Flood than after it, it is likely that all the continents would have been joined by dry land. (There is no reason to assume that pre-Flood continents had any close resemblance to those of today.) Therefore, it would have been possible for all the creatures that God intended to board the Ark to have traveled across land, all the way to the Ark.

It seems unrealistic to hold that all the creatures who came aboard the Ark traveled spontaneously from all around the world, but it can reasonably be deduced from Scripture (*Gen.* 6:19-7:9) that God intervened in the laws of nature and led the creatures to the Ark. (As recorded in *Gen.* 8:17, these creatures would repopulate the Earth after the Flood.) There should be no difficulty involved for Christians to believe that God could do this, because He is omnipotent and could have suspended the creatures' normal behavior patterns.

Since God created the Universe anyway, He could do any number of works that so pleased Him. Bearing in mind that many of the various

---

23. See Steve Austin, video *Mount St. Helens: Explosive Evidence for Catastrophe!* (N. Santee, CA: Institute for Creation Research).

species in a particular "kind" are thought by creationists and some evolutionists to have descended from one original pair (e.g., dogs/dingoes/wolves/coyotes), only a relatively small number would have had to be accommodated on the Ark.

Huge giraffes need not have been taken on board—young ones like those seen in zoos today would have sufficed. Similarly, dinosaurs could have been housed there, the large animals as young creatures. And of the many species of insects—ants, etc.,—most would have been able to survive either in the Flood waters or upon vast quantities of floating rafts of debris from trees and other vegetation. (Such debris rafts occurred after the 1980 Mt. St. Helens explosion.)

Regarding the problem of survival for both saltwater and freshwater fish during the period of a global flood, it is estimated that it is easier for fish to adjust themselves to excess water than to excess salt. Since all fish are thought to be adaptable to a certain range of salinities, it is not unreasonable to suggest that enough individuals of each kind of fishes would survive the gradual mixing of waters and gradual change in salinities during and after the Flood.

The Bible informs us that "God remembered Noe, and all the living creatures, and all the cattle which were with him in the ark." (*Gen.* 8:1). This passage supports the possibility that air-breathing creatures who were accommodated on the Ark may have been kept by God in a docile state of aestivation, rather like hibernation. Otherwise, the task of controlling the creatures and feeding them would have been quite impracticable for Noah and his family, and the noise and stench of the animals would have soon become unbearable. Unless these creatures were in a docile state, how would the human inhabitants get any sleep amidst the turmoil?[24] This idea should not present any problems for Christians, since it is known that God intervened each night to sustain the Israelites with *manna* at one stage of their journeying.[25]

Scripture also indicates that order prevailed aboard the Ark. Note the hint of order even during the entry of the passengers into the Ark: "They and every beast according to its kind, and all the cattle in their kind, and every thing that moveth upon the earth according to its kind, and every fowl according to its kind, all birds, and all that fly, went in to Noah into the ark, two and two of all flesh." (*Gen.* 7:14-15). The scene depicted is that of tranquility and order.

---

24. In a recently published book, *Noah's Ark: A Feasibility Study*, John Woodmorappe argues that hibernation would not have been necessary.
25. See the reference to *manna* in *Exodus* 16.

The mysterious extinction of the dinosaurs is often attributed to the impact of huge meteorites that crashed into Earth and adversely affected the environment, and there may be much truth in this speculation. But their extinction can also be attributed to another cause: they may have been killed in great numbers by the Flood events.

If there were greater atmospheric pressure before the Flood, the lower pressure afterwards would have adversely affected the huge dinosaurs. The few juvenile types of such dinosaurs that survived aboard the Ark were quite conceivably unable to cope with undersized lungs in the changed environment after the Flood. Though smaller dinosaurs may have survived for several centuries after the Flood, as suggested by the widespread existence of dragon legends, the larger types may have succumbed to environmental pressures.

It will be argued later that Earth's atmospheric environment seems likely to have been markedly changed by the Flood events. As the Flood subsided, the reemerging Earth was then subjected for the first time to the full ravages of the Sun, with a very hot equator, freezing cold polar regions, strong winds and cloud systems. The reshaped Earth, with new high mountains, cloud systems and larger and deeper oceans, would have in a short time rapidly frozen over at the polar regions and frozen solid all the animals already entombed there in sediment, muck and ice. This explains the preservation of tens of thousands of frozen creatures found entombed in the Arctic regions of Siberia and Alaska.

With respect to the Ice Age and the means by which creatures could have reached all the continents, a possible explanation can be offered: It is conceivable that, for a certain period after the year-long inundation with water, there would have been extra evaporation from the oceans (due to the higher water temperature, as most of it came from below the ground). In the mid-latitudes, the result would have been more rainfall, but closer to the poles there would have been increased ice and snowfall.

For perhaps several hundreds of years there would have been land bridges connecting virtually all the continents, as the continental shelf areas indicate, while the excess ice melted. In addition, it seems likely that large rafts of trees and foliage debris would also have been afloat for many years after the Flood. During this period animals, birds and human beings could have traveled to all the habitable parts of Earth, until the excess amount of polar ice melted and the oceans stabilized to their present levels, thus isolating various continents.

This process of gradual isolation of continents may have finally stabilized in the time of Phaleg, possibly explaining why "in his days the earth was divided." (*Gen.* 10:25). During the process, deserts would also have come about because of lack of moisture as the Earth stabilized to its present state. Certain inland seas and lakes would have evaporated until little or no water was left, leaving large salt deposits or soil salinity conditions.

The turmoil of a year-long global flood would explain why much of the Earth's surface, including the tops of mountains, is covered in sedimentary rock. It explains why many fossils of fish are found in high mountain areas and why large numbers of fossilized jellyfish, of all things, are found in the inland desert area of Australia.

The sheer numbers of fossils trapped and buried in sediment all around the Earth suggests the action of rapid catastrophic events. (An estimated five million mammoths would be required for the half million tons of tusks buried along the Arctic Coast in Siberia and Alaska.) The idea that the geological column was formed by very slow processes seems implausible. On the contrary, stupendous forces were required, and it *must* have happened fairly quickly:

> It is impossible to account for most of the important geological formations according to uniformitarian principles. These formations include the vast Tibetan Plateau, 750,000 square miles of sedimentary deposits many thousands of feet in thickness and now at an elevation of three miles; the Karoo formation of Africa, which has been estimated by Robert Broom to contain the fossils of 800 billion vertebrate animals; the herring fossil bed in the Miocene shales of California, containing evidence that a billion fish died within a four-square mile area, and the Cumberland Bone Cave of Maryland, containing fossilized remains of dozens of species of mammals, from bats to mastodons, along with the fossils of some reptiles and birds—including animals which now have accommodated to different climates and habitats from the Arctic region to tropical zones. Neither has the uniformitarian concept been sufficient to explain mountain building nor the formation of such vast lava beds as the Columbian Plateau in northwest United States, a lava bed several thousand feet thick covering 200,000 square miles. . . . The fossil record, rather than being a record of transformation, is a record of mass destruction, death, and burial by water and its contained sediments.[26]

---

26. Gish, *Evolution: The Fossils Say No!*, p. 61.

Formation of the Continents

Another aspect which requires comment is the theory of plate tectonics, sometimes referred to as "Continental Drift." The passage of immensely long periods of time is considered essential to the concept.

According to this theory, Earth has a series of huge plates carrying the continents. The theory holds that these huge plates are considered to have broken up from an original super-plate covering the Earth, and that ever since this break-up, the plates have been gradually shifting into new formations. But much uncertainty surrounds the nature of the mind-boggling force required to keep moving them.

Another explanation for the existence of fault-lines on Earth is that they were brought on by the stresses associated with the Flood events. Prior to the Flood, it seems likely that mountains were much lower and oceans were much shallower. But some mountains rose up to great heights, and ocean valleys sank down to great depths, as the Earth's crust sought a new *isostatic* balance following the catastrophic events associated with the Flood.[27]

In contrast to the commonly supposed Continental Drift idea, Walt Brown has proposed an alternative model—the *Hydroplate Theory*.[28] He suggests that when God caused the Flood events to occur, intense pressure under the outer granite crust unleashed a gigantic fracture which went racing around the Earth. This fracture in turn unleashed violent forces of subterranean water up through the widening crack and caused a colossal amount of erosion to spill over the Earth as sediment (about 65% from the granite crust and 35% from the lower basalt crust), quickly engulfing creatures and plants in the process.

Brown envisages that the forces unleashed were so immense that great sections of the Earth's outer plates very likely were forced up, slid sideways and collided, forcing great mountains to be pushed up, as well as great ocean basins to sink down; as a result, there is some instability in the rocks which now comprise the Earth's crust. He points out that the enormous flooded Earth would have experienced successive unimpeded tidal waves impacting upon a vast amount of newly eroded sediments and giving rise to a "sorting" effect in deposition.

Brown suggests that a phenomenon known as *liquefaction* occurred

---

27. See Dr. Andrew Snelling and David Malcolm, "Earth's Unique Topography," *Creation Ex Nihilo* (Australia: Creation Science Foundation Ltd., December, 1987), Vol. 10, No. 1.
28. Walt Brown, "The Fountains of the Great Deep," *In the Beginning* (Phoenix, AZ: Center For Scientific Creation), pp. 73-149.

during the Flood events. After the initial sedimentary deposits had spilled out, the action of each huge tidal wave further compressed the sediments. Water and lighter particles then moved upwards, giving rise to vertical water vents and huge vertical plumes, which sometimes spilled out to form domes, the most immense of which may be Uluru (Ayers Rock) in central Australia. (This huge mound consists of compressed [sharp-edged, thus "fresh"] sand particles which are found in *vertical strata*. It rises 1,140 feet high, and its base perimeter measures over five miles). Brown cites quicksand as an example of liquefaction; lighter particles move up and denser particles move down, giving rise to buoyancy when an object falls into quicksand.[29]

Rapid formation explains why some strata can be found in otherwise strange contour shapes (such as folded mountains, which indicate that sideways compression must have occurred while the sediment was still soft) and why fossils of fishes have been found at mountain-top level. It also explains why earthquakes are a legacy of the Flood.

Brown contends that the main fracture line is actually the Mid-Atlantic Ridge and the Mid-Oceanic Ridge, which today dominate the seascape under the oceans. He further suggests that the general shape of the continental shelves on each side of the Atlantic Ocean fit quite well the shape of the Mid-Atlantic Ridge, having been both eroded away and forced sideways from it when the Flood events took place.

However, since the very nature of research into the events which took place at the time of the Flood is fraught with difficulty, it may be too soon to know whether the hypothesis proposed by Walt Brown is entirely valid. Nevertheless, his detailed proposal provides a neat model for further research.[30]

The Entombed Evidence

The vast amount of evidence buried worldwide cannot be ignored when considering Flood credibility. That the Earth suffered a fairly sudden but immense catastrophe can be observed in the permafrost regions, where hundreds of thousands of various types of animals lie entombed in sediment, volcanic ash and ice—frozen as hard as flint.

In 1901, a mammoth was dislodged from the permafrost near Beresovka in Siberia and was examined by a team of scientists.

---

29. See Liquefaction Theory—Walt Brown, pp. 138-149.
30. For a comprehensive discussion of catastrophic plate tectonics, see *"What About Continental Drift?" The Answers Book* (Australia: Creation Science Foundation Ltd., 1990), pp. 27-41.

Twenty-four pounds of undigested vegetable matter were found in the stomach, and a comprehensive list was compiled of the wide range of shrubs, herbs, meadow grasses, mosses and plants found in the food. The color of the leaves of one plant was still intact, as if freshly picked. Another mammoth unearthed in 1908 was found lying on green grass frozen with the carcass.

The data on the Beresovka mammoth shows the northern polar region underwent a rapid dramatic change of climate and has since stayed that way. The type of food found in the mammoth's stomach cannot grow in that area today; the climate must have been much milder before the change. Further, since the mammoth lacked oil-producing glands in the skin and hair-erector muscles, it could not survive in a very cold environment. The climate must once have been quite temperate.

There seems little doubt that the creature was suddenly overwhelmed in the middle of summer, within a half-hour of eating the food, and instant death occurred, followed by rapid deep-freeze. The sudden death is proved by the unchewed bean pods, still containing the beans, that were found between its teeth; the deep freeze is suggested by the well-preserved state of the stomach contents and the presence of edible meat. Some of the carcass was thawed and fed to dogs in the expedition, who ate it with great relish and with no after-effects.

The preservation of entombed creatures in the permafrost is quite intriguing. If the change of climate had been only gradual, the animals could easily have moved away southwards to a warmer climate. If the ground had already been frozen, the remains of the creatures could not have penetrated the rock-hard earth or would have suffered damage if crushed in a crevasse. If the ground were soft, as burial requires, then the temperature must have been warm, and the remains should have rotted away. This shows that immediately after burial the ground was quickly frozen and has remained so ever since.

The Beresovka mammoth is not an isolated phenomenon. Some idea of the vast extent of burial can be gauged from the very large amount of usable tusks unearthed in northern regions. The fact that much of it is still good enough for ivory turning indicates rapid burial in what was thereafter permanently frozen ground. Normally, animal matter in tusks tends to rot away soon after death.

In *The Waters Above,* Joseph C. Dillow outlines the likely chain of events which led to the entombment of the mammoths and the onset of the Flood. The sheer enormity of the catastrophe is evident:

> As the day of the Deluge approached, Earth was already in growing instability, and in some areas, violent storms with blizzards having temperatures lower than -40°F were present. Other parts of the Earth, however, were relatively peaceful and, except for the growing cloud cover and apparent cooling, sheltered from the violence of other geographical areas. . . . Suddenly, after the northern temperatures had already dipped into the low 40s due to the cloud cover, as the Beresovka mammoth was having his last chilly lunch, a 200-mile-per-hour storm front drove raging blizzards of ice, vegetable matter, mud, rain, and snow across the tundras. . . .
>
> On the opening day of the Flood, the atmospheric instabilities that had been building up for a year or more suddenly precipitated into a global instability, and the "windows of heaven were opened." Simultaneously, the fountains of the deep spewed forth, and the Earth was overwhelmed with rainfall and deluge. Some of the mammoths were caught in raging blizzards at -175°F temperatures and frozen in their tracks. Other "Pleistocene" animals were swamped in a rain of freezing mud and buried in tons of frozen vegetable material along with trees, shrubs, and gravel that was driven in front of the great winds . . . in the northern areas this entire mass of mud, trees, shrubs, and animals became frozen to depths of hundreds of feet into what is called the permafrost today.[31]

Walt Brown's hypothesis may be somewhat different in detail from Dillow's, but nevertheless it exemplifies the general modern research being conducted by scientists into the type of events thought likely to have occurred before, during and after the Flood. Since *Genesis* gives only a partial revelation of the Creation events, there is no need to seek exact information about these events. Nevertheless, it may be possible gradually to discern a clearer picture of what probably took place. In discussing the Flood Phase, Brown envisions the likely events which led to the burial of mammoths and other creatures:

> As the crack raced around the Earth, the ten-mile-thick "roof" of overlying rock opened like a rip in a tightly stretched cloth. The pressure in the subterranean chamber immediately beneath the rupture suddenly dropped to almost atmospheric pressure. Water exploded with great violence out of the ten-mile-deep "slit," which wrapped around the Earth like the seam of a baseball.
>
> All along this globe-circling rupture, a fountain of water jetted supersonically into and above the atmosphere. The water fragmented

---

31. Dillow, *Waters Above*, pp. 414, 415.

THE CONCEPT OF SPECIAL CREATION 75

into an "ocean" of droplets that fell to the Earth great distances away. This produced torrential rains such as the Earth had never experienced—before or after. Some jetting water rose above the atmosphere where the droplets froze. Huge masses of extremely cold, muddy hail fell at certain locations where it buried, suffocated, and froze many animals, including some mammoths.[32]

Large hills of "rock ice," or *yedoma*, still exist where frozen mammoths were found in Siberia, and they are yielding powerful evidence. The rock ice consists of compacted hail, which probably fell to Earth in great quantities after the "fountains of the deep" had burst forth and flung water and fine particles high into the atmosphere, literally suffocating the mammoths and giving rise to instantaneous snap freezing at somewhere around -175°F. This snap freezing *must* have occurred, otherwise their internal organs would have suffered decomposition.

Closely intermingled with the rock ice is *loess*, a fertile soil rich in carbonates but which lacks internal layering. It is also found extensively in formerly glaciated regions, but is thought to have originally formed simultaneously with the rock ice:

> Much of the resulting ice fell in a gigantic "hail storm." Some animals were suddenly buried, suffocated, frozen, and compressed by tons of cold, muddy ice crystals. The mud in this ice prevented it from floating as the flood waters submerged these regions after days and weeks. . . . After the flood waters drained off the continents, the icy graves in warmer climates melted, and their contents decayed. However, many animals, buried in what are now permafrost regions, were preserved.[33]

The Canopy/Firmament Puzzle

The sheer extent of buried tropical vegetation and animals which could only live in temperate conditions shows that Earth *must* once have had a worldwide temperate climate, and a sudden freezing of Polar regions *must* have occurred. What is the explanation?

Joe Brenner, an American pilot with some eight million miles of air travel (mostly in Boeing 747's), has made some most interesting points affecting pre-Flood climate theory. In an unpublished manuscript, *The Pilot's Guide to the Universe,* he points out that

---
32. Brown, p. 88.
33. *Ibid.*, p. 119.

- Unfossilized stumps and trunks of five groves of coastal redwoods, which only grow in a warm climate, have been found as near as 550 miles from the North Pole. Some of these are still rooted to the ground in which they originally grew. Also, the remains of warm water corals, tropical animals and plants, and swamp-dwelling dinosaurs have been found in Polar regions.
- As well as the flash frozen mammoths of Siberia and Alaska, the ground is frozen to a depth of 2,250 feet near Point Barrow.
- So great is the amount of ice in Antarctica that some of it sits on the sea floor 5,000 feet below sea level. At one time, Earth's northern ice cap covered 17 million square miles and was 6,000 feet to 12,000 feet deep. Certainly, along with the rain, an awesome amount of ice must have formed in association with the Deluge.
- To make ice caps of only 6,000 feet in thickness would require an amount of snow 60,000 feet thick, but most snow clouds are less than 40,000 feet high. Also, the action of currents and saline water prevents much ice forming from snow falling on Polar seas.
- The air above Earth is very cold (at 30,000 feet it is about -60° F) Until a height of 87,000 feet is reached, at which point the temperature rises to +84° F. Higher still, in the Thermosphere, the temperature rises greatly, to several thousand degrees F.
- There are incredible numbers of sharp-edged, jagged mountain ranges around the world which are indicative of a young Earth; erosive forces should have worn them down long ago.

Various creationists have suggested that when God created the Earth, He lifted up sufficient water (*Gen.* 1:7) to expose the desired extent of dry land and then suspended it high in the atmosphere.[34] They suggest that God instantly turned this water into vapor (super-heated transparent steam) and established it in a pressure-temperature distribution that would not require miracles to maintain.

It is argued that a greenhouse effect would have ensued, with pleasant temperatures at all parts of the globe, allowing a state of tranquillity on Earth. This in turn would have enabled lush vegetation to grow in all areas. The atmospheric pressure would have been about 2.18 times that of today's atmosphere, thus facilitating giant forms of life to exist. It would also have aided longevity.[35]

---

34. For a comprehensive exposition of the Canopy Theory, see Joseph C. Dillow, *The Waters Above*.
35. Dillow, *The Waters Above*, p. 102.

Brenner points out that the extremely hot air of the Thermosphere could have held a prodigious amount of water in transparent vapor form, thus creating a canopy and giving the entire Earth a greenhouse atmosphere. The weight of this additional water would result in a greater sea level air pressure. Since the lungs of larger dinosaurs were undersized for their mass, this hyperbaric pressure was necessary to supersaturate the dinosaurs' hemoglobin with oxygen, otherwise they could not have grown as large as they did.

Thus, before the advent of the Flood, the Earth very likely would have had an entirely different environment from today, with low level mountains. (Had they not sinned, Adam and Eve would have felt no need to wear clothes out of shame; the pleasant and tranquil environment would not have required the use of clothes.)

Since only gentle winds would have prevailed before the Flood catastrophe, the now-extinct giant flying creatures could readily have coped, whereas today's violent winds would have caused them to crash.

If the idea of a canopy and greenhouse effect seems fantastic, one can point to an atmospheric canopy in existence today on the planet Venus. The temperatures at the poles and equator there are similar. If this canopy theory is valid, the higher atmospheric pressure provides a clue as to why Noah got drunk after he had re-established his family on the Earth after the waters had subsided. Perhaps he was caught off guard. With the canopy gone, the rate of formation of the alcohol in the wine would have speeded up, and Noah would not have anticipated there being more alcohol in the wine.

It must be made clear, however, that the water vapor canopy theory still remains scientific speculation; other creationists are not convinced it even existed and have argued against it. Nevertheless, although not yet understood in detail, there *must* have been some mysterious mechanism in place above Earth which kept a worldwide temperate environment in place.

In view of serious doubts raised against the canopy theory, especially at a time when the integrity of *Genesis* has been assailed by many liberal theologians, it is becoming ever more important to attempt clarification of the meaning of the "firmament" in *Genesis* 1. (The Hebrew *raqia* is variously translated as "firmament," "expanse" or "vault.")

What was, or is, the baffling "firmament"? What were the "waters under" and the "waters above the firmanent"? (*Gen.* 1:7). How is "heaven" or "heavens" defined? Did a canopy of some sort really exist above Earth? As far as this writer is aware, there are no agreed answers to such questions, and the Catholic Church has not given a specific definition. Therefore, any speculation must remain tentative.

Russell Humphreys in *Starlight and Time* (1994) opposes the water vapor canopy theory and proposes instead that the firmament is not atmospheric space above Earth, but rather is interstellar space. He suggests that the "waters above" the firmament now form a boundary around the Universe and the "waters below" the firmament are all the waters under and within the Earth's atmosphere.

On the other hand, Walt Brown in *In the Beginning* (1995) rejects the idea that waters originally created around Earth now form a boundary around the Universe (p. 179). He maintains that the amount of available water would be insufficient to form a substantial surface boundary on the immense distances around the perimeter of the Universe. He argues that, assuming the edge of the Universe is only eight billion light years away, this would result in water molecules being spread out one mile apart around the boundary.

Brown also advances arguments against a purely water vapor or ice water canopy *per se*. But he also advances a novel idea about the firmament. He speculates that it is actually the crust of the Earth itself; the "waters below" the firmament are underground waters below Earth's crust and the "waters above" are those waters above ground. In discussing the Hebrew term, *raqia,* Brown notes that "originally it probably meant something solid or firm that was spread out." (p. 178).

Both Humphreys and Brown are eminent Origins researchers, and both argue that vast distances of the Universe are explicable in six 24-hour creation days. Yet their theories of the firmament could hardly be more different. Other Christians have held different views: St. Augustine thought it applied to upper atmospheric areas.

We know from *Genesis* 1:6-7 that the firmament came into existence when God separated the waters on Earth. Perhaps the firmament did refer to the atmospheric area above Earth. But what explains the worldwide temperate climate? And did the Earth's magnetic field itself somehow facilitate such a continuing climate? Humphreys points out that

> As a younger creationist, one of the great attractions of the canopy model for me was that its logical consequences would provide an explanation of several scientific problems for creationists. The most important of these consequences were: (1) a "greenhouse" effect to make the warm, uniform pre-flood climate indicated by the fossils; and (2) a shielding of cosmic rays to reduce carbon 14 in the pre-flood world, thus explaining "old" radiocarbon dates.
>
> However, we now have very good scientific reasons to think that the

amount of carbon dioxide in the pre-flood atmosphere was many times greater than today. That would produce a strong greenhouse effect, a warm climate, and as a bonus, stimulate plant growth to produce the large amount of plant life we find in the fossils. The additional ordinary carbon in the biosphere would dilute carbon 14, so that the pre-flood $^{14}C/^{12}C$ ratio would be considerably lower, due to that effect alone, thus explaining the "old" post-flood radiocarbon dates. In addition, *we have evidence suggesting that the earth's magnetic field was at least ten times greater before the flood than now. That would enable the geomagnetic field to be a very effective shield for cosmic rays, thus greatly reducing the production of carbon 14, making the pre-flood world a healthier place, and further explaining post-flood radiocarbon dates.* Thus we have alternative scientific explanations for the main things the canopy model was supposed to explain.[36]

The actual mechanism of Earth's magnetism is still debated among scientists, but it is known that the magnetosphere (Earth's magnetic field) traps the majority of radioactive particles from the solar wind which blows against it and thus protects Earth's inhabitants from much harmful radiation; the trapped particles remain in the Van Allen radiation belts. Humphreys has shown elsewhere, from investigation of actual magnetism in rock strata, that Earth's magnetic field reversed itself in the past, and could have done so various times during and soon after the Flood events. His published predictions—that evidence of such rapid reversals would be found—have since been confirmed.

Consider the following tentative scenario: On Day 2, God activated a very strong magnetic field around the Earth and simultaneously separated the waters below from the waters above. The "waters below" were both underground water and sea water at ground surface level. The substantial "waters above" may have existed in the form of very fine vapor moisture, in a fairly transparent state, through which one could see the stars, and perhaps supplemented by evaporation, because "a spring rose out of the earth, watering all the surface of the earth." (*Gen.* 2:6).

Evaporation could also have been facilitated by the fact that river systems were in operation, apparently flowing up from underground. If the higher magnetic field of Earth itself were the primary factor which somehow, mysteriously, impacted upon and controlled atmospheric pressures and thus facilitated a worldwide tranquil greenhouse effect, then such fine moisture high above Earth would not have been func-

---

36. D. Russell Humphreys, *Starlight and Time: Solving the Puzzle of Distant Starlight in a Young Universe* (Colorado Springs, CO: Creation-Life Publishers, Inc., 1994), p. 62. Emphasis added.

tioning as a canopy *per se*. But the mysterious "waters above" referred to in *Genesis* would nevertheless have been in place above Earth.

Most importantly, the "firmament," "expanse" or "vault" would not have been kept in place by continuous Divine intervention, but rather by *secondary causes* set in motion during Creation events and which gave rise to a naturally continuing greenhouse environment.

The effect of the strong magnetic field would have meant that Earth's atmospheric area appeared to early human beings as being rather like a "dome" or "vault," and we know that ancient writers depicted the idea of a dome standing on columns. This tranquil atmospheric area, lacking violent winds, storms and rain, truly would have been an almost invisible "expanse" giving the *appearance* of something quite solid. Joseph Dillow comments on the "firmament":

> Frequently in Hebrew the meaning of a noun can be more clearly understood by an examination of the meaning of the verb form of that noun. The noun *raqia* ("expanse") is derived from the verb *raqa*. This verb is used eleven times in the Masoretic text of the Old Testament. It is used of "spreading out" in the sense of "pounding" (*2 Sam.* 22:43) or "stamping" (*Ezek.* 6:1; 25:6); in fact, this is its basic meaning: "to spread out, stamp, beat out." [37]
>
> Because the verb is sometimes used of hammering out strips of beaten metal (see *Ex.* 39:3, *Num.* 16:39; *Jer.* 10:9), some have postulated that the noun *raqia* "harks back to the conception of the sky as a mirrorlike surface," as in Homer's "brazen heaven" (*Iliad*, 5.504). Because of this possible verbal meaning, Moses has been accused of believing in the primitive, upside down, domelike celestial vault of ancient mythology. Sometimes *Job* 37:18 is cited as proof of this: "Thou perhaps hast made the heavens with him, which are most strong, as if they were of molten brass?" Here the verb form *raqa*, "to spread out," is used and not the noun *raqia* of *Genesis* 1:6-7. Supposedly God is here presented as stamping out the metallic dome of the firmament. But the noun receiving the action in the Job passage is not the firmament (*raqia*) but "sky" (*shahaq*), to which the lexicon ascribes the major meaning of "dust" or "thin cloud"...
>
> The point of similarity between the molten mirror and clouds is obviously not solidarity but susceptibility to spreading or expansion. . . . So it is clear that the verb *raqa*, on which the noun *raqia*, "firmament," is built, simply means to expand or spread out. The result of the action of the verb does not necessarily produce an object that is solid. If metal is being spread out, then a solid would result. If

---

37. Brown, Driver, Briggs, *A Hebrew & English Lexicon of the Old Testament*, p. 955.

THE CONCEPT OF SPECIAL CREATION

clouds or atmosphere are being spread out, then an open expanse would result. Something solid is not in the meaning of the verb.[38]

Prior to the global Flood, stars would have been seen from the Earth through the high, transparent moisture canopy and the night view, free from pollution, would probably have been quite spectacular. In addition, it seems likely that Earth's axis tilt was in existence from the creation of Earth to facilitate the stars being used for signs and seasons.

This tentative scenario allows for birds to fly at all levels of atmospheric regions, "under the sky's vault"—as high as oxygen would allow them to fly. Seen from the ground level, birds flying at 25,000 feet would *appear* to fly across the "firmament of the heavens" but would actually fly under it.

When the Flood events occurred—if this speculation has merit—the "windows of the heavens" were opened when underground water jets were flung high into the atmosphere and fell through the thin moisture band high above Earth's atmosphere (giving rise figuratively to "windows"), and then came crashing down as *geshem* (Hebrew for violent rain involving huge hailstones—see Brown p. 175). Earth's magnetic field began somehow to dissipate in strength—perhaps related to the escape of vast underground waters during the Flood and/or reversals of Earth's magnetic field.

What triggered the actual catastrophic events that produced the Flood? No definite answer is known; there is only conjecture. Brown speculates that perhaps God brought on great instability underground, giving rise to gargantuan explosive forces, but Brenner speculates that meteoric bombardment was brought on by God, and Earth was bombarded with tens of thousands of large meteorites all at once, punching holes in the canopy:

> When meteorites punched their way into Earth's atmosphere at 40,000 to 60,000 mph, they didn't push the air aside but bunched it up ahead of the meteorites. This left large holes in the atmosphere: the meteorites opened "windows in the heavens." When the surrounding water-laden air rushed in to fill these holes, it suddenly expanded, and this expansion was a cooling process that formed immense amounts of extremely cold ice and snow, which also fell from the sky.[39]

---

38. Dillow, *The Waters Above*, p. 44.
39. Joe Brenner, *The Pilot's Guide to the Universe* (Scotsdale, AZ: unpublished manuscript, 1993), Chap. 33.

Brenner envisions that meteoric impact could have cracked Earth's crust, heating rocks to melting point and breaking open large subterranean water chambers. The combined forces of meteoric impact, downward pressure from overlying strata and tremendous heat which vaporized water, giving an enormous increase in pressure, all combined to unleash the underground water and give rise to the great "fountains of the deep" so graphically depicted by Walt Brown. Once the underground chambers were empty, further collapse of strata then occurred.

The source of the meteorites is unknown, but the sheer extent of impact craters all throughout the solar system indicates that an incredibly large planetary object must have exploded, thus giving rise to the meteorites. Some astronomers have even speculated that there once existed a giant planet next out from Mars, in the position now occupied by the asteroid belt of rocks which orbits the Sun between Mars and Jupiter. If this planet did exist and exploded immediately prior to the Flood events, it would explain the rapid bombardment which impacted upon the upper level of Earth's strata and battered all the hard surface planets and moons. Whether it would explain the impact craters upon asteroids themselves is unclear.

The "windows of the heavens" closed up when underground fountains finally subsided and the tranquil atmospheric state ceased to exist. For the first time, differences in the four seasons were now very pronounced and fully experienced on Earth. Cloud systems began to function as a result of the Earth's being exposed to vastly differing degrees of heat and cold and such things as great changes in high mountains and sea currents. The "rainbow" covenant also took place.

The firmament was described in *Genesis* as *heaven* (translated in Knox as *sky*) and everything above the firmament was described in plural: *heavens*. Perhaps the different types of galaxies in the Universe, about which we know very little, can also be defined as "highest heavens." Both the RSV Version "firmament of the heavens" and the Knox "vault of the sky" seem to indicate the existence of interstellar space.

The earthly firmament is described in *Genesis* as *heaven,* and interstellar space can also be described as *heavens*. If Earth's atmospheric space really was/is the firmament, then the "waters above" (whether pre-Flood high in the thermosphere as transparent vapor or post-Flood as high clouds) could qualify as both above and within the firmament.

References to *heaven* and *heavens* are used almost interchangeably in Scripture, and it seems hard to place too much weight on either singular or plural usage. This makes possible the alternative use of "*waters above the heaven*" or "*waters above the heavens*" or even "*waters in the heavens*"—all of which are recorded in Scripture. Perhaps a qualifying standard for *heaven* could be defined as *space*.

The foregoing hypothesis contains much speculation, but nevertheless, one is struck by similarities in the fascinating account given in Scripture. Is it related, somehow through typology, to the Flood events?

> The earth shook and trembled: the foundations of the mountains were troubled and were moved, because he was angry with them. There went up a smoke in his wrath: and a fire flamed from his face: coals were kindled by it. He bowed the heavens, and came down: and darkness *was* under his feet. And he ascended upon the cherubim, and he flew; he flew upon the wings of the winds. And he made darkness his covert, his pavilion round about him: dark waters in the clouds of the air. At the brightness *that was* before him the clouds passed, hail and coals of fire. And the Lord thundered from heaven, and the highest gave his voice: hail and coals of fire. And he sent forth his arrows, and he scattered them: he multiplied lightnings, and troubled them. (*Psalms* 17:8-15).

Mankind after the Flood

When the Flood subsided and Noah had gone forth from the Ark and had burnt offerings on an altar, "And the Lord . . . said: I will no more curse the earth for the sake of man: for the imagination and thought of man's heart are prone to evil from his youth: therefore I will no more destroy every living soul as I have done." (*Gen.* 8:21). God then established the rainbow covenant with Noah that "neither shall there be from henceforth a flood to waste the earth." (*Gen.* 9:11).

If there were no clouds prior to the Flood, rainbows would have only come into existence afterwards (a rainbow is only formed upon water droplets greater than 0.30 millimeters in clouds) and the *Genesis* passage is likely to have a quite literal meaning: "And when I shall cover the sky with clouds, my bow shall appear in the clouds: And I will remember my covenant with you, and with every living soul that beareth flesh." (*Gen.* 9:14-15).

Genesis records that the sons of Noah who went forth from the Ark were Shem, Ham, and Japheth, together with their wives, and the whole Earth was repopulated.

Subsequently, the Earth again became increasingly sinful, and God found it necessary to introduce a number of languages into mankind to confound their arrogance and to scatter them all over the Earth. In the

Genesis passage referring to the Tower of Babel, it is recorded that "the earth was of one tongue, and of the same speech . . . And he [the Lord] said: Behold, it is one people, and all have one tongue: and they have begun to do this, neither will they leave off from their designs, till they accomplish them in deed. Come ye, therefore, let us go down, and there confound their tongue, that they may not understand one another's speech. And so the Lord scattered them from that place into all lands, and they ceased to build the city." (*Gen.* 11:1, 6-8).

This would explain the existence of a dozen or so markedly different language groups or families (e.g., Indo-European, Sino-Asiatic, etc.) which diverged over the subsequent centuries, resulting in the many different languages of today. It would also shed light on the so-called "primitive" tribes as being those who perhaps drifted away from the more learned sections of mankind. These do not owe their origin to evolution from apelike ancestors; rather, they would seem to have degenerated culturally into primitive consciousness. Regarding this explanation for the mentality of the primitive tribes, it is noteworthy that many versions of the Creation and Flood legends are found on all continents. These may be perversions of early mankind's memory of events which actually happened.

The authors of *God's Promise to the Chinese* have pointed out that, for some 4,000 years without interruption, until the overthrow of the Manchu dynasty in 1911 A.D., each Chinese Emperor offered an annual eastern border sacrifice in praise of the "God above." Their study of ancient Chinese pictographs (3,500 years old) incised on tortoise shell plastrons has revealed a remarkably similar Creation-Fall-Redemption account to that given in *Genesis*. The "border ceremony" pointed to a coming Saviour, and a bull was sacrificed to ShangDi, whom the authors contend is none other than Jesus Christ.

The authors believe that this ancient Chinese reverence for ShangDi originated at much the same time as the Tower of Babel event, when God introduced new languages upon Earth. The eastern border location seems to have been very significant, because the pictographs depict Adam and Eve, after the Fall, worshipping God and seeking forgiveness at the east gate of the Garden of Eden.[40] (There is a long tradition within the Catholic Church concerning priests "facing the East" when celebrating the Mass).

The remarkable similarity of the 4,000 year old Chinese account and

---

40. See Ethel R. Nelson, Richard E. Broadberry and Ginger Tong Chock, *God's Promise to the Chinese* (Dunlap, TN: Read Books, 1997).

THE CONCEPT OF SPECIAL CREATION 85

the *Genesis* account supports the belief that *Genesis* was intended by God, the Principal Author of Scripture, to be understood as primarily historical, containing *real* history, rather than as a religious mythology.

Apparently the concept of "the Fall of man" can also be found in all ancient theistic theologies. The common theme is that God created human beings with intelligence and freedom, but man abused his freedom by rejecting God and was then punished. And so the resulting alienation of man from God explains the present human condition— one of suffering, guilt and death.

Scripture informs us that Adam and Eve had other children after Cain and Abel: "And the days of Adam, after he begot Seth, were eight hundred years: and he begot sons and daughters." (*Gen.* 5:4). Thus Adam lived for 930 years, Noah for 950 years (*Gen.* 9:29) and Shem, the first-born son of Noah, 600 years. (*Gen.* 11: 10-11).

Is this data about Patriarchs living for hundreds of years to be accepted in the literal, as-given sense and as reliably revealed to us by the infallible God, who is the Principal Author of Scripture, or are these texts little more than unbelievable mythology, now rejected by modern man?

Origins researchers believe it is quite possible and reasonable to believe that early human beings lived for such a long time. The hundreds of years attributed to some personalities in *Genesis* may seem fantastic to modern man, but why should a literal, as-given understanding be rejected out of hand? After all, it had not been intended by God that human beings would die, and so the aging process probably would not have taken place if Adam had chosen obedience.

Apart from accident or murder, death would primarily have been caused by degenerative conditions due to aging, and aging would have required a long time to take effect in order to reduce the average life-span down to that with which we are now familiar. Man's pre-Flood longevity could also have been influenced by the environment, as people were probably living in conditions optimum for health before the Flood occurred.

Since Adam and Eve were created in a state of high perfection, with the full complement of genetic material, this would have allowed substantial variety to arise quickly among human beings, who were free from inherited defects. Thus it would have taken a long time for the gene pool within mankind to become contaminated enough to prohibit intermarriage between close relatives.

Although incest was later forbidden by God at the time of Moses (*Leviticus* 18), procreation must have occurred, by definition, between the earliest human beings. Marriage between closely related human beings could not have been regarded as sinful by God at that time.

To hold otherwise is to assume there must have been other human beings who did not descend from Adam and Eve. Not only does this assumption of *polygenism* beg the question of where the others came from (and it does not address the problems in genetics if they are supposed to have evolved), but for Christians this position cannot be held because it places doubt upon the revealed truth about Original Sin. (It will be argued in Chapter 13 that polygenism is really *impermissible*, because it clashes with Catholic doctrine.)

A fascinating argument to consider, from within Scripture itself, is the possibility that Shem and the priest-king Melchisedech were in fact the same person. *Shem is recorded in Scripture as having lived for about 600 years.* Scott Hahn argues strongly that "Melchisedech," king of Salem (known afterwards, most significantly, as Jerusalem—later the scene of Christ's Crucifixion), was actually a priestly title rather than a personal name.[41] Since Melchisedech is not mentioned until some ten generations after the time of Shem (*Gen.* 14:18), this indicates—if the very detailed argument assembled by Hahn is valid—that the 600 years attributed to Shem (*Gen.* 11:10) were indeed factual.

(Shem was, of course, aboard the Ark during the Flood events, so the truth of this aspect of the Origins controversy is important also to the discussion about the historical reality of the global Flood of Noah.)

Genesis 14 and *Hebrews* 7 record that Abraham defeated four kings (who in turn had defeated five kings), but then he deferred to the king of Salem and paid tithes to Melchisedech, *who brought out bread and wine and blessed Abraham* (significantly, the first blessing recorded since Shem was blessed by Noah, some five hundred years earlier). According to biblical chronologies, Abraham was about 150 years old when Shem died.

Hebrews 7 also records that the mysterious *man* Melchisedech "who first indeed, by interpretation, is king of justice, and then also king of Salem, that is, king of peace: Without father, without mother, without geneology having neither beginning of days nor end of life, but likened unto the Son of God, continues a priest forever." (*Heb.* 7:2-3).

Since Melchisedech has been mentioned in the Latin rite of Mass for

---

41. Scott Hahn, Tapes 3 and 4 of audiotape series on *St. Paul's Letter to the Hebrews* (West Covina, CA: St Joseph Communications). See also Fr. James L. Meagher D.D., *How Christ Said the First Mass* (Rockford, IL: TAN , 1984, originally published 1906).

many centuries, one would think that he must have been an outstanding person. Many scholars down the years have held that Melchisedech was in fact Jesus Christ, appearing ahead of time. A brief appearance ahead of time is one thing, but reigning on Earth for years as king of Salem, hundreds of years before being born to Mary, seems out of the question. In addition, the man known as Melchisedech *must* have had human parents and also some form of genealogy. Therefore, the detailed case assembled by Hahn has coherence.

Understood as a priestly title, "Melchisedech" of course had no human parents or genealogy and foreshadowed the institution of the Catholic priesthood, ultimately to be brought into reality by Jesus Christ. In contrast to the serious limitations which existed in the Levitical priesthood (a man could only be a priest between the ages of 30 and 50 and then only after passing stringent ancestry regulations), the Catholic priesthood would incorporate the essence of personal sacrifice instituted by the perfect Redeemer. In the light of this true priesthood later instituted, the blessing bestowed by Melchisedech upon Abraham was of no small significance.

If a strong case exists from within Scripture itself that one of the Patriarchs lived for 600 years, why not accept the biblical report that the other Patriarchs also lived for hundreds of years? It is not a question of being overly literalistic, but of granting credibility to the true historicity of *Genesis*. In fact, the real problem for modern biology is not whether such long life is possible, but why multi-celled creatures age and die in the way they do. Our cells seem to carry genetic "switches," which make them stop replicating after a time, so that organs can no longer regain or rejuvenate themselves. Genes for longevity can be bred into and out of fruit flies. After the Flood, some such genes may have been rapidly lost due to "drift" in such a small human population. In any event, within only 10 generations—from the Flood of Noah (who died at 950 years) to Abraham (who died at 175 years)—the lifespan of man had declined, for whatever reason or reasons, to that approaching the life of man in our own times.

The Effects of the Fall

The Fall most certainly brought death to mankind, and it also seems beyond doubt that it profoundly affected the whole of Creation. As St. Paul teaches in his letter to the Romans, "the creature also itself shall be delivered from the servitude of corruption. For we know that every creature groaneth and travaileth in pain, even till now." (*Rom.* 8:21-22).

Fr. Valentine Long wrote about the effects set in motion by the sin of Adam, and how Satan instigated it:

> We live in a damaged world. It had been all good at creation, but "sin came into the world through one man and death through sin," as St. Paul explains, and evil has ever since made its presence felt. Adam in disobeying God turned the world disobedient to himself. He lost his unsullied dominion over it. The harmony between the elements and Adam, between the lower forms of life and Adam, suffered a break, and remains disrupted, so that mankind thereafter has lived under the threat of disease and in danger of the beast, the fangs of the serpent, the sting of the insect, and indeed has had to maintain a livelihood against great odds. The mishaps of drought and floods and quakes and storms and raging fires, and the very obstinacy of the soil, render it difficult to make a living on an Earth no longer paradise. The Bible early records God's curse on the world. And the Bible, as God's word, could not leave the devil out of the account. He instigated the tragedy. With him, contrary to the fuzzy Pelagian notion, the whole divine story of Original Sin opens. The material world, still a magnificent symphony of wonders acclaiming its Creator, owes whatever disorders it has known, and the human race may attribute its own consequent ills, not to Adam alone, nor Eve, but to Satan more than either.[42]

Certain passages in *Genesis* where God speaks to Adam suggest that both the Earth and man's relationship with the animal world were adversely affected: "cursed is the earth in thy work; with labour and toil shalt thou eat thereof all the days of thy life" (*Gen.* 3:17), "And let the fear and dread of you be upon all the beasts of the earth, and upon all the fowls of the air, and all that move upon the earth." (*Gen.* 9:2). *The Catechism of the Council of Trent*, citing *Gen.* 3:17, recognized that great disorder resulted from the disobedience of Adam:

> Our condition, therefore, is entirely different from what his and that of his posterity would have been, had Adam listened to the voice of God. All things have been thrown into disorder, and have changed sadly for the worse.... The dreadful sentence pronounced against us in the beginning remains.[43]

---

42. Fr. Valentine Long O.F.M., *Upon This Rock* (Chicago: Franciscan Herald Press, 1982), p. 28.
43. John A. McHugh, O.P. and Charles J. Callan, O.P., *The Catechism of the Council of Trent* (Rockford, Illinois: TAN Books & Publishers, Inc. 1982).

Adam's body was naturally subject to death, but he was given the preternatural gift of immortality. Had he chosen to obey God, the fate of human beings to die would have remained over-ridden by grace. And quite possibly the instinctive *behavior* of wild creatures would have been over-ridden, diseases suppressed, and perhaps even obnoxious plants would have remained harmless.

By way of analogy, it is as if the Creator set up two computer programs within creation. The choice of which program would be implemented rested upon the decision made by Adam. Obey God, and harmony will prevail thereafter; disobey God, and chaos will be unleashed as the unavoidable consequence of disobedience.

One can only speculate as to how things would have been in God's plan if Adam had chosen obedience. Perhaps the predatory nature of wild beasts and reptiles would have been over-ridden by God's grace, so they and human beings could have lived happily together as vegetarians. The size of the animal does not alter the argument. Dinosaurs with six-inch teeth embedded only two inches into the gums seem designed to be vegetarians, and the large teeth of other beasts can be quite useful for ripping plants (e.g., the teeth of the gorilla are designed for vegetarian use).

Ample food would have been available from the lush vegetation. There would have been no need for animals to kill each other for survival. Despite the fact that some animals are clearly designed as predators, man in this situation would have lived in harmony with the animals, and the animals now considered as wild would have lived in harmony with each other. (The death of plants is not a problem. No bloodshed is involved, and Scripture indicates that God intended plants to be eaten.)

The reproduction rate in the animal world would very likely have functioned in perfect harmony. In some mysterious way, not only animals but birds, reptiles, insects and even sea creatures would very likely have lived in peace. Fishes generally may have fed on plant organisms such as plankton, and perhaps an abundance of small sea herbs would have satisfied the appetites of otherwise predatory sea creatures, which thus did not feel the need to attack other sea creatures or human beings.

Is it possible that the intrinsic *nature* of wild beasts, snakes, vicious sea creatures, etc. was not changed by the Fall, but rather that their *behavior* was affected adversely, in consequence of the disobedience of Adam? Today, the highly complex chemicals in snakes' venom can attack the victim's central nervous system to arrest breathing, but perhaps snakes were not intended by God to behave primarily this way.

Just as mankind was wounded "in his natural powers" by the Fall, so also were the now-ferocious wild animals and sea creatures wounded too in the natural powers proper to them, and their behavior thus disordered?

These speculative comments are not immaterial to the Origins debate, for the evolutionary idea of *red in tooth and claw* does not sit easily with the notion of man living in an earthly Paradise. If Original Sin had not occurred, would human beings have lived in a state of paradise while all around them other creatures were killing each other, with vultures preying on the spoils and disease rampant? In an earthly Paradise, how could human beings have gone swimming among Great White sharks, piranhas or huge crocodiles, unless a state of tranquillity prevailed? Even if such creatures ignored mankind, how could human beings tranquilly swim in sea water which was being bloodied by a great shark devouring seals in a frenzied attack?

This discussion of the Fall highlights a conceptual problem in Theistic Evolution and Progressive Creation. (See Chapters 12 and 16). Both hold that mankind was created much later than other creatures, and both challenge the Scriptural message that everything was created good and only degenerated because of and *after* Adam's sin.

We know that God can somehow use suffering and evil from which to bring about good. But it seems paradoxical that Jesus Christ who, on hearing of the distress caused by the death of Lazarus, "groaned in the spirit, and troubled himself . . . And . . . wept" (*John* 11:33, 35) could, as the Second Person of the Divine Trinity, have intentionally devised *survival of the fittest* in which violence, bloodshed and death would have been the norm whether the Fall occurred or not.

Any speculation about the earthly Paradise is obviously fraught with problems, and we are hampered by the constraints of seeing things through the eyes of fallen man. We do not even know the length of time that Adam and Eve lived in the Garden of Eden before their disobedience, or whether any deaths of animals took place there.

Regarding the death of animals, since they do not possess a rational soul and thus are not necessarily destined for immortality, it seems valid to allow the possibility that God did intend for them to die at some point of time in the earthly Paradise. But what would have been the process of their death? In the world of today, animals will flee from danger and avoid death—but in an earthly Paradise would this also be applicable? It is a mystery which awaits explanation.

Another aspect to be considered here is that of entropy. (Entropy refers to the increasing loss of useable energy and the tendency for

THE CONCEPT OF SPECIAL CREATION   91

things to go to disorder and to decay.) Entropy would still have prevailed, even if Adam had chosen to obey God. Some examples of this can be given: The very existence of the Sun would mean that its useable energy would have been decreasing slowly. Simply walking along Earth's surface would involve some friction and also a slow decrease in useable energy. As plants shed their fruit and died, they would decompose and return to the soil in the form of organic matter.

If the process of decay in matter was taking place, would disease also be involved, or rather, would the net tendency toward maximum entropy have been over-ridden by the powers of the good Angels?

Fr. Long records how most of the Church Fathers and the ancient exegetes of the Torah have down the years held that the good Angels are very active in the natural world and actually operate the Universe. (Perhaps they were also aboard the Ark, taking care of things for Noah.) It seems that they have been delegated power from God to supervise the laws of nature and to interrupt them if necessary:

> Their power over nature has an impressive record. An angel kept a shipwrecked St. Paul from drowning by not letting the turbulent waters sink the boat. An angel prevented the fire from so much as touching the three youths in the raging furnace. An angel restrained the lions in the den from harming Daniel. The angel, who would roll back the boulder from Christ's sepulchre, announced his flight to earth by having it quake. There seemed to be nothing of physics, or of chemistry, or of anatomy, that the angels of Scripture did not know. An angel, by just touching Jacob's thigh, threw it out of joint. One angel after another, sometimes more than one, even [a] "company of destroying angels" could bend the laws of nature to their purpose of wreaking God's vengeance on the guilty. Thus [all] of a sudden in Israel, from the northernmost city of Dan to the southern extremity of Canaan, seventy thousand died of pestilence and when the Lord "said to the angel who was working destruction among the people, 'it is enough, stay your hand,'" the angel did. And the pestilence was gone.[44]

The incident of Daniel in the lion's den is particularly relevant to this discussion about the effects of the Fall. "My God hath sent his angel, and hath shut up the mouths of the lions, and they have not hurt me." (*Dan.* 6:22).

The fact that wild beasts have sometimes been restrained by some

---

44. Long, *Upon this Rock*, p. 158.

unseen force from attacking martyrs has been recorded by Warren Carroll. In describing a savage persecution of Christians which was carried out in the year 177 A.D. at Lyons, he notes that

> It was not simply a matter of throwing grown men to wild beasts in the arena. A woman, Blandina, was tortured until "her whole body was broken and opened" . . . and when at last Blandina was offered to the beasts, she was hung on a stake for them like a haunch of meat—but not one of the savage, half-starved animals would touch her. As often happened in these hellish scenes, the beasts felt the presence and power and goodness of their Maker in their victims, when men would not.[45]

The Premise of Special Creation

The foregoing account shows that Special Creation offers a coherent basis for understanding the earliest events and for understanding how mankind came to be in a state of confusion and distress. It also takes note of the unbounded love and compassion that is encompassed by God. One could argue that this account contains much that is convenient speculation which suits those who want to believe in Special Creation, but there are quite different basic premises involved in the creationist or evolutionist viewpoints, and these must be understood.

The Evolution view postulates that biological Evolution must have occurred in the past, but the rationale for it must constantly keep changing ground as the theory encounters more and more difficulties—and meanwhile, the elusive mechanism of Evolution still remains missing.

The Creation view accepts, on the basis of faith in the revealed God, that Scripture must be free from error and that empirical science will never discover any data which can conclusively contradict Scripture; in addition, finite man can also know something about the infinite, can deduce the existence of an unseen Designer who possesses awesome intelligence, and can discern absolute principles.

The evolutionist viewpoint offers only a Naturalistic explanation of the material Universe, whereas the creationist view goes further and acknowledges the existence of an unseen Designer and a supernatural dimension of existence. Since the Creation model alone can explain such things as the existence of coded information, order in design and

---

45. Warren H. Carroll, *A History of Christendom, Vol. 1: The Founding of Christendom* (Front Royal, VA: Christendom College Press, 1985), p. 463.

the laws of nature, it can fairly lay claim to a more comprehensive concept of science than that of the Evolution model.

—PART II—

THE DISCOVERIES OF SCIENCE

—Chapter 5—

THE DISCOVERIES OF SCIENCE

Modern theories of Evolution, as promoted by Naturalistic scientists, rely on the simultaneous occurrence of such conditions as random beneficial mutations and natural selection, without the involvement of a Creator. Accordingly, there can be no question of any master plan being involved in the proposed process of Evolution. On the other hand, many Christians also believe in "Evolution," while still believing in God.

If the model of Special Creation is a reasonably accurate explanation of what God actually did when creating the Universe, then we should expect to find that the discoveries made in modern science would tend to verify this explanation and to confound the predictions associated with evolutionary models.

Despite the religious beliefs of individuals, a crucial consideration must be addressed: What actual scientific discoveries have been made concerning matters relating to Origins? Where does Evolution now stand in the spectrum of scientific truth?

A vast array of data has now been analyzed and documented across a range of disciplines. The weight of arguments points strongly toward design and Creation, rather than Evolution, as being the most likely explanation of the reality of the Universe. In addition to the scientific material already presented, the following chapters will consider further specific aspects. Nevertheless, they represent only a part of the whole story—no attempt has been made to compile a complete catalog. Discoveries are still emerging, and the explanation of how the Universe was created is becoming more refined as Origins researchers continue to develop the overall picture.

The case for Creation is based on logical inference from scientific observations. When the same standards are applied to Evolution Theories, it becomes clear that Evolution is devoid of confirming evidence and is destined to remain a series of conflicting hypotheses plagued with profound conceptual weaknesses.

Empirical science alone cannot answer ultimate questions of existtence, but scientific inquiry about Origins and about the Universe leads inevitably to consideration of metaphysical concepts and theological reality. The material presented in Part II of this book thus has to be seen in light of arguments explored in Part III.

—Chapter 6—

THE FOSSIL RECORD

Uniformity or Catastrophe?

As noted earlier, Evolution has usually been linked to the concept of *uniformitarianism*, the idea that physical features of the Earth and the Universe were formed very slowly by natural forces.

According to Evolution Theory, the path leading up to man can be traced as follows: Invertebrates evolved into fish, fish into amphibians, amphibians into reptiles, reptiles into mammals, some of the mammals into primates, primitive primates into apes, and a common progenitor of ape and man evolved into human beings.

If Evolution did occur, the fossil record should provide ample evidence of transitional forms (i.e., intermediate stages) of different creatures. This would demonstrate how the various phyla, classes, families, orders and the like have evolved through many stages. Darwin himself noted that transitional forms are crucial to Evolution Theory:

> By the theory of natural selection all living species have been connected with the parent-species of each genus, by differences not greater than we see between the varieties of the same species at the present day; and these parent-species, now generally extinct, have in their turn similarly connected with more ancient species; and so on backwards, always converging to the common ancestor of each great class. So that *the number of intermediate and transitional links, between all living and extinct species, must have been inconceivably great. But assuredly, if this theory be true, such have lived upon this Earth.*[1]

In view of the multitude of fossils now discovered, and the completeness of the fossil record, there should no longer be problems in

---

1. Darwin, *The Origin of Species,* p. 293. Emphasis added.

finding ample buried evidence showing an evolutionary progression from simple forms of life to more complex. If there is a problem, it should be one of identifying the basic "kinds" of creatures. The model of Special Creation, however, supports the idea that the features of the original Earth were formed rapidly by God, and later were subjected to great changes during the Flood of Noah. Thus there should be no trace of true transitional forms found among the fossilized remains smothered within sedimentary rocks.

The Buried Evidence

After 150 years of excavations, millions of fossils have now been discovered, an ample quantity to provide a thorough analysis of the fossil record. The buried evidence points against the evolutionary explanation. Three-quarters of the Earth's land area has a colossal amount of sedimentary (i. e., water-borne) rock strata covering it, ranging in depth down to an enormous 10,000 meters deep. Within the Cambrian strata are found most of the fossils—known as the "Cambrian explosion."

At some stage in the past this colossal amount of sediment was eroded and dispersed all over Earth. But how could all this have come about in such vast quantities through slow uniformitarian processes, or even by intermittent localized catastrophes? The forces involved must truly have been immense.

As Gary Parker points out,[2] some geological formations are spread out over vast areas of a whole continent. He cites as examples the Morrison Formation, which covers much of the mountainous western United States, and the St. Peter's Sandstone, an expanse of glass sand that stretches from Canada to Texas and from the Rockies to the Appalachians. (St. Peter's Sandstone is about 500,000 square miles of mostly pure quartz.) While sediment does build up slowly at the mouth of rivers, such as the Mississippi Delta, slow build-up could not produce such massive deposits, such broadly consistent sedimentary and paleontological features, as seen in the Morrison and St. Peter's formations.

Many fossilized trees have been found around the world, often standing vertically through various strata deposits, generally showing evidence of having been uprooted and washed into position. The trees *must* have been buried rapidly to avoid decay and provide the mineral

---

2. Morris & Parker, *Creation Science?*, p. 138.

environment necessary for them to be uniformly fossilized. The strata *must* have formed around them in less time than it takes for trees to rot away—not over eons of time.

In Arizona, the Petrified Forest National Park covers some 50,000 acres and contains six separate "forests" of petrified coniferous tree trunks. Some are almost two meters in diameter and thirty meters in length. Examples of fossilized trees on such a massive scale as this run counter to the slow layer-by-layer deposition envisaged by Uniformitarian Theory.

The vast coal and oil deposits found around the world have formed from previously living matter; the pressure of sedimentary strata apparently facilitated the conversion of plant life into coal, and marine organisms into oil. (Laboratory evidence suggests that a short-term, moderate heating event, in the absence of pressure, will suffice to form coal from wood, if acidic clay is present as a catalyst.)

In the Latrobe Valley region of Victoria, Australia, there exist large brown coal deposits which rest on a clearly defined clay base, which is devoid of roots. Many large tree trunks have been found lying through the coal strata, and there are clearly defined layers of pollen, one foot deep, residing between various layers of coal, denoting a sorting process during water-borne deposition. Coal is supposed to be formed from swamps and was thought to have slowly accumulated as peat and then very slowly changed into coal. Yet the trees in this Australian coal are pine trees, of varieties known today which do not tolerate swampy conditions.

But what of the existence of tree fossils, with trunks extending vertically through many layers of coal? The swamp idea does not explain satisfactorily the existence of these fossils. The type of plants involved in the coal (especially layers of tree bark), the state of their preservation and the texture of coal layers indicate that another explanation is required. Rather than stagnation over vast amounts of time (which would mean that the tree trunks should have rotted away before the coal layers could accumulate), the state of deposition indicates the action of rushing water and fairly rapid deposition.

The geological column is, of course, a theoretical model. Nowhere on Earth is the complete sequence found. Even the walls of the Grand Canyon include only five of the twelve major systems (one, five, six and seven, with small portions here and there of the fourth system, the Devonian). Nevertheless, there is a tendency for fossils to be found in groups in a certain vertical order.

It seems highly likely that the strata were formed rapidly. The authors of *Scientific Creationism* suggest the following sequence of formation during the Flood events:³

- Each stratum must have formed rapidly, since a set of hydraulic factors are involved that cannot remain constant very long.
- Each succeeding stratum in a formation must have followed rapidly after its preceding stratum, since its surface irregularities have not been truncated by erosion.
- Therefore the entire formation must have been formed continuously and rapidly. This is further confirmed by the fact that its rock type required rapid formation, and its fossil contents required rapid and permanent burial.
- Although the formation may be capped by unconformity [i.e., physical break], there is no world-wide unconformity, so that if it is traced out laterally far enough, it will eventually grade imperceptibly into another formation, which therefore succeeds it continuously and rapidly without a time break at that point.
- The same reasoning will show that the strata of the second formation were also formed rapidly and continuously, and so on to a third formation somewhere succeeding that one.
- Thus, stratum-by-stratum and formation-by-formation, it is possible to proceed through the entire geological column, proving the whole column to have been formed rapidly and continuously.
- The merging of one formation into the next is further indicated by the well-recognized fact that there is rarely ever a clear physical boundary between formations. More commonly, the rock types tend to merge and mingle with each other over a zone of considerable thickness.⁴

Indeed, fossil evidence is consistent with sudden burial; huge death pits containing vast numbers of creatures have been discovered worldwide. The fossil record is truly a vast museum of death. As noted in Chapter 4, the far northern hemisphere regions provide ample evidence

---

3. Experiments supervised by French scientist Guy Berthault have further refined the likely processes involved in the behavior of sedimentary particles under the action of fast running water.
4. *Scientific Creationism,* edited by Henry M. Morris (San Diego: Creation-Life Publishers, 1974), pp. 115, 116.

of the immense *catastrophic* extinction of hundreds of thousands of animals, frozen in mid-motion, overcome by violent flood and wind.

The very presence of the frozen muck demonstrates a sudden and permanent temperature drop. The muck is full of plant and animal remains to depths of several thousand feet. In numerous cases, the remains are in relatively fresh condition. Hippopotamuses, sabre-tooth tigers, elephants and other low-latitude animals are found buried in the tundras, freshly preserved. If the muck were frozen at the time these animals lived, the animals could never have been thrust into the rock-hard frozen ground.

After all, fossils do not form when plants and animals simply die and rot away on the surface of the ground or on the bottom of the sea. To have any chance of being preserved as a fossil, a plant or creature must be buried rapidly under a heavy load of sediment, which must also harden relatively rapidly to exclude oxygen and bacteria. Otherwise, scavengers or forces of erosion and decay would destroy the specimen. How long would jellyfish survive?

The colossal amount of eroded sediment must have been dispersed fairly rapidly all over Earth and not over eons of time. The field evidence is fully consistent with the idea of an enormous Flood of worldwide dimensions, and Evolution Theory is confounded by the evidence.

The Links Are Missing

If Evolution really did occur down through the ages, an ample number of transitional creatures should by now have been found among the immense number of fossils now unearthed. But, back in the mid-19th century, Charles Darwin himself realized that there was already trouble in finding transitional forms in the fossils, and he sought to find reasons which might yet validate his naturalistic theory. However, by Darwin's own candid recognition of the gravest objection to his theory, Evolution Theory stands exposed as completely mistaken:

> As natural selection acts solely by the preservation of profitable modifications, each new form will tend in a fully-stocked country to take the place of, and finally to exterminate, its own less improved parent or other less-favoured forms with which it comes into competition. Thus extinction and natural selection will, as we have seen, go hand in hand. Hence, if we look at each species as descended from some other unknown form, both the parent and all the transitional varieties will generally have been exterminated by the very process of

formation and perfection of the new form. But, *as by this theory innumerable transitional forms must have existed, why do we not find them embedded in countless numbers in the crust of the earth?* I believe the answer mainly lies in the record being incomparably less perfect than is generally supposed; the imperfection of the record being chiefly due to organic beings not inhabiting profound depths of the sea, and to their remains being embedded and preserved to a future age only in masses of sediment sufficiently thick and extensive to withstand an enormous amount of future degradation; and such fossiliferous masses can be accumulated only where much sediment is deposited on the shallow bed of the sea, whilst it slowly subsides. These contingencies will concur only rarely, and after enormously long intervals. Whilst the bed of the sea is stationary or is rising, or when very little sediment is being deposited, there will be blanks in our geological history. The crust of the Earth is a vast museum; but the natural collections have been made only at intervals of time immensely remote.[5]

The numbers of intermediate varieties which have formerly existed on the Earth [must] be truly enormous. *Why then is not every geological formation and every stratum full of such intermediate links? Geology assuredly does not reveal any such finely graduated organic chain; and this, perhaps, is the most obvious and gravest objection which can be urged against my theory.* The explanation lies, I believe, in the extreme imperfection of the geological record.[6]

No longer can the possible imperfection of the geological record be cited in excuse of Evolution Theory. Thousands of fossils unknown to Darwin have since been discovered and about a quarter of a million fossil *species* have now been cataloged,[7] but the gaps have still not been filled. In addition, Darwin's explanation is much too convenient. Having been assured that the numbers of transitional forms must have been *"inconceivably great,"* we are asked to accept that few will be found in the fossil record. All this supposedly over many millions of years.

Since millions of fossils have now been found, it is reasonable to expect that large numbers of unmistakably clear transitional forms should also have been unearthed. In theory, one should expect to find whole chains of them in series between one major kind and another—but then, how would anyone know which ones were the major kinds?

---

5. Darwin, *The Origin of Species*, p. 206.
6. *Ibid.*, p. 292. Emphasis added.
7. Davis *et. al., Of Pandas and People*, p. 94.

Vast numbers of fossils identical to creatures alive today have been found, but not a trace of transitional forms. The fossil record is devoid of "missing links" grading up from simple to more complex creatures. Even those thought to be among the oldest are very complex in composition, with no ancestors leading up to them. All the major phyla appear suddenly, without grading up from other, ancestral, phyla.

This absence of transitional forms is a major problem, and some evolutionary scientists are quite candid in their comments that the fossil record provides little support for Evolution:

> This regular absence of transitional forms is not confined to mammals, but is an almost universal phenomenon, as has long been noted by paleontologists. It is true of almost all orders of all classes of animals, both vertebrate and invertebrate. A fortiori, it is also true of the classes, themselves, and of the major animal phyla, and it is apparently also true of analogous categories of plants.[8]

> Darwin . . . was embarrassed by the fossil record . . . we are now about 120 years after Darwin and the knowledge of the fossil record has been greatly expanded. We now have a quarter of a million fossil species, but the situation hasn't changed . . . We have even fewer examples of evolutionary transition than we had in Darwin's time.[9]

Even the widely published genealogical tree of the evolution of horses has been rejected by some evolutionary scientists, on the ground that it is a contrived argument.[10]

The documentation regarding the fossil record is now too extensive to ignore; in addition, there are many difficult questions confronting both uniformitarianism and Evolution Theory. For example, where do the organic deposits of coal come from in such vast quantities if swamps cannot be shown to be associated with these deposits? Why are there no transitional creatures leading up to the giant dinosaurs? Why have vast numbers of fossilized sea creatures been found, sometimes in areas high above sea level, when normally a fish would be eaten upon death or else quickly decay? (There are several classic fossil examples of one fish half swallowing another fish, indicating rapid fossilization.)

---

8. George Gaylord Simpson, *Tempo and Mode in Evolution* (New York: Columbia University Press, 1944), p. 107. As quoted by Walt Brown, *In the Beginning*, p. 46.
9. David M. Raup, "Conflicts Between Darwin and Paleontology," *Field Museum of Natural History,* Vol. 50, No. 1 (January, 1979), p. 22.
10. Niles Eldredge, *Discovery* (Melbourne, ABC Television), July 29, 1982.

The best evidence for Evolution is, according to Richard Leakey in his TV series, *The Making of Mankind*, a series of snails which have differences in the roundness of their shoulders.[11] Hardly the decisive evidence one would want to produce in a court of law!

As will be discussed in Chapter 7, there seems little doubt that phenotypes intermediate between the major phyla did not exist. In other words, the missing links are not missing—they never existed!

Human/Hominid Fossils

As to whether man and other primates have evolved from a common progenitor, Marvin Lubenow shows that the supposed hominid fossils leading up to modern man have turned out to be either fully human, fully animal or fake. Some six thousand hominid fossil individuals have now been unearthed but, once again, we encounter the *illusion* of Evolution:

> We have all seen pictures of the impressive sequence allegedly leading to modern humans—those small, primitive, stooped creatures gradually evolving into big, beautiful you and me. What is not generally known is that this sequence, impressive as it seems, is a very artificial and arbitrary arrangement because (1) some fossils are selectively excluded if they do not fit well into the evolutionary scheme; (2) some human fossils are arbitrarily downgraded to make them appear to be evolutionary ancestors when they are in fact true humans; and (3) some nonhuman fossils are upgraded to make them appear to be human ancestors. . . . The fact that objects can be arranged in an "evolutionary" sequence does not prove that they have a relationship or that any of them evolved from any of the others.[12]

Fossils which are indistinguishable from modern humans trace back 4.5 million years,[13] according to the supposed "long-ages" time-scale. In addition, footprints found in 1978 at site G, Laetoli (northern Tanzania), closely resemble those of habitually unshod modern humans. The strata above them has been dated at 3.6 million years and the strata below them at 3.8 million years. Such evidence is contradic-

---

11. Richard Leakey, *The Making of Mankind* (Melbourne: ABV2 Television, Episode One), April 24, 1983.
12. Marvin L. Lubenow, *Bones of Contention: A Creationist Assessment of Human Fossils* (Baker Book House Company, 1992), p. 21.
13. See *ibid.* for Composite Fossil Chart, p. 171.

tory to Evolution beliefs, but the public is rarely informed about such discoveries. Walt Brown reports:

> Bones of many modern-looking humans have been found deep in rocks that, according to evolution, were formed long before man began to evolve. Examples include: the Calaveras skull, the Castenedolo skeletons, Reck's skeleton, and many others. Other remains present similar problems, such as: the Swanscombe skull, the Steinheim fossil, and the Vertesszollos fossil. These remains are almost always ignored by evolutionists.[14]

Media reports about new "missing link" discoveries continue to be made constantly, and this reinforces in the mind of the casual observer the unquestioned belief that genuine transitional forms do exist. But a crucial factor which can be easily overlooked is that drawings of transitional creatures largely depend upon the imagination and presuppositions of the artist. For example, the amount of body hair is impossible to determine from a skullbone and can be anything that the artist cares to assign to the creature. Similarly, the general body features can be made to fit whatever end result is desired.

Any similarity between man and other primates, or any similarity between any other creatures, can also be explained by there being the same designer involved. Much is made of the similarity between man and the apes, but many substantial differences are often overlooked. The feet of other primates, for example, function very much like hands and are radically different from the feet of human beings.

Regarding brain sizes, gibbons' brains are about 100 cubic centimeters, chimpanzees' about 400cc, gorillas' about 500cc, and Australopithecines' were about 600cc. The various anthropoid apes thus average about 500cc and do not exceed 600cc. The so-called *Homo erectus* was about 900cc, and this fits in the lower scale of modern human brain sizes. Neanderthal was larger on average, but also fits within modern human brain sizes.

This marked difference in brain size is most important in any discussion of "missing links." In contrast to the apes, the modern human *Homo sapiens'* brain varies in size from about 700cc to about 2200cc and with no difference in ability or intelligence. Human beings exist in very small and very large sizes but, as Lubenow points out, differences in cranial capacity are not as important as the relationship of brain size

---

14. Walt Brown, *In the Beginning*, pp. 11, 12.

to brain organization. The theory of an increasing brain size leading up to human beings is false.

The belief that Australopithecines evolved into *Homo erectus* and then later evolved into *Homo sapiens* has no support in genetics or in the fossil record. Australopithecines appear much too late in the fossils, in the supposed long-ages timescale, and it is by no means certain they were bipedal, anyway. (*Homo habilis* should not be defined as a separate taxon [biological group] because some specimens are australopithecines, while others are very modern.)

Homo erectus sites show evidence of the extensive use of tools and of controlled fire. One rock tool closely associated with *Homo erectus* sites is the Acheulean hand ax. This fascinating tool has been discovered around the world in large numbers nearby ancient rivers and lakes. It is rather narrow in cross section, yet almond or tear-shaped, generally with a knife-sharp edge around its perimeter, and without a handle. Researchers now suspect these tools were used by early human beings and thrown like a discus when hunting herds of game. They could be flung some one hundred feet and would then usually fall point down upon large beasts, inflicting great damage.[15] Such tools surely require human hunters.

Java Man (*Pithecanthropus 1*) was claimed by Eugene Dubois as a large gibbon contender for the "missing link," but it now seems very likely that the specimen is *Homo erectus*. The Selenka Expedition of 1907, in fact, found that humans with wide morphological diversity had lived together in Java. Similarly, while the fragments of Peking Man disappeared during World War II, these also are now thought to be those of *Homo erectus*. Cro-Magnon Man (famous for cave paintings in France) and the Heidelberg Man *(Germany)* jawbone are now both considered modern *Homo sapiens*.

Neanderthal Man has been established as fully human, but of rather large size and probably of great strength. Curiously, as long ago as 1872, Rudolf Virchow diagnosed from Neanderthal specimens that rickets gave rise to bone deformation in Neanderthal children and adults, on a worldwide scale.

Various other contenders for Evolution fame have credibility problems: *Ramapithecus*, deduced initially only from teeth and jaw fragments, was only a baboon which Richard Leakey acknowledges did not stand erect.[16] Another one, "Lucy"—an Australopithecine found by

---

15. Lubenow, *Bones of Contention*, pp. 142 and 143.
16. Richard Leakey, *The Making of Mankind*.

Donald Johanson—has been ruled out by fellow evolutionists Richard and Mary Leakey. Piltdown Man was cited as proof of Evolution Theory for forty years until found in 1953 to be an elaborate hoax. Nebraska Man consisted of only one tooth, which turned out to be that of a peccary (a type of pig).

Lubenow argues, regarding brain size, that *Homo erectus* is on the lower end of a continuum which includes *Homo erectus*, archaic *Homo sapiens* and Neanderthal.[17] Because of morphological compatibility with *Homo erectus*, and contemporaneous dating right throughout the fossil record with both australopithecines and *Homo erectus*, modern human beings (*Homo sapiens*) cannot be the product of Evolution.

The "Sudden Leap" Evolution Theory

The fossil record is far less Darwinian than commonly supposed, and some evolutionists believe that a series of dramatic events occurred in the past which resulted in the extinction of multitudes of creatures. One estimate is that as much as 96% of all life forms were destroyed in a series of extinctions.[18] This estimate is remarkably similar to that expected from a global flood. Indeed, a series of catastrophes is not necessarily in conflict with Evolution Theory. However, if only 4% survived, their survival can only be attributed to luck and not to superior adaptations. (As Hitching also points out, instead of *survival of the fittest*, you may only get the *survival of the luckiest*.)

The absence of phylogeny [development, evolution] in the fossils has now been established beyond reasonable doubt, and some scientists have rejected Darwinian gradualistic Evolution in favor of another concept: long periods of time, punctuated by very short episodes of rapid *upward* change—in effect, each time a sudden leap in Evolution. This concept of *punctuated equilibrium* was proposed by Stephen Jay Gould and Niles Eldredge and represents, at least in some of its more extreme manifestations, a partial return to the "hopeful monster" idea.[19]

This hypothesis was first proposed in 1952 by Richard Goldschmidt. The idea is that the origin of new groups cannot be explained by a series of small changes; therefore, rapid changes in small isolated com-

---

17. Lubenow, *The Bones of Contention*, pp. 138, 178-180.
18. David M. Raup, *Conflicts Between Darwin and Paleontology*, as quoted by Francis Hitching in *The Neck of the Giraffe*, p. 166.
19. Stephen Jay Gould, *Everyman*, September 2, 1982 (Melbourne: ABC TV).

munities must have been involved in between long intervals of time. When a new phylum, class or order appears in the fossil record, there follows a rapid diversification and a fairly quick jump to a new and more complex creature. Thus, there would be few transitional forms found in the fossil record. But punctuated equilibrium only explains the *absence* of transitional forms; a major weakness of the hypothesis is that there is no evidence for it and no conceivable mechanism for it, and so it remains hypothetical.

While agreeing that this may explain microevolution, Michael Denton (writing as a non-creationist) also outlines, in *Evolution: A Theory in Crisis*, why it is unlikely to validate macro-scale changes:

> The gaps which separate species: dog/fox, rat/mouse etc. are utterly trivial compared with, say, that between a primitive terrestrial mammal and a whale or a primitive terrestrial reptile and an *Ichthyosaur*; and even these relatively major discontinuities are trivial alongside those which divide major phyla such as molluscs and arthropods. Such major discontinuities simply could not, unless we are to believe in miracles, have been crossed in geologically short periods of time through one or two transitional species occupying restricted geographical areas . . . To suggest that the hundreds, thousands or possibly even millions of transitional species which must have existed in the interval between vastly dissimilar types were all unsuccessful species occupying isolated areas and having very small population numbers is verging on the incredible!

The advent of the theory of punctuated equilibrium and the associated publicity it has generated have meant that, for the first time, biologists with little knowledge of paleontology have become aware of the absence of transitional forms. After this revelation of what Gould has called "the trade secret of paleontology," it seems unlikely that we will see any return in the future to the old comfortable notion that the fossils provide evidence of gradual evolutionary change.[20]

Another obvious problem for the punctuated equilibrium concept is the question of mutations. Interestingly, where one was once asked to believe in the existence of transitional links, against all the evidence at hand, now the request is to believe rather in the existence of beneficial, "uphill" mutations, against all the evidence at hand.

---

20. Denton, *Evolution: A Theory in Crisis,* pp. 193, 194.

Inherent Problems in Transitional Forms

Conceptual problems plague Evolution Theory at every turn, and there are major inherent problems involved in the very idea of transitional forms. For example, if a creature which once could run swiftly, and thus survive, began to grow wings for flight, it then would have to pass through a period of Evolution during which it would be very clumsy and not good at either running or flying. It would thus likely become easy prey for predators and die out.

Bats are a good example of this inherent problem of intermediate stages. Unless the forelimb is substantially developed, the bat would not be able to fly or walk properly. Since the bat's pelvic girdle is rotated 180° compared to other mammals—which enables it to fly— what advantage would there be in the only 90° rotation that would occur at an intermediate stage of development? Also, unless the milk teeth of a baby bat are fully turned inward, the young creature would not be able to cling to its mother's hair, high above the ground.

Evolutionists tend to gloss over hard questions such as these. Consider, for example, *Time-Life's* textbook account of the first fish that dragged itself up onto land and decided to stay:

> A fish that differed significantly from all its forebears appeared in the rocks laid down about 390 million years ago. The front of its skull could be raised and lowered slightly, a change that would ease the shock when the jaws snapped shut. Its teeth were sharply pointed and well adapted to grasp prey. A single bone articulated the fins with the structural girdle. Such a single bone and related structures were to become familiar in later ages as the leg bones of land-dwelling animals. These unusual fish, called *crossopterygians* (lobefins), had made an even more vital change. In addition to gills, they were developing lungs. "The *crossopterygians* are to us perhaps the most important of fishes," said Colbert, "they were our far distant but direct forebears."
>
> Some 365 million years ago, some of these *crossopterygians* ventured out on the land. It is a plausible guess that they lived in streams which dried in the drought of summer into a few scattered pools. Did the fish struggle and flop from one drying pool to another with more water? No one knows, but those fish that were able to stay out of the water for a longer time certainly would have been the survivors and would have left behind progeny with their own greater ability to breathe in the air. In eastern Greenland, fossil hunters have found a creature more advanced than the most advanced *crossopterygians*; it was one of the primitive amphibians. The *Ichthyostega* combined a

fish tail with lungs and well-developed legs and feet. With their lungs and "walking legs," these early fish-out-of-the-water had a whole new source of food open to them. They could crawl along the banks of streams and snap up the insects which were beginning to swarm there. And the Earth lay open before them—for no other vertebrates were there to contest it.[21]

It all sounds so straightforward. However, in one of the major creationist textbooks, *Evolution, The Fossils Say No!* Duane Gish noted many anatomical weaknesses in evolutionist assertions.[22]

With respect to the above discussion about fish, he points out that no transitional forms have been found showing an intermediate stage between the fin of the *crossopterygian* and the foot of the *ichthyostegid*. The limb and the limb girdle of *Ichthyostega* were already of the basic amphibian type, showing no vestige of a fin ancestry. There is a basic difference in anatomy between all fishes and all amphibians not bridged by transitional forms. In all fishes, living or fossil, the pelvic bones are small and loosely embedded in muscle. There is no connection between the pelvic bones and the vertebral column. None is needed. The pelvic bones do not and could not support the weight of the body.

Also, it seems true that there are no walking fishes, including the "walking catfish" of Florida. These latter fish do not walk, but slither along on their belly, using the same type of motion they use in the water. In tetrapod amphibians, living or fossil, on the other hand, the pelvic bones are very large and firmly attached to the vertebral column. This is the type of anatomy an animal must have to walk. It is the type of anatomy found in all living or fossil tetrapod amphibians, but which is absent in all living or fossil fishes.

Unbelievably, having somehow overcome all the incredible difficulties in the transition from life at sea to life on land, the vague processes of natural selection supposedly resulted in some creatures returning to the sea. It really is unbelievable. Walter ReMine points out the mind-boggling problems involved in the transition:

> A classic case is the whales, beautifully adapted to an aquatic life. Allegedly, the whales evolved from four-legged mammals who returned to the sea. The waters were already full of advanced

---

21. *Evolution*, p. 112.
22. Gish, *Evolution, The Fossils Say No!*, p. 79.

predators and competitors. Imagine the difficulties as these land-roving animals lost their legs and underwent a total body transformation. The transformation would change their pelvis, their thermal insulation, their eating, breathing, sight, hearing and navigation apparatus.[23]

These organisms—whales, porpoises, seals, penguins, and ichthyosaurs—are significant. (1) They are dramatic examples of so-called "convergence." (2) In each case they fail to provide a clear lineage. (3) Their abrupt appearance into the [fossil] record (otherwise called "adaptive radiation") occurs *after* the seas contained numerous highly refined predators and competitors.[24]

Yet more conceptual problems can be tabled. Consider the following examples of problems for Evolution via natural selection:

> All mammals, living or fossil, have a single bone, the dentary, on each side of the lower jaw, and all mammals, living or fossil, have three auditory ossicles or ear bones, the malleus, incus, and stapes. In some fossil reptiles the number and size of the bones of the lower jaw are reduced compared to living reptiles. Every reptile, living or fossil, however, has at least four bones in the lower jaw and only one auditory ossicle, the stapes. There are no transitional forms showing, for instance, three or two jaw bones, or two ear bones. No one has explained yet, for that matter, how the transitional form would have managed to chew while his jaw was being unhinged and rearticulated, or how he would hear while dragging two of his jaw bones up into his ear.[25]
>
> The origin of the amniotic egg and the amphibian-reptile transition is just another of the major vertebrate divisions for which clearly worked out evolutionary schemes have never been provided. Trying to work out, for example, how the heart and aortic arches of an amphibian could have been gradually converted to reptilian and mammalian condition raises absolutely horrendous problems.[26]

Regarding the evolution of flight, nothing exists to support the evolutionist hypothesis other than the controversial, extinct creature *Archaeopteryx*, which is alleged to be fossil evidence of the transition from reptiles to birds. Archaeopteryx is described in *Time-Life's* textbook on Evolution as being a bird (p. 114) which possessed teeth. Thus

---

23. ReMine, *The Biotic Message*, p. 143.
24. *Ibid.*, p. 422.
25. Gish, *Evolution, The Fossils Say No!*, p. 85.
26. Denton, *Theory in Crisis*, p. 219.

it had wings and feathers, and it flew—it was not transitional.

Michael Denton has also pointed out other profound problems in the idea of transitional forms. A good example of such problems, typical of the conceptual difficulties which make Evolution untenable, is that of the evolution of birds. Regarding flight problems he notes,

> It is not easy to see how an impervious reptile's scale could be converted gradually into an impervious feather without passing through a frayed scale intermediate which would be weak, easily deformed and still quite permeable to air. Take away the exquisite coadaptation of the components, take away the coadaptation of the hooks and barbules, take away the precisely parallel arrangement of the barbs on the shaft, and all that is left is a soft pliable structure utterly unsuitable to form the basis of a stiff impervious aerofoil. The stiff impervious property of the feather which makes it so beautiful an adaptation for flight, depends basically on such a highly involved and unique system of coadapted components that it seems impossible that any transitional feather-like structure could possess even to a slight degree the crucial properties.[27]

As if this were not enough to contend with, there is also the profound problem of the avian lung and respiratory system:

> In all other vertebrates the air is drawn into the lungs through a system of branching tubes which finally terminate in tiny air sacs, or alveoli, so that during respiration the air is moved in and out through the same passage. In the case of birds, however, the major bronchi break down into tiny tubes which permeate the lung tissue. These so-called parabronchi eventually join up together again, forming a true circulatory system so that air flows in one direction through the lungs.
>
> ... Just how such an utterly different respiratory system could have evolved gradually from the standard vertebrate design is fantastically difficult to envisage, especially bearing in mind that the maintenance of respiratory function is absolutely vital to the life of an organism to the extent that the slightest malfunction leads to death within minutes. Just as the feather cannot function as an organ of flight until the hooks and barbules are coadapted to fit together perfectly, so the avian lung cannot function as an organ of respiration until the parabronchi system which permeates it, and the air sac system which

---

27. Denton, *Theory in Crisis*, p. 209.

guarantees their air supply, are both highly developed and able to function in a perfectly integrated manner.[28]

For a non-flying creature to be able to fly, a great number of modifications would be necessary. And yet flight is supposed to have evolved separately in insects, birds, mammals (bats) and reptiles (the extinct pterosaurs). The fact that Archaeopteryx is so regularly cited by evolutionists shows just how tenuous are their arguments.

Regarding the possession of teeth, it is true that birds of today tend not to have teeth, but some extinct birds did possess them. The same can be said for other creatures. Some reptiles have teeth, while others do not. Some amphibians have teeth, others do not. In fact, this is true throughout the entire range of the vertebrate subphylum—fishes, Amphibia, Reptilia, Aves and Mammalia, inclusive.[29]

The uniform absence of transitional forms in the fossil record confirms the validity of *typology*, the idea that there are quite distinct classes of creatures which cannot be bridged by transitional forms:

> It is a remarkable testimony to the almost perfect correspondence of the existing pattern of nature with the typological model that, out of all the millions of living species known to biology, only a handful can be considered to be in any sense intermediate between other well defined types. The lungfish is a classic example. It has fins, gills and an intestine containing a spiral valve like any fish but lungs, heart and a larval stage like an amphibian. Another classic example of an intermediate type is the egg-laying mammals, the monotremes such as the duck-billed platypus. In laying eggs the monotremes are reptilian, but in their possession of hair, mammary glands, and three ear ossicles they are entirely mammalian.
>
> ... although the lungfish betrays a bewildering mixture of fish and amphibian character traits, the individual characteristics themselves are not in any realistic sense transitional between the two types ... The biology of the monotremes is similar ... Instead of finding character traits which are obviously transitional, we find them to be either basically reptilian or basically mammalian.[30]

The emergence of phenetics and cladistics, in modern studies of biological patterns, are also presenting problems for Evolution Theory:

---

28. *Ibid.*, pp. 210-212.
29. Gish, *Evolution, The Fossils Say No!*, p. 92.
30. Denton, *Theory in Crisis*, p. 109.

It has only been over the past two decades, with the adoption of new methodologies which have subsequently revitalized and popularized the science of classification, that the conflict between hierarchy and evolution has re-emerged and come to the attention of significant numbers of biologists. The re-emergence of the conflict is evidenced today, not only in the increasing skepticism being expressed by some of the more radical cladists over many aspects of evolution theory, but also in the increasing resemblance that is developing between the modern cladistic framework and the non-evolutionary perception of pre-Darwinian biology.[31]

While on the subject of phylogeny, mention must be made of Ernst Haeckel's infamous (but still influential) *recapitulation theory*, known also as *"ontogeny recapitulates phylogeny."* This theory, also called the Biogenetic Law, was known to be false in the late 19th century, but it made a profound impact upon both Sigmund Freud and Carl Jung, and it lingers still in popular evolutionary mythology. (Haeckel was rebuked openly by his peers for blatant fraud in faking his documentation.)[32]

This theory holds that the human embryo, supposedly like all other embryos, takes on during its early stages the appearance of its so-called evolutionary ancestors in their adult form. The human embryo is said to have gill slits during the early stages of development. However, the pouches in the neck region are not really gill slits and develop instead into various glands, the lower jaw, and structures in the inner ear. The proper name for these "gill slits" is *pharyngeal arches*. An important weakness of recapitulation theory has been noted by Duane Gish:

> If the human embryo recapitulates its assumed evolutionary ancestry, the human heart should begin with one chamber and then develop successively into two, then three, and finally four chambers. Instead, the human heart begins as a two-chambered organ which fuses to a single chamber which then develops directly into four chambers. In other words, the sequence is 2-1-4, not 1-2-3-4 as required by the theory. The human brain develops before the nerve cords and the heart before the blood vessels, both out of the assumed evolutionary sequence. It is because of many similar contradictions and omissions that the theory of embryological recapitulation had been abandoned by embryologists.[33]

---

31. *Ibid.*, p. 137.
32. Russell Grigg, "Ernst Haeckel: Evangelist for Evolution and Apostle of Deceit," Creation Ex Nihilo (Creation Science Foundation Ltd., March, 1996).
33. Gish, *Evolution, the Fossils Say No!*, p. 180.

In conclusion, in view of lack of fossil evidence and multitudes of conceptual weaknesses in the idea of transitional forms, it seems clear that the fossil record cannot be cited in support of Evolution Theory.

—Chapter 7—
GENETICS

The Complexity of Cells

The science of genetics is crucial to the credibility of Evolution Theories. In considering this vital aspect of the Origins debate, it is worth pondering briefly the sheer complexity in the design of cells.

The cell is the basic structural unit of a plant or creature. It is built up from molecules, which in turn are comprised of atoms. Atoms have a nucleus (of a varying number of protons and neutrons), around which the electrons revolve, not unlike planets around the Sun. Atoms are incredibly small. In a single drop of water there are billions of atoms. If an apple could be magnified to the size of the Earth, each atom in the now enlarged apple would be about the size of the original apple.

The cell contains many separate components which work together harmoniously. Living things have dead cells in some parts of the body (e.g., the outer surface of the skin and fingernails), but internally there are millions of living cells in many shapes and sizes, performing all sorts of functions. Cells tend to be so small they can only be seen clearly through powerful electron microscopes.

Put simply, the main parts of a living cell are the outer cell membrane, which has the ability to keep some substances out of the cell and to keep others in; the nucleus, which is surrounded by its own membrane; and the cytoplasm, a jelly-like complex filling the space between the nucleus and the outer cell membrane and which distributes nourishment in the cell and gets rid of waste. The cell may exist as a tiny unit or in groups comprising tissue, and may combine with other tissues to form organs.

The nucleus of the cell is the control center which regulates most of the operations of the cell. It carries vital information about the sex of the plant or creature, the shape of the body, color of hair and eyes and the like. This information is recorded in the chromosomes, which are made up of nucleic acids (deoxyribonucleic acid, or DNA) in strings of tiny genes. The genes control an immense variety of characteristics and thus are the basic units of heredity.

In the human cell there are 46 chromosomes arranged in 23 pairs. Unlike some basic forms of life which reproduce asexually, new human life begins at conception when 23 chromosomes from the male sperm and 23 chromosomes from the female egg combine to form the first cell of the new human person, and then the "spiral staircase" of DNA uncoils and replicates itself.

Because of the immense number of possible genetic combinations derived from the parents, no two human beings are alike. *Time-Life International*, in its textbook on Evolution, describes something of the amazing reality of the DNA structure:

> In the 1950s . . . Francis Crick and James Watson fashioned a wire model which portrayed DNA as a helix, looking like a spiral staircase. The sugars and the phosphate were the framework and the four nucleotides, Adenine, Thymine, Cytosine and Guanine, were strung around it like four kinds of repeated steps. The 46 human chromosomes, H. J. Muller estimated, contain some four thousand million of these bases, or steps. If each one were written down as a single alphabetical letter, they would fill 100 large dictionaries—a sort of code, defining man. The order of such steps is different for each living thing; it is the endless variety of their order that explains the limitless variety of the living world. The long coils of DNA have a property uniquely their own—their capacity for reproducing themselves. At the right time for self-reproduction, the staircase divides down the middle. From free nucleotide units in the cell nucleus, each half-step, or base, picks up another unit complementary to itself, and a new coil is formed.[1]

All living organisms have in their genes an information storage and retrieval system which can be likened to an alphabetical code. The system in all living organisms consists of the four "letters" which are of chemical nature. The sequences of these four letters in their arrangements on the double helix of the DNA molecule determine the message they store and transmit. These sequences constitute the chemical information the organism needs to synthesize cell components.

While the chemical letters in themselves do not mean much (just as alphabetical letters mean little on their own), when arranged in

---

1. *Evolution*, p. 94.

sequences, they take on specific meaning. They contain information for specific protein synthesis. The intricate complexity of the information on the genetic code, along with its simple concept of four "letters" of chemical molecules, all packed into such a tiny size, overwhelmingly suggests design by an intelligent Designer. For this to arise by chance processes, impossibly high odds would be involved. As Denton points out, there is no evidence to support the idea that cells have evolved:

> Molecular biology has shown that even the simplest of all living systems on Earth today, bacterial cells, are exceedingly complex objects . . . far more complicated than any machine built by man and absolutely without parallel in the non-living world.
>
> Molecular biology has also shown that the basic design of the cell system is essentially the same in all living systems on Earth from bacteria to mammals. In all organisms the roles of DNA, mRNA and protein are identical. The meaning of the genetic code is also virtually identical in all cells. The size, structure and component design of the protein synthetic machinery is practically the same in all cells. In terms of their basic biochemical design, therefore, *no living system can be thought of as being primitive or ancestral with respect to any other system, nor is there the slightest empirical hint of an evolutionary sequence among all the incredibly diverse cells on earth.* For those who hoped that molecular biology might bridge the gulf between chemistry and biochemistry, the revelation was profoundly disappointing.[2]

The structure of DNA and its RNA replicate was only properly understood as recently as 1953, and a problem for Evolution Theory has emerged: how did the first replicating system get started?

As noted in *Scientific Creationism*,[3] higher organisms contain a tremendous number of specialized cells, and within each cell is an intricate complex of specialized protein molecules. Each protein molecule is a specially organized structure consisting of about twenty different amino acids. Each amino acid is made up of the four elements: hydrogen, oxygen, nitrogen and carbon, and in two cases a sulphur atom is also present. These complex systems are all, in the case of every known organism, reproduced and assembled on the basis of the instructions built into the DNA molecular system.

The DNA molecule not only has the information required for syn-

---

2. Denton, *Evolution: A Theory in Crisis*, p. 250. Emphasis added.
3. *Scientific Creationism*, p. 47.

thesis of the specific protein molecules required by the cell, but also that which is needed for its own replication. Thus, reproduction and inheritance depend directly on this remarkable molecule, which is organized differently and specifically for each kind of organism.

DNA can only be replicated with the specific help of certain protein molecules, which in turn can only be produced at the direction of DNA. Each thus depends on the other, and both must be present for replication to take place. But that is not all:

> There is much more to the cell than the "mere" origin of the protein synthetic apparatus. In fact, the protein synthetic mechanism cannot function in isolation but only in conjunction with other complex subsystems of the cell. Without a cell membrane the components of the protein synthetic apparatus could not be held together. The integrity of the cell membrane, however, depends on the existence of a protein synthetic apparatus capable of synthesizing the protein components of the membrane and the enzymes required for the synthesis of its fat components. However, the protein synthetic apparatus consists of a number of different components and can only function if these are held together by a membrane: two seemingly unbreakable interdependent systems. To continue, the protein synthetic apparatus also requires energy. The provision of energy depends on the coherent activity of a number of specific proteins capable of synthesizing the energy-rich phosphate compounds—proteins which are themselves manufactured by the protein synthetic apparatus. A further couple of interdependent cycles! As we have seen, the information for the specification of all the protein components of the cell, including those of the protein synthetic apparatus, is stored in the DNA. However, the extraction of this information is dependent on the proteins of the protein synthetic apparatus—yet again another set of interdependent cycles.[4]

So how did it all get started, other than by a Designer in possession of awesome intelligence? The supposed primordial cell was portrayed as being probably very simple and to have arisen by chance, but how could the first living cell have emerged by itself from non-living matter? Even if this were true, however, no other life was in existence which could have then served as nourishment—so how would it have survived?

All forms of nourishment must involve a high degree of enzyme sys-

---

4. Denton, *Evolution: A Theory in Crisis*, p. 268.

tem complexity. Any nourishment by means of reduction of carbon dioxide, solar energy and chlorophyll is ruled out, due to the complex chemistry involved. And any simplified forms of photosynthesis are also most unlikely to have been in existence, since all known types of this kind of synthesis require a high chemical order. Not surprisingly, the composition of the Earth's early atmosphere has long been a vital area of dispute:

> Chemists categorize the chemical effect of atmospheres into a range, from oxidizing, to neutral, to reducing. These are terms for the ability of atmosphere to remove (oxidize) or add (reduce) electrons to an atom, ion, or molecule. Our atmosphere contains abundant free oxygen. We have an oxidizing atmosphere. By the 1920s, Oparin and Haldane knew our oxidizing atmosphere would forbid the naturalistic origin of life. So, they dutifully suggested that the primitive earth atmosphere contained *no* free oxygen. . . . They assumed a strongly reducing atmosphere, since this is the most favorable to formation of organic molecules. Thus was born the myth of the primitive reducing atmosphere [with problems]:
>
> The lack of free oxygen would have left the upper atmosphere without an ozone layer to filter out ultraviolet rays from the sun. These rays would directly destroy most exposed organic matter since they can penetrate tens of meters beneath the ocean. . . . Ultraviolet rays would also have a destructive effect indirectly. The ultraviolet sunlight converts surface minerals into materials that will destroy organic molecules even more effectively than will oxygen gas.[5]

Once again, the odds are heavily stacked against random Naturalistic Evolution. Formation of the first living cell would have involved processes of molecular self-organization unknown to science, and the complex mechanism of cellular reproduction must have evolved in the brief period of viability of the first cell. Since there is no known way for this to occur, acceptance of it as a natural process requires a great act of faith by the evolutionist:

> Regardless of how primitive and simple a primordial cell may have been, it must have been complicated enough to reproduce itself, carry on metabolism and to excrete waste products. In short, it must have been complex enough to live. All living chemical processes require stored but retrievable information to direct them. How did the stored

---

5. ReMine, *The Biotic Message,* pp. 70, 71.

information arise *before* the actual cell chemistry itself became functional? If this question is answered as many establishment biologists do, usually by mumbling the word "chance," one comes into conflict with all information science. . . . For these and other reasons, the evolution of a single primordial cell by chance to a high degree of complexity, as demanded by present biological thought, is untenable.[6]

The "tree" and "bush" concepts of phylogeny are wishful thinking:

> The world of living organisms appears to be much more like a forest than a single tree. The many kinds of life which exist in this "forest of life" all have their own separate roots of origin. . . . Although they stand and "grow" next to each other, the one did not grow out of the other, but rather separately and next to the others. One thing still remains certain in serious biology—the genetical relationships between the protozoa, the metazoa and the metaphyta are still a mystery. The phyla all grow next to each other more like trees in a forest rather than like branches growing out of one root. A family tree in which the protozoa are below and the others "above" is today out of the question.[7]

Differences in Chromosomes

Another major problem for Evolution Theory lies in the differences in chromosomes between man and the other primates.

The late French geneticist Jerome Lejeune (who discovered the chromosomal abnormality for Down's Syndrome) compared the distinctions in chromosomes of the orangutan, gorilla, chimpanzee and man. All of the others have 24 pairs of chromosomes, while man has 23 pairs; the apes have two pairs of chromosomes, called 2p and 2q, while man has only the single pair No 2. In addition, there are very profound differences between each type of ape. Lejeune also rejected, as definitely debarred, the possibility of accumulation of small genetic changes. This finding stands in contrast to the supposed *Modern Synthesis* (a term coined by Julian Huxley in 1942) of evolutionary biology, which holds that point mutations are a vital part of evolutionary progress.

---

6. Arthur E. Wilder-Smith, *A Basis for a New Biology* (Telos-International, 1976), p. 129.
7. *Ibid.*, p. 90.

Whatever may be the process actually employed by nature, close consanguinity over several generations or speciation by a single couple in one generation only, the laws of the chromosomal mechanism require absolutely that a new species appear in an extremely small branch which almost immediately detaches itself from the root stock. To put it another way, the classic neo-Darwinian hypothesis of an imperceptible accumulation of small genetic changes (point mutations) causing a whole population to evolve very slowly toward a new type, is definitely debarred. Starting from a common ancestor, whose karyotype[8] we can reconstruct very closely, it is not possible to produce a continued chimpanzification, an indiscernible gorillization, an imperceptible orangutanization or again a slowly progressing hominization.[9]

Lejeune held that each species traces back to an original pair, and the only possibility for evolutionary progress, if it happened, is that of sudden detachment from the parent species. He also noted that twelve chromosomal transformations would be necessary to pass from one primate species to another, in a sudden emergence of new species.

Thus, it is very hard to envisage how "upward, higher" *Evolution speciation* could take place. How could truly new genetic information be transmitted from life forms which do not already possess it?

A further observation made by Lejeune is worth consideration. Noting that two scientists, King and Wilson (1975), had assembled all the available data and compared the proteins of man and of the chimpanzee by three methods (amino-acid sequence, immunological properties and electrophoretic properties), Lejeune then summarized their conclusions. *The very order of information is of vital importance*:

> For all the proteins analyzed, more than 99% of the amino-acids are identical in man and in the chimpanzee. These same authors have compared, by molecular hybridization methods, the constitution of nucleic acids of the two species. For a segment of DNA comprising 3000 nucleotides, they estimated that only 33 nucleotides were different between the two species. . . . As King and Wilson remark, the genetic differences between man and the chimpanzee appear extremely tenuous. For example, the differences observed between neighbouring species of squirrels are notably greater. Between two

---

8. Karyotype: The appearance, number and arrangement of the chromosomes in the cells of an individual.
9. Jerome Lejeune, "On the Mechanisms of Speciation," Proceedings of the Sessions of the Biological Society, Paris, Abstract 9, Volume 169, No. 4 (1975), p. 828.

species of frogs, they are thirty times more significant. It is necessary therefore to abandon definitely the view that small mutations can alone explain the difference between man and the chimpanzee. In other words, between the two species, the genetic vocabulary is common, except for less than 1% variations, whereas the structure of the chromosomes is notably different. . . . *It all seems to happen as if the genetic instructions being almost identical, the order in which they are expressed makes all the difference.*[10]

Mutations

The question of mutations is of central importance to Evolution. No-one argues that mutations do not occur, but the crucial question, rather is this: Do beneficial mutations ever occur which add new *information*, and, if they do, by what process? Upon the resolution of this question stands the credibility of Evolution Theory. The perennial problem for theorists is to demonstrate how a species could "jump" to a new and more complex type. How do the additional characteristics come about? How do simpler forms of life give rise to much more complex types?

Evolutionists are thus at pains to establish the mechanism involved. Genetic variation is unlikely to account for it, since each species has a unique DNA code structure. For example, a horse (with 64 chromosomes) and a donkey (with 62 chromosomes) can interbreed and produce offspring, but this hybrid is known to be always sterile and thus a one-generation phenomenon only. (Perhaps both were derived from one horse kind, but vital information has been since lost.)

Artificial selection and breeding have definitely established, after decades of research involving a great number of experiments, that there is a limit beyond which no further change is possible. And the breeder is always left with the same basic species, whether it be fruit flies, chickens, cattle or vegetables.

When mutations occur, these "copying mistakes" may be caused by various factors, such as ionizing radiation. The sum of these deleterious genes in the gene pool is called the *genetic load*. To date, scientists have established that the vast majority of mutations either have no effect on function or are deleterious and can result in mental and physical abnormalities. Sometimes a loss or corruption of information can be "beneficial"—a beetle on a windy island that has a mutation which

---

10. Lejeune, "On the Mechanisms of Specialization." Emphasis added.

causes all its descendants to lose wings will give rise to a "favored" race which is less likely to be blown into the sea. But additions of useful, complex information are another matter altogether.

Yet again, Evolution Theory is up against the problem of immense odds, and vast periods of time are of little help. As discussed earlier, if beneficial mutations do exist, they are thought to occur only once in perhaps 10,000 mutations. In addition, in order to be inheritable, a beneficial mutation must occur in the genes of the germ cells. The germ cells make up only a tiny fraction of all the cells of an organism and are generally relatively well protected from the environment. The question of odds cannot be ignored, as Gary Parker observes:

> The mathematical problem for evolution comes when you want a *series* of *related* mutations. The odds of getting two mutations that are related to one another is the product of the separate probabilities: 10^7 x 10^7, or 10^{14}. That's a 1 followed by 14 zeros, a hundred trillion. Any two mutations might produce no more than a fly with a wavy edge on a bent wing. That's a long way from producing a truly new structure, and certainly a long way from changing a fly into some new kind of organism.[11]

The problem with beneficial mutations is not so much whether they are theoretically possible or impossible. Rather, the crucial problem is that the price of good mutations is too high. To postulate biological Evolution by the gradual selection of beneficial mutations, one must also explain how the thousands of harmful mutations are overcome by the few good ones. Natural selection cannot weed them all out, since harmful mutations are often only expressed if inherited from both parents. In other words, the defects will still be "carried" by healthy individuals. Gary Parker shows how mutations actually point in favor of intelligent design rather than Evolution:

> Mutations are "going the wrong way" as far as evolution is concerned. Almost every mutation we know of is identified by the disease or abnormality that it causes. Creationists use mutations to explain the origin of parasites and disease, the origin of hereditary defects, and the loss of traits. In other words, time, chance, and random changes do just what we normally expect: tear things up and make matters worse. Using mutations to explain the *breakdown* of existing genetic order (creation) is quite the opposite of using muta-

---

11. Morris & Parker, *What Is Creation Science?* p. 63.

tions to explain the *build up* of genetic order (evolution). Clearly creation is the most direct inference from the effects of mutations that scientists actually observe.[12]

Modern science has shown that each species has its own specific DNA code structure and complementary protein molecules. Each has an inherent specification which ensures that the basic kind remains unique. Adaptations of diseases resistant to antibiotics, or of weeds resistant to herbicides, are not evidence of positive mutations, because information is lost and truly new, "higher," species do not arise. The genetic system offers resistance to extreme changes, with a tendency to revert back toward the basic type. Thus, the possibility of human beings (with 46 chromosomes) and the other primates (48 chromosomes) emerging from a common progenitor seems definitely ruled out.

Organs—Could They Have Evolved?

Further problems for Evolution Theory can be seen when one reflects upon the incredible complexity of systems within living things. Gary Parker raises consideration of the complexity of organs:

> Perhaps the clearest anatomical evidence of creation is "convergence." The classic example is the similarity between the eyes of humans and vertebrates and the eyes of squids and octopuses. Evolutionists recognize the similarity between the eyes easily enough, but they've never been able to find or even imagine a common ancestor with traits that would explain these similarities. So instead of calling these eyes homologous organs, they call them examples of "convergent evolution." That really means that we have another example of similarity in structure that cannot be explained as evolutionary descent from a common ancestor.[13]

Consider the human eye. Light from outside enters the eyes, and messages about the light are carried to the brain by optic nerves, involving hundreds of millions of electrical impulses every second. The brain then recognizes the nerve messages and so the person comprehends what is being seen. Without the brain, the eyes would be of little use.

The human eye has been likened to a small, ball-shaped camera. It fits into the eye socket within the skull and can be swiveled around by

---

12. *Ibid.*, p. 65.
13. *Ibid.*, p. 22.

various small muscles. The pupil of the eye contracts or enlarges in accordance with the amount of light available, and light passes through a transparent lens. The light rays are bent as they pass through the lens and they focus on the retina, a layer situated at the back of the eye. When other interrelated parts are considered, such as eyelashes, tears, blood vessels and the ability to refocus almost instantly, the eye can only be regarded as an incredible accomplishment. If the slightest thing goes wrong, the operation of the eye may be greatly impaired.

Not only does the very existence of the eye point toward a Creator, but it poses immense difficulties for Evolution. Sylvia Baker pointed out the dilemma:

> All the specialized and complex cells that make up our eyes are supposed to have evolved because of advantageous mutations in some more simple cells that were there before. But what use is a hole in the front of the eye to allow light to pass through if there are no cells at the back of the eye to receive the light? What use is a lens forming an image if there is no nervous system to interpret that image? How could a visual nervous system have evolved before there was an eye to give it information?[14]

Underground creatures may not need eyes, of course, and poor vision seems much better than none, but such reasoning is little consolation for Evolution Theory. The crucial question relating to the eye is this: How could the incredibly complex "system of systems" involved in the eye evolve, since *irreducible complexity* is involved—and no-one has yet shown conceptually how such interdependent systems involving irreducible complexity can have intermediate stages.

How could the haphazard, random process of beneficial mutations bring about such a magnificent masterpiece as the eye? Yet sight (like flight) is supposed to have evolved concurrently and independently many times within all types of creatures. *Encyclopaedia Brittanica's* comment on Evolution purports to address the problem of the eye but, like Darwin's earlier and similar "explanation,"[15] it is no answer at all:

> The eye was not produced out of nothing: ancestors of animals with eyes had simpler eyes, even single cells sensitive to light such as are found in the most primitive organisms. Improvement of function

---

14. Sylvia Baker, *Bone of Contention* (Evangelical Press, 1981), p. 17.
15. Darwin, *The Origin of Species*, pp. 231, 232.

confers increased survival value from the start, and the same is true of all other cases subjected to rigorous analysis; electric organs, the origin of flight in birds, the parasitic habit of the cuckoo, the adaptation of whales to life in deep water, and the evolution of color-vision. In each case, new functions and new organs were developed out of old ones. In color vision, the ability to distinguish color is a by-product of two other improvements: accuracy of sight, and sight at low thresholds of light. Gradations can be made out, even in the evolution of such an organ as the eye. This is important because adaptations are not perfect and leave room for further improvement.[16]

This statement purports to explain the evolution of the eye, but it resorts to the use of *illusion* and effectively begs the question. It does not address seriously the detailed problems involved, and simply asserts that there are no real problems.

Another aspect of Evolution Theory which requires comment is that of vestigial organs. These are parts of the human body which are supposed to be no longer functional, as they once were in distant ancestors. At one stage some 200 vestigial organs were cited by evolutionists, but the list has been reduced to the vanishing point. It is one thing not to know what the function of an organ is; it is quite another to establish that it has no function. Human organs once thought to be vestigial include the appendix, tonsils, coccyx, pituitary gland and the thyroid gland.

An inherent weakness in the idea of vestigial organs is the absence of nascent organs. These are organs which should be seen to be evolving but which have not completed their evolution. If vestigial organs are phasing out, then nascent organs should be phasing in. It seems clear, however, that no such organs are found in the body.

In conclusion, it is fair to say that the modern science of genetics has brought to light insoluble problems for the possibility of Evolution.

---

16. *Macropaedia,* Vol. 7, p. 14.

—Chapter 8—

ENTROPY

Entropy refers to (among other things) the amount of unusable energy in the Universe. Its significance has been deduced from study of the laws of thermodynamics, which deal with relationships involved in the conversion of heat and other forms of energy into work.

The first and second laws of thermodynamics have been experimentally tested many times and are universally accepted as established laws of nature. The first law is a statement of the principle of conservation of energy; it states that no matter or energy is now either being created or completely destroyed. The second law states that every system left to its own devices always tends to move from order to disorder, its energy tending to be transformed into lower levels of availability, finally reaching a state of complete randomness and unavailability for further work.

There are several ways to describe the second law, all of them equivalent and interchangeable.[1] In physical systems the second law has been expressed in three ways:

- As a measure of the increasing unavailability of the energy of the system for useful work (classical thermodynamics).
- As a measure of increasing disorder, randomness or probability of the components of the system (statistical thermodynamics).
- As a measure of the increasingly confused information in the transmission of the coded message through a system (informational thermodynamics).

In each case, entropy is a measure of the lost usefulness of the system. In *classical thermodynamics*, the useful energy is measured which has had to be converted into nonuseable heat energy to overcome fric-

---

1. A comprehensive treatment of entropy is given in *Thermodynamics and the Development of Order*, Ed. Emmett Williams (Norcross, Georgia: Creation Research Society Books, 1981).

tion and keep the system running. In *statistical thermodynamics*, the probability of the structured arrangement of the system is measured, with a state of complete disorganization being most probable. In *informational thermodynamics*, the amount of garbled information that accompanies the transmission of information by the system is measured. The same mathematical equations can be shown to apply to all three types of situations, so that all three are equivalent to each other.

The behavioral tendency of matter is always in a *downward* direction, never *upward* in the absence of any preprogramming (e.g., coded machinery such as characterizes living things). Without such programming, matter does not transform itself into higher and higher levels of organization. For example, buildings and equipment always tend to deteriorate unless maintained adequately. Another example: Heat will only flow from a hot object to a cold object, until a temperature equilibrium is reached in both.

Thus, entropy is a measure of the tendency toward decreasing arrangedness or decreasing complexity in the Universe. All observed systems display such tendencies toward disorder. As bodies such as the Sun dissipate their energy, the entropy of the Universe increases (however slowly the process), and this points to a future state of equilibrium, one of maximum entropy. The concept can be extended further: In biological systems, the phenomena of sickness, death and extinction represent outworkings of the second law. Far from being limited to, say, the study of heat engines, the laws of thermodynamics illustrate broad categories of phenomena which affect man.

The Implications for Evolution Theory

The process of Evolution requires energy in various forms, and thermodynamics is the study of energy movement and transformation. The two fields are clearly related. Scientific laws that govern thermodynamics must also govern Evolution.

According to Evolution Theory, matter transformed itself on Earth from simple gases, through spontaneous natural processes, into an extremely complex living cell. Evolution is also held to be an irreversible process which necessarily leads to greater variety and increasing complexity of organization. But this is quite against the observable tendency of matter.

Evolution Theory is therefore inconsistent with the findings of modern science with respect to entropy. Sean O'Reilly commented on the implications of entropy as it affects Creation/Evolution:

> The first law speaks to the finite nature of the universe of matter. If we listen to the implications of its finite nature and all that science has to tell us about the high degree of order which it has exhibited, we should conclude that its existence and the evident conservation of its finite mass-energy are not likely to be self-explanatory, or to be explained by science, since it too is limited, is finite. The second law contains a direction, an "arrow of time," aimed at the ultimate heat death of the Universe, with its total mass-energy unchanged in quantity, but totally unavailable for further work. It also clearly implies that the Universe cannot be infinitely old since if it were, it would already be dead. The second law also directly contradicts evolutionary theory; if language has any meaning, both cannot be true. Evolution theory requires a universal principle of upward change; the entropy law is a universal principle of downward change. The latter has been proved to apply in all systems tested so far; the former cannot even be tested scientifically.[2]

The reality of entropy suggests that the Universe must have had a beginning. As the differences of energy level throughout it are being gradually ironed out, eventually there will come a time when a dead level of energy will have been produced and the process of change will end. The Universe has been likened to a clock which is running down and not being rewound. Such a clock must have been wound up at some point of time at a measurable distance from that at which it runs down; hence the conclusion that it must have had a beginning.

Thus, the second law requires the Universe to have had a beginning, and yet the first law precludes its having begun by itself. It seems likely that it came into existence through the work of a transcendent Cause "beyond" the Universe. But there is nothing within the present observable space-mass-time framework which is an adequate Cause. Therefore the Cause must be either an evolutionary process beyond observable space or prior to observable time, or else a transcendent Creative Force which brought space and matter and time into existence concurrently.

The idea that matter evolved into its present structure far out in non-observable space is called the "steady-state" theory. To offset the tendency toward universal decay, it has been suggested that new matter, in the probable form of hydrogen gas, is continually evolving into existence somewhere in outer space. Alternatively, the idea that matter evolved into its present structure far back in non-observable time is

---

2. Sean O'Reilly, *Bioethics and the Limits of Science* (Front Royal, VA: Christendom Publications, 1980), pp. 56 and 57.

called the Big Bang Theory—a gigantic explosion gave rise to everything in the Universe. Some speculate that the explosion perhaps was caused by a previous collapse into a super-dense state.

Open Systems

But what of life on Earth—does this not contradict the process of entropy with respect to the possibility for Evolution to occur?

Evolutionists tend to argue that the Earth is an open system, and that the energy of the Sun would be sufficient to offset any losses through entropy. However, Origins researchers have pointed out the weakness of this idea. The Sun's energy is a vital factor, but alone it is insufficient to overcome the thermodynamic barrier to the origin of complex chemicals and complex systems on the Earth. A programmed conversion mechanism, such as photosynthesis, is also required to direct the energy. Further, an influx of indiscriminate energy is more likely to increase the tendency toward disorder and to work against evolutionary processes. This problem is illustrated by the oxygen-ozone discussion:

> If the layer of ozone that surrounds the Earth were removed, allowing all the radiant energy from the Sun to reach the surface of the Earth, all life, from micro-organisms to man, would rapidly be snuffed out. This is because the shortwave, highly energetic, highly destructive portion of the radiant energy from the Sun is absorbed by ozone. The removal of this protective shield of ozone would allow this deadly energy to reach the Earth and destroy all living things.
>
> According to the evolutionary scenario that must be suggested by evolutionists, there could have been no protective shield of ozone in the hypothetical primordial atmosphere. Ozone is composed of tri-atomic oxygen. Ultraviolet light from the Sun converts ordinary diatomic molecular oxygen into ozone. If there were no oxygen in the primordial atmosphere, then there could have been no ozone. Evolutionists must, of necessity, exclude oxygen from their primitive Earth scenario, however. If oxygen were present, all organic substances, such as amino acids, sugars, etc would be oxidized to carbon dioxide, water, and other oxidized substances.
>
> Oxygen is thus incompatible with an evolutionary scenario, but then so is its absence! No oxygen, no ozone. No ozone, no protection from the deadly destructive shortwave ultraviolet light that is rapidly fatal for the existence of amino acids, proteins, DNA, and RNA. Again, the evolutionist is caught between the horns of a dilemma.[3]

---

3. *Thermodynamics and the Development of Order*, p. 78.

In commenting on the failure of the *Viking* spacecraft to find signs of life on Mars in 1976, Michael Denton observed that

> Significantly, the absence of organic compounds in the Martian soil has been widely attributed to just such a strong ultraviolet flux which today continuously bombards the planet's surface. What we have then is a sort of "Catch 22" situation. If we have oxygen, we have no organic compounds; but if we don't have oxygen, we have none either. There is another twist to the problem of the ultraviolet flux. Nucleic acid molecules, which form the genetic material of all modern organisms, happen to be strong absorbers of ultraviolet light and are consequently particularly sensitive to ultraviolet-induced radiation damage and mutation. As Sagan points out, typical contemporary organisms subjected to the same intense ultraviolet flux which would have reached the Earth's surface in an oxygen-free atmosphere acquire a mean lethal dose of radiation in 0.3 seconds.[4]

Evolution theorists also face the added problem of explaining how photosynthesis came into being—how did this controlling mechanism of the Sun's energy come about, quite against the observable tendency toward lower levels of organization?

> It is not sufficient to say, as the evolutionist does, that the Sun's energy is great enough to maintain the process of evolution. The essential, and unanswered, question is: "*How* does the Sun's energy maintain the process of evolution? What is the specific mechanism of 'evolutionary photosynthesis' that converts solar energy into the transformation of particles into atoms, then into molecules and stars and galaxies, complex molecules into replicating molecules, simple cells into metazoan life, marine invertebrates into reptiles and birds and men, unthinking chemicals into conscious intelligence and abstract reasoning?"[5]

There are apparent exceptions which seem to run counter to the process of entropy, but in fact, for every real process the net entropy of the Universe (a system plus its surroundings) must always increase.

The chemical reactions proceeding in cells do so by way of entropic laws. But by its intrinsic automation, a living entity perfects itself internally. It takes in food and breaks it down to its simplest components,

---

4. Denton, *Evolution: A Theory in Crisis*, p. 261.
5. Henry M. Morris, *The Troubled Waters of Evolution* (San Diego: Creation-Life Publishers, 1980), p. 19.

then builds it up to a higher order of being, to make its structural proteins, fats, bone, etc. The decay tendency seems suspended for such an operation, unique to things containing program machinery (or during the operation of intelligence). Nevertheless, the net entropy of the organism and its surroundings still increases. But when a living thing dies, it then proceeds to decay. The second law tendency is thus fully expressed because the matter is no longer living.

When viewed against the general reality of increasing entropy, this apparent suspension for a time of the second law can be seen as a temporary aberration. It does however touch upon an important point: What is it that constitutes life? What animates the many separate particles within a plant or creature?

Most Origins researchers today argue that the program machinery—the information contained in living things which enables them to temporarily overcome this decay tendency—is the key. Such information is never observed to arrive spontaneously from raw matter, thus it had to arise initially from mind, and be imposed upon the matter. Living things, of course, subsequent to the original Creation, have been able to pass this information to their descendants by means of the complex preprogrammed processes of reproduction.

In view of the difficulties facing Evolution Theory with respect to the reality of entropy, it may fairly be claimed that the Creation model explains the observable data far better than does Evolution Theory.

— Chapter 9 —

HOW OLD IS THE UNIVERSE?

With respect to the age of the Universe and when the Creation events took place, there are several distinct sets of beliefs now competing for acceptance within Christianity:

- God created the Universe billions of years ago and used Evolution and/or divine intervention in the creation of life forms and subsequent changes *beyond* kind.
- God created space, time and matter only about six thousand years ago and specially created each kind of life form in fairly rapid succession, and He set genetic variation into operation to allow subsequent changes *within* kind.
- God created the Universe billions of years ago and much later specially created each kind of life form with repeated divine intervention, at great intervals of time apart, and set genetic variation into operation to allow subsequent changes *within* kind.

Since each belief set is exclusive of the others, only one of these possibilities can be true. Evolution or Special Creation? Is the Universe billions of years old, or only about six thousand years old (in time as measured now on Earth)? The answer is unlikely to be found by focusing only on empirical science, in isolation from aspects drawn from theology and exegesis. (See Chapter 15 for consideration of these aspects and related conceptual arguments.)

Unfortunately, the range of arguments concerning the age of the Universe is not as clear-cut as those which can be advanced for other aspects of the Origins debate, such as those drawn from the science of genetics. How could they be, when there is so much that is not yet understood about the gargantuan and mysterious Universe?

In addition to this, most people—including many who are disposed favorably to other aspects of the Special Creation position—tend to baulk at the idea that the Universe may be only about 6,000 years old.

Thus, anyone who does not share the prevailing "orthodoxy" in favor of billions of years faces a difficult task in being taken seriously.

There is no proof whatsoever in favor of Evolution, and where is the proof beyond doubt that the Universe is billions of years old? We live in a world where complex, complementary *interdependence* exists between atmosphere, vegetation and all sorts of creatures. Many plants, birds and insects depend on mutual co-existence; an obvious example is the cross-fertilization process. Like the DNA/RNA relationship, how could all this come about, if not quickly?

To "force" God to implement the Creation, intermittently, over billions of years is really to indulge in convenient *anthropomorphism*, a label long placed on "young-age" creationists, who are often unfairly labeled as "disruptive." Implicit in this pejorative labeling is the idea that there exists somewhere "a noble group of unbiased scientists" who can be relied upon absolutely in the search for objective truth about the Universe, despite whether or not they believe in God.

Consider what has occurred during the last two centuries. Swayed by mistaken 19th century scientific arguments in favor of evolutionary descent, many scholars have become convinced that an age of many millions of years was necessary to accommodate the process of Evolution. That the Universe *must* be billions of years old became conventional wisdom across modern society and remains largely so today, despite the formidable case which now exists against Evolution.

Must the Universe be billions of years old? Why? Since Evolution cannot occur, there is no need to expect an age of billions of years for the Universe. The quest which began in the 19th century to seek evidence for billions of years of time to enable Evolution to have occurred, can now be seen as quite pointless anyway, for no amount of time can "save" Evolution. The fossil record shows no evidence whatsoever of evolutionary descent—and Naturalistic Evolution cannot occur, because DNA is designed to allow only variety *within* kind.

Dating Method Problems

The modern urge to explain all things in terms of the Universe being billions of years old traces from the concept of *uniformitarianism*, made popular by James Hutton and Charles Lyell in the early 19th century. The idea that various features of the Earth's strata, including the fossil record, only formed slowly over many millions of years by the same processes that can now be observed at work upon the Earth, became known as "the present is the key to the past" theory.

Assumptions were accepted much too soon, long before sufficient field data was known about strata and fossils. Arguments began to arise across the spectrum of scientific disciplines, seeking to justify the "long-ages" view. Findings supporting a "young-age" hypothesis could easily be discarded, by definition, as incorrect. The various dating methods have tended to justify the goal of proving long ages and to convince the public that a young-age view is "unscientific" and unworthy of serious consideration.

As discussed in Chapter 2, the validity of extrapolating from current process rates back millions of years has by now been challenged by many Origins researchers working in various scientific disciplines. One of the major difficulties involved in examining the question of "Age" is that of finding reliable dating methods. Often the age of an object dated by one means will be confounded by another dating method.

The *Carbon 14 method* (C-14) is applicable only to tens of thousands of years! This method, proposed by Willard Libby, is perhaps the one most popular in the public mind, but it was never intended as a proof for the millions-of-years age theory. Carbon-14 is continually entering the atmosphere (and into the carbon cycle) by the action of cosmic rays, which convert nitrogen (N-14) into C-14. And it is continually leaving the system by its radioactive decay back to N-14. When a plant or animal dies, the C-14 atoms which "decay" are not replaced by new ones from outside, and the amount of C-14 in the remains gradually reduces, and therefore *theoretically*, it should be possible to establish the time period since death.

However, if the Earth's magnetic field was once of much greater strength than at present, there probably would have been far less C-14 formed before the Flood, and this would now give a misleading reading, indicating an extremely old age to the Earth. The present formation rate of C-14 is thought to be significantly higher than the disintegration rate, and this suggests that equilibrium has not yet been reached. Since it is reckoned that equilibrium should have been reached after only 30,000 years, the C-14 method gives little support to the "old-Earth" idea.[1]

Other difficulties in dating methods can be cited. The idea that stalagmites and stalactites in caves are a reliable indication of extreme age has been overturned. In the Carlsbad Caverns, a bat has been found entombed in a stalagmite. If slowly operating process rates applied, the

---

1. For discussion of the C-14 dating method, see "What About Carbon-14 Dating?" *The Answers Book*, pp. 43-50.

bat should have rotted away long before the stalagmite could have formed.

Precious *opals* are dug out from underground in central Australia and assumed to be millions of years old. But an Australian opal miner, Len Cram, has discovered the secret of "growing" opals in glass jars in natural opal dirt at Lightning Ridge in New South Wales—in only a matter of weeks! Under an electron microscope, they look identical to natural opals. Andrew Snelling describes the surprising discovery:

> All it takes is an electrolyte (a chemical solution that conducts electricity), a source of silica and water, and some alumina and feldspar. The basic ingredient in Len's "recipe" is a chemical called tetraethylosilicate, which is an organic molecule containing silica. The amount of alumina which turns to aluminium oxide determines the hardness of the opal.
>
> The opal-forming process is one of ion exchange, a chemical process that involves building the opal structure, ion by ion. (An ion is an electrically charged atom, or group of atoms [molecule]). This process starts at some point and spreads until all the critical ingredients, in this case the electrolyte, are used up. Within a matter of weeks of this initial formation, the newly forming opal has beautiful color patterns, but it still has a lot of water in it. Slowly over months, further chemical changes take place, the silica gel consolidating as the water is "squeezed" out.[2]

An opalized marine reptile—"Eric" the pliosaur—was excavated from underground at Lightning Ridge in the 1980's and was reckoned to be 100 million years old. But, as Snelling points out, after the burial by catastrophic deposition of the pliosaur, the most likely explanation of the pliosaur bones' preservation via opalization is therefore the same replacement (ion-exchange) process which has been demonstrated to happen in only a period of months.

Such discoveries highlight doubts about presuppositions inherent in Uniformitarianism. Scientists tend to examine the physical phenomena currently prevailing and then extrapolate back in time to try to establish the age of the Universe and of the Earth. But is it valid to extrapolate from currently prevailing process rates and assume that present rates have always been constant? The evidence suggests rather that the present rates have not always applied and that worldwide catastrophic

---

2. Andrew A. Snelling, "Creating Opals," *Creation Ex Nihilo*, Vol. 17, No. 1, December, 1994.

events *must* have occurred in the past.

Discerning the real age of an object from the apparent age can be very difficult. For example, regarding the method known as potassium-argon (K/Ar)—since as much as 1% of the Earth's atmosphere consists of argon, and rocks can easily absorb this gas, how is one to know how much of the isotope Argon 40 has come by decay from potassium and how much from the atmosphere? Various methods of dating have been developed, but all are open to question. If constant *radioactive decay rates* actually varied in the past, the presently calculated ages may be misleading and invalid. Duane Gish gives the creationist perspective:

> It should be realized that there is no *direct* method for determining the age of any rock. While very accurate methods are available for determining the *present* ratios of uranium-lead, thorium-lead, potassium-argon, and other isotope ratios in mineral-bearing rocks, there is, of course, no direct method for estimating the *initial* ratios of these isotopes in the rocks when the rocks were first formed. Radiochronologists must resort to indirect methods which involve certain basic assumptions. Not only is there no way to verify the validity of these assumptions, but inherent in these assumptions are factors that assure that the ages so derived, whether accurate or not, will always range in the millions to billions of years (excluding the Carbon 14 method, which is useful for dating samples only a few thousand years old). Recent publications have exposed weaknesses and fallacies in radiometric dating methods, while some recent publications have described many reliable chronometers, or "time-clocks," that indicate a young age for the Earth.[3]

The method of radioactive dating suffers from the problem that it is impossible to know the exact isotopic composition of a particular object as it was in times gone by. And it does not necessarily follow that the present level has come about in one way only:

> It is possible that within a radioactive mineral, one lead isotope might be changed into another by a process known as "neutron-gamma" reaction. Lead-206 could be changed into lead-207, or lead-207 into lead-208. . . . The scientist dating the rock would not know whether the lead-207 had come by decay from uranium-235, or whether it had come from lead-206 by the neutron-gamma reaction.[4]

---

3. Gish, *Evolution, The Fossils Say No!*, p. 63.
4. Sylvia Baker, *Bone of Contention*, p. 22.

"Recent" Earth Features

Around the world, "older" strata have been found lying above "younger" strata, requiring mountains to have moved upwards and across country, in some places for many miles, by natural forces.

A profound problem is that, while it is possible for rock overthrusting to occur in a limited way (which would show abundant associated evidence of broken rocks and gouge marks), on a massive scale the idea is untenable. Beyond a certain mass, the friction resistance would exceed the rock's cohesive strength, and the rock would break up. Instead of resulting in the formation of great mountains, such great forces would tend rather to have a leveling effect. Rather than slow, intermittent formation, a more reasonable explanation is that the sedimentary features of mountains with a "swirling" effect *must* have been shaped that way quickly, before they hardened into rock.

Another difficulty is that many creatures appear abruptly, already in very complex composition, in Cambrian strata. Tiny multi-cellular creatures—fishes, rodents, even dinosaurs—all appear suddenly, with no trace of earlier transitional forms. The sheer extent of vast death pits around the Earth indicates that the geological column formed very quickly. *If the geological column in fact formed quickly, this does not prove a "recent" Creation, but it does undermine a central assumption in the idea that Earth must be billions of years old.*

Quite apart from these vast death pits, the sheer extent of *dinosaur* findings in Utah and Montana alone suggests rapid entombment on a massive scale. The uniformly scattered bones of some 10,000 maiasaur dinosaurs have been found entombed in one huge Montana deposit, all with their noses pointing in the same direction, apparently choked by volcanic ash and later swept by immense water-borne forces. In addition, there have been fascinating findings of human-like footprints alongside dinosaur footprints in cretaceous rock strata along the Paluxy River, near Glen Rose, Texas, and in various parts of North America.

> Human footprints, both normal size and giant size, sometimes side by side with dinosaur prints, have been found in Mexico, Arizona, Texas, Missouri, Kentucky, Illinois and in other US locations. ... Some believe that the prints are frauds ... the following counter-arguments suggest otherwise: (a) The tracks are widely distributed; (b) The impressions are usually only exposed by flood erosion or bulldozers; (c) Two paleontologists are on record as having pronounced them genuine; (d) Strings of up to 15 to 23 right-left tracks have been uncovered; (e) Upon sawing through the tracks, the rock

particles found underneath the impressions are more compressed than the particles surrounding the prints; (f) The associated dinosaur fossil tracks are accepted as valid; (g) Some prints have ridges of mud pushed up around them.[5]

Also, fascinating findings of "fresh" juvenile Hadrosaur (duck-bill) dinosaur bones have been found along the Colville River on the Alaskan North Slope.[6] The buried sections of bones are not yet mineralized into rock, thus indicating a "recent" age, much less than the widely assumed extinction of dinosaurs 65 million years ago.

The vast strata evidence suggests that a water-borne catastrophe occurred on a global scale, rather than a series of smaller localized catastrophes. (One coal deposit, in Montana, is immense—hundreds of miles long, about fifty miles wide and several hundred feet deep.) The 1976 discovery of a large fossilized baleen whale in a Lompoc, California diatomite quarry provides yet another indication of the sheer size of the Flood forces. The bones are lying in an inclined position, encompassing various strata, indicating (through their excellent state of preservation when compared to whale skeletons found on today's sea floor) that the whale must have been entombed rapidly.

Henry Morris has pointed out that each "layer" in the geological column in fact merges into others, with no worldwide time break between the end of one "age" and the beginning of the next.[7] And recent experiments on sedimentology, carried out under the direction of Guy Berthault, add further weight to the strong likelihood that strata were formed rapidly and thus cannot be cited in support of the "vast ages."[8] As discussed in Chapter 2, these experiments help explain how the sedimentary strata formed and were deposited while in a soft state, rapidly engulfing all sorts of creatures and plants. A huge upheaval would then have occurred, giving rise to great mountains and enormous sea canyons, after which the sedimentary deposits hardened in their present position.

A further compelling argument in favor of a rapidly occurring global Flood is that of *polystrate fossilized tree trunks*. Stripped of branches and showing evidence of water-borne deposition, they have been found

---

5. Wysong, *The Creation-Evolution Controversy*, p. 373.
6. Such "recent" dinosaur findings have been closely studied by the *Creation Research Science-Education Foundation*.
7. Henry Morris, *The Troubled Waters of Evolution*, p. 95.
8. See video *Fundamental Experiments on Stratification*, Pierre Y. Julien and Guy Berthault.

standing vertically through various strata, including layers of coal. If the strata were deposited over millions of years, the tree trunks would have decayed long before the uppermost strata could have been deposited. For such specimens to have become fossilized, the strata *must* have been deposited rapidly—in less time than it would take for the tree to have rotted away.[9]

It is well known that dating methods associated with fossils tend to involve circular reasoning. Fossils generally are dated by the supposed age of the strata in which they are found, and the strata are dated by the age of the fossils found there. Nothing is proved in this manner.

Sometimes creatures deemed to have become extinct millions of years ago are discovered to be still alive. A famous example is the *coelacanth*, a type of fish which was thought to be long extinct, its fossil remains being dated at about 70 million years old, with none having been found in "younger" rocks. However, about one hundred now have been caught off the coast of Madagascar. The coelacanth is now designated, curiously, as a "living fossil." (The idea of *"living fossils"* raises an obvious problem. If a creature has not changed in 70 million years, in what way does it support Evolution Theory?) [The huge sea creature hauled up by Japanese fishermen off New Zealand in 1977 and featured that year on Japanese stamps—thought originally to be a *plesiosaur*—is now known to have been the rotting corpse of a Basking Shark.]

Further arguments suggest that the geological column formed rapidly. When new *oil fields* are tapped, the underground reservoir is under such great geostatic pressure that an oil geyser results. The high pressures found within oil and natural gas beds require sudden deep burial. The phenomenon of the oil fields indicates formation in the period up to 10,000 years ago. If these oil beds were really ancient, one would expect the permeability of the rocks surrounding them to have allowed the pressure to dissipate and bleed off long ago into the adjacent rocks. Formation of oil does not need vast periods of time; it can be produced quickly from certain plants and waste products.

The age of trees provides little evidence of an ancient Earth. The oldest known living trees, the bristlecone pines of California and Nevada, are commonly held to be about 4,900 years old. All dendro-chronological extrapolations to earlier wood are based on very tentative evidence.

---

9. For evidence of present-day formation of vertical tree trunks in sedimentary layers as well as compelling evidence of rapid formation of strata and canyon erosion, see Steve Austin, video *Mount St. Helens Explosive Evidence for Catastrophe!* (N. Santee, CA: Institute for Creation Research).

And what is known beyond doubt about the length of time that mankind has lived on Earth? It seems that almost every week an item appears in the media to the effect that yet another ancient fossil finding has been made, linking mankind back many hundreds of thousands of years. Typical of such findings is the skull known as "KNM-ER 1470" found in Kenya (1972). Estimated by Richard Leakey to be 2.61 million years old, it has a strikingly modern appearance, and this is inconsistent with Evolution/old-ages theory. As discussed in Chapter 6, Marvin Lubenow has shown that specimens of *homo sapiens* trace back to the finds of australopithecines dated at the earliest. (Cf. p. 109, par. 1.)

The Discoveries of Archaeology

Charles Darwin's proposition that mankind developed through a slow evolutionary process gave rise to the idea that man's cultural development could be traced by analysis of the objects found in the various strata upon the Earth. Archaeologists have now refined a theoretical time scale of the history of mankind. Our cultural development has been classified into approximate periods of history:

- *Iron Age* (1500 B.C. to 500 A.D.): Spread of cities, Greek and Roman eras.
- *Bronze Age* (3500 B.C. to 1500 B.C.): The first cities, metalwork, writing.
- *New Stone Age "Neolithic"* (10,000 B.C. to 3500 B.C.): Permanent settlements, pottery, ground stone tools, clay figurines, domestication of animals.
- *Middle Stone Age "Mesolithic"*: An intermediate period in parts of Europe and Africa.
- *Old Stone Age "Palaeolithic"* (Before 10,000 B.C.): Hunters, crude tools, cave paintings.

The Neolithic period is markedly different from earlier periods:

> Before the tenth millennium B.C., we know only of hunters, fishers, collectors of edible fruits and roots, of people living in caves and in temporary shelters. . . . But from roughly 9000 B.C. onwards we find the first real settlements in the Near East, as people gradually became keepers of goats, sheep or cattle, and learned to cultivate grain crops. From this Neolithic phase, a series of ancient towns has come to light in

Palestine, Syria, Anatolia and Mesopotamia. Oldest Jericho became in time a walled township ten acres in extent with massive watchtowers and round houses, for a population of perhaps 2000 people. The sheer mass of the stone-built defenses, the economy based partly on local cultivation of irrigated ground and partly on trade, and the general material layout and quality of life—all suggest a well-organized community under effective leadership able to muster the common resources for major undertakings—and fearing jealous foes against whom defense was thought needful. . . . Far north in Anatolia, remarkable towns grew up at this early epoch (7th millennium if not earlier), at Hacilar (old settlement) and especially Catal Huyuk—a townsite of 32 acres, thrice as large as Jericho. Life here was enlivened with some of the world's earliest pottery.[10]

Two aspects here are noteworthy. Firstly, as Kitchen observes, mankind's cultural history appears to have a sudden and very sophisticated beginning no more than about 10,000 years B.C. Secondly, much cultural evidence assigned to the earliest human beings, apart from tool use, relates to cave paintings.

In discussing ancient cave paintings, G. K. Chesterton noted the crucial point that man differs from the brutes in kind—and not in degree. He pointed out that such art is really mankind's *signature*.[11] The point which comes to mind is this: Man being what he is, one would expect to find his "signature" right from the earliest time of his existence upon Earth. The inquisitive and creative nature of human beings would have left behind a highly significant impact. We know that brilliant individuals have made their presence felt in every generation—surely such human beings would have left behind ample and varied evidence of their inventiveness, from the beginning of mankind. There should be much more than cave paintings and artifacts.

It seems extremely unlikely that all human beings would have been content to live only a simple hunting and fishing lifestyle for some 40,000 years. It is quite reasonable to expect that some would have taken an interest in constructing buildings or other structures, leading soon to substantial accomplishments, which would show clear evidence of human presence. (Alternatively, if one proposes that early human beings were intellectually inferior to those of today, then tangible evi-

---

10. K. A. Kitchen, *The Bible in Its World: The Bible and Archaeology Today* (The Paternoster Press, 1977), pp. 20, 21.
11. G. K. Chesterton, *The Everlasting Man* (Image Books, 1955), p. 34.

dence will have to be provided to verify the claim.)

The presence of the human *signature* suggests the need for and use of human *language* with which to express abstract ideas. While the earliest known writing dates back only to about 3,100 B.C. (the Sumerians of Mesopotamia),[12] the very origin of language is quite a puzzle for modern scholars. If Evolution Theory were true, one would expect to find some evidence of the gradual development of language, perhaps from very crude grunting sounds to intelligible communication. If man's cultural history really dates back tens or hundreds of thousands of years, there should be ample evidence of this available—but such evidence is missing. According to *Funk & Wagnalls New Encyclopaedia*,

> Many contemporary primitive languages are actually more complex than modern developed languages. Many students of language today consider the problem of the origin of human language to be implicitly insoluble, since the earliest evidence of language yet discovered is necessarily only in written form, and possibly dates back no further than 6000 B.C.[13]

Since the very use of language demonstrates that human beings have rational minds, why is there is no trace of languages developing upwards from very crude to very complex forms? Did such languages all vanish without trace? Rather, the "links" of language are also missing:

> Many other attempts have been made to determine the evolutionary origin of language, and all have failed . . . why is there no trace of languages. . . . Even the people with least complex cultures have highly sophisticated languages, with complex grammar and large vocabularies, capable of naming and discussing anything that occurs in the sphere occupied by their speakers. . . . The oldest language that can reasonably be reconstructed is already modern, sophisticated, complete from an evolutionary point of view.[14]

Languages tend, it seems, to devolve from complex to simpler forms,

---

12. Kitchen, *The Bible in Its World*, pp. 15, 16.
13. *Funk & Wagnalls New Encyclopaedia*, 1978, Vol. 15, p. 43.
14. George Gaylord Simpson, "The Biological Nature of Man," *Science*, Vol. 152, April 22, 1966, p. 477.

but the very existence of language highlights an even more fundamental question for Evolution Theory. Where did speech come from? And how is it possible that infants can master a complex arrangement of mouth and tongue movements to produce intelligible sounds in so short a time?

Evolution Theory declares that man dates back at least 50,000 years, but mankind's "signature" appears to be effectively missing beyond, at best, about 10,000 years B.C. So where is the proof beyond doubt that mankind and the supposed hominid ancestry date back hundreds of thousands of years?

Radioactive Halos

The very idea that the Earth took millions of years to form and cool down is under direct challenge. Were the immense bedrocks of the Earth formed originally as hot magma, which slowly cooled over a very long time? A serious challenge to this possibility has arisen.

In countries all around the Earth, there have been found various types of minute extinct radioactive "halos" in rocks, generally granite, in quantities reaching into the trillions. Of particular interest are those radiohalos which contain only polonium atoms.

Radiohalos refer to a radiation-damaged area within a mineral, resulting from either alpha or, more rarely, beta emission from a radioactive center such as Uranium-238 (which decays in a chain of steps through atoms of different types, with lead being the final stable product). The alpha emitters produce discolored spheres, which in cross-section appear as concentric rings (the ring diameters within each sphere correspond to the ranges of the various alpha emitters). For the layman, the shape of these concentric spheres can be thought of as resembling somewhat the rings of an onion.

The importance of these halos for the question of "Age" is quite profound. Robert Gentry contends that they constitute evidence of direct, virtually instantaneous Creation—less than three minutes—of the foundation rocks of the Earth. The radiohalos denoting only polonium should not be present if the rocks had formed slowly. The radioactivity responsible for these halos had such a fleeting existence that it should have disappeared long before the magma could have had time to cool and form the rocks.[15]

---

15. Robert V. Gentry, *Creation's Tiny Mystery* (Earth Sciences Associates, 2nd Edition, 1988). See also *The Young Age of the Earth* video, Alpha Productions, 1994.

Three types of polonium elements are part of the Uranium-238 chain: polonium-210 (half-life = 138.4 days), polonium-214 (half-life = 164 microseconds) and polonium-218 (half-life = 3 minutes). Since polonium elements are only supposed to come into existence as part of the decay process of Uranium-238, the discoloration caused by these atoms should only be found in the Uranium-238 radiohalos. However, abundant examples have been found which exhibit only the polonium halos, thus indicating they were not derived from Uranium-238.

The radiocenters of these polonium halos apparently contain a type of lead not previously known (greatly enriched in the isotope 206 Pb), and it seems the polonium radiocenters have come into existence independently of Uranium-238.

> His [Gentry's] specialty is the study of minute halos in mica and biotite crystals and, more recently, in coalified wood from uranium-bearing sands in the Colorado Plateau and the Chattanooga Shale. The halos are created by alpha-particles of differing energies emitted by such substances as uranium, thorium, polonium, and other radioactives. . . . The polonium halos, especially those produced by polonium-218, are the center of a mystery. The half-life of the isotope is only 3 minutes. Yet the halos have been found in granitic rocks . . . in all parts of the world, including Scandinavia, India, Canada, and the United States. *The difficulty arises from the observation that there is no identifiable precursor to the polonium; it appears to be primordial polonium. If so, how did the surrounding rocks crystallize rapidly enough so that there were crystals available ready to be imprinted with radiohalos by alpha particles from polonium-218? This would imply almost instantaneous cooling and crystallization of these granitic minerals, and we know of no mechanisms that will remove heat so rapidly; the rocks are supposed to have cooled over millenia, if not tens of millenia.*[16]

Rather than mislead the reader, it should be stated that the issue may not be fully settled because of debate concerning the nature of the various types of rocks in which the radiohalos have been found. Some of the rocks are pegmatites, which are regarded as intrusive cross-cutting rocks. This means that they should be younger than the rocks that they intrude, and this implies a certain length of time.

Nevertheless, the essential point to note is that the polonium radio-

---

16. Raphael G. Kazmann, *Geotimes* (*EOS*, January 9, 1979), as quoted in *Creation's Tiny Mystery,* p. 61. Emphasis added.

HOW OLD IS THE UNIVERSE?

halos are there in the rocks—and they exist in enormous numbers. As Gentry points out, they provide powerful evidence that upon the creation of the Earth, many specks of specially created polonium rapidly began to decay and interacted with the primordial matter, and then the bedrocks were almost instantly cooled and "frozen" solid.

In his efforts to research other sources of polonium halos which might invalidate the instantaneous Creation Theory, Gentry instead discovered powerful evidence supporting the Flood and compounding the problems facing Uniformitarianism.

Polonium radiohalos from a secondary source had in fact already been found by others (in coalified wood), but only with the longest-life 210 isotope. Curiously, these polonium halos were found in both circular and elliptical shapes, and some had a circular shape superimposed directly on an elliptical shape. The very existence of these secondary source polonium radiohalos is itself quite remarkable, defying the odds, and has resulted in a further crucial discovery:

> We must realize that the formation of secondary polonium halos required an extraordinarily complex, interrelated series of geological events. The basic ingredients were: (1) water, (2) uprooted trees as the source of the logs and smaller wood fragments, (3) a rich uranium concentration near the wood, and (4) a compression event occurring after the uranium solution invaded the wood, but prior to its becoming coalified. The gel-like condition of the wood suggests only a short time had elapsed since the trees had been uprooted. At the very time the wood was in this special condition, it had to be infiltrated by a solution that had recently dissolved uranium from a nearby deposit. Note that if the water had contacted the uranium deposit *after* infiltrating the wood, there would have been no radioactivity in solution, and hence no possibility of forming secondary halos. The same is true if the wood had already turned to coal before contact with the uranium solutions.[17]

The polonium-210 atoms (half-life =138.4 days) would have lived long enough for them to be captured from the infiltrating uranium solution before they decayed away; whereas the other two types would have decayed away very quickly. The absence of polonium-218 (half-life = 3 minutes) is significant because it is so prevalent in granites.

It has been objected that the presence of the polonium radiohalos in

---

17. Gentry, *Creation's Tiny Mystery*, p. 56.

granites is because of infiltration (i.e., the polonium has precipitated from hot circulating fluids that have moved along grain boundaries and cracks in the granite). Yet compared to coalified wood, granite is a very tight rock into which infiltration is very slow.

The coalified wood represents optimum conditions for infiltration, and yet the key polonium-218 radiohalo did not make it under those circumstances. So, if it could not make it under ideal circumstances in coalified wood, it could not have made it by infiltration into granites.

As Gentry points out, the fact that these radiohalos were found in three different geological formations, supposedly millions of years apart in age, is most important and must be emphasized. It lends great support to the historical occurrence of a global Flood:

> The evolutionary scenario requires that the complex sequence of events described above must have been repeated more than ten million years later in the same geographical location. *That this scenario would occur a third time, again in the same area about fifty million years later, seems improbable....*
>
> Those formations represent three geological periods: Triassic, 180 to 230 million years ago; Jurassic, 135 to 180 million years ago; and Eocene, 35 to 60 million years ago. The occurrence of the elliptical secondary polonium-210 halos in specimens from all of these formations is evidence par excellence that the wood in all of them was in the *same gel-like condition* when infiltrated by a uranium solution. These data fit the flood model perfectly....
>
> If deformation of the wood occurred within just a few years after the introduction of the uranium, then only one polonium-210 halo could have been compressed because only one (from polonium-210) had been formed. Several years later another circular halo could develop (as 210 Pb-lead decayed to polonium-210) and superimpose on the elliptical halo ... From this sequence a very relevant conclusion emerges: only a few years elapsed from the introduction of the uranium to the time when the wood was compressed. These data very specifically support the flood model, which includes considerable readjustment and deformation of freshly deposited sedimentary rocks in the years after the flood waters receded.[18]

Problems in the Solar System

In addition to the information given in Chapter 4 about recent find-

---

18. *Ibid.*, pp. 56, 57. Emphasis added.

ings in the solar system, a wide range of problems has been raised against the idea that the solar system is billions of years old. Consider the difficulties concerning the continued existence of comets:

> Comets travel around the Sun and are believed to be of the same age as the solar system. Each time a comet orbits, it loses a certain amount of its mass through gravitational forces, tail formation, meteor stream production and radioactive forces. There are numerous comets, both short and long period, traveling around the Sun, but no source of new comets is known. If the Universe is billions of years old, then these comets would have traveled many thousands of times around the Sun and lost a huge amount of mass. Considering the size of comets today, their loss of mass per orbit, and extrapolating backwards billions, millions or even hundreds of thousands of years, their original mass would have had to have been several times that of the Sun—in which case the Sun would have been orbiting the comets! Therefore, the existence of hundreds of comets in our solar system with closed elliptical and random aphelia [maximum orbital distance from the Sun], proving that they are not being added to by a particular source outside the solar system—suggest a youthful solar system.[19]
>
> The lifetimes of the short-period comets have been estimated variously by different investigators. By lifetime is meant the time elapsed from the origin of the comet to its destruction. The estimates have ranged from R. A. Lyttleton's calculation of ten thousand years to Fred Whipple's calculation that on the average a short-period comet will make two hundred trips around the Sun during its lifetime. Thus, from Whipple's estimate, 1,400 years would be the lifetime of an average short-period comet, since the average period of these comets is seven years. Obviously these estimates are far apart, but several thousand years seems to be the best estimate of the life time of a short-period comet. *This seems then to put an upper limit on the age of the solar system*, if the comets came into existence at the same time as the solar system. Thus, the solar system is quite young . . . The failure to find a mechanism to resupply comets or to form new comets would seem to lead to the conclusion that the age of the comets and hence the solar system is quite young, on the order of just several thousand years at most.[20]

---

19. R. L. Wysong, *The Creation-Evolution Controversy* (Midland, MI: Inquiry Press, 1980), pp. 159-179. This book cites 33 arguments which point toward a "young" Universe.
20. Harold S. Slusher, "Age of the Cosmos," *ICR Technical Monograph,* No. 9 (San Diego: Institute for Creation Research, 1980), pp. 45, 53. Emphasis added.

Consider also the argument relating to the *Poynting-Robertson Effect*. This refers to the solar drag force exerted upon micrometeoroids in the solar system, which causes the particles to spiral into the Sun. The Sun is vacuum-sweeping space at the rate of about 100,000 tons per day. If the solar system is really billions of years old, there should no longer be any significant quantities of micrometeoroids because there is no known source of adequate replenishment (the Kuiper belt notwithstanding). In reality, however, the solar system is abundant in micrometeoroids.[21]

Further problems against an age of billions of years for the Universe can be cited: the high abundance of hydrogen in the Universe (Hydrogen is constantly being converted to helium throughout the Universe, but hydrogen cannot be produced in large enough quantity through the conversion of other elements; if the Universe were extremely old, there should not be much hydrogen left), the existence of spiral-arm galaxies which should each be almost wound up into circular shape if the Universe is billions of years old, and star clusters which are not bound by gravity and should have dispersed in less than tens of thousands of years.

Regarding the possibility that the Sun is shrinking, much caution is warranted about such a difficult matter, but it is worthwhile to note the comment made by astrophysicist John Eddy:

> I suspect . . . the Sun is 4.5 billion years old. However, given some new and unexpected results to the contrary and some time for frantic recalculation and theoretical readjustment, I suspect that we could live with Bishop Ussher's value for the age of the Earth and Sun [about 6,000 years]. I don't think we have much in the way of observational evidence in astronomy to conflict with that.[22]

Consideration must be given to data concerning the magnetic field of Earth, which behaves like a giant magnet. Its magnetism is thought by some to result from electrical currents which are generated in the core by the revolution of Earth on its axis. The magnetic field protects Earth's inhabitants from solar radiation. As the solar wind "blows" from the Sun, it confines the magnetic field to a limited space around the Earth. Measurements over the last 150 years indicate that the

---

21. Wysong, *The Creation-Evolution Controversy*, p. 171.
22. John A Eddy, quoted in R. G. Kazmann, "It's About Time: 4.5 Billion Years," *Geotimes,* Vol. 23 (September, 1978), p. 18.

strength of the magnetic field is decaying. Even using the extrapolation basis favored in uniformitarian arguments, the data indicates that Earth is likely to be quite young:

> Based on figures from 1835 to 1965, the half-life of the Earth's magnetic field has been computed to be 1,400 years. Calculating the rate of decay of the magnetic field, and extrapolating backwards, the strength of the Earth's field at any point can be computed. If we extrapolate to just 20,000 years, the Joule heat generated would probably liquefy the Earth. If we go back just one million years, the magnetic field would be 3×10^{215} Tesla, which is greater than the magnetism of all objects in the Universe and would generate so much heat that the Earth would vaporize! Therefore, according to this dating method, the Earth cannot be millions or even tens of thousands of years old. Rather, the decay of the Earth's magnetic moment speaks to (1) a creation of the Earth; and (2) an age for the Earth of less than 10,000 years. (We might also add that the magnetic field of the Earth influences cosmic ray influx. The greater the magnetic field, the less the cosmic ray influx. Since the magnetic field was greater in the past, the cosmic ray dependent C-14 method would show ages for the past greater than what they are in reality.)[23]

Barnes' contention that Earth's magnetic field was caused by a decaying electric current in Earth's metallic core stands in opposition to the self-sustaining dynamo model favored by evolutionists. Russell Humphreys has more recently proposed that rapid reversals of the magnetic field could have been brought on by the plunging of tectonic plates at the onset of the global Flood, with hot fluids rising and cold fluids sinking. He proposed that magnetic reversals should be found in rocks known to have cooled in days or weeks. As Jonathan Sarfati has pointed out, rapid reversals and fluctuations would result in even faster decay. Subsequent confirmation, by others, of Humphrey's predictions lends great support to the likelihood that Earth cannot be older than about 10,000 years.[24]

It may be objected that creationists should not use arguments which

---

23. Wysong, *The Creation-Evolution Controversy*, p. 161. See also Thomas G. Barnes, "Origin and Destiny of the Earth's Magnetic Field," *ICR Technical Monograph* (San Diego: Creation-Life Publishers Inc., 1973), No. 4.
24. See Jonathan Sarfati, "The Earth's Magnetic Field: Evidence that the Earth Is Young," *Creation* magazine (Acacia Ridge, DC, Qld 4110 Australia, Answers in Genesis, 20[2] March-May, 1998).

involve extrapolation back in time from the process rates in operation in today's world, since they criticize evolutionist scientists for using processes of extrapolation. But the use of such arguments only illustrates the inherent weaknesses in the slow, uniformitarian concept assumed in dating methods which are called on to support evolutionary hypotheses. Due consideration ought to be given to arguments for and against extrapolation from current process rates.

In his book *The Crumbling Theory of Evolution* (an important work which drew attention to the unacceptability of Evolution to Catholicism), Wallace Johnson noted the creationists' arguments which confront the idea that the Sun and planets formed from a swirling cloud of gas and dust following the supposed Big Bang:

- The cloud would not condense; force of gravity would be insufficient. Gravitational attraction would not be effective until the particles were the size of the moon.
- Particles would not accumulate to form planets. Modern science knows of no process whereby grains of dust will stick together, and accumulate to a size where gravity will take effect.
- The Sun is rotating much too slowly. The Sun's mass is more than 99% of the total mass of the solar system. If it had condensed from a cloud, the Sun should have 99% of the rotational momentum of the solar system. Instead, we find the lazily turning Sun has less than 1% of the rotational momentum, while the insignificant planets have more than 99%. How could the Sun have transferred almost all its rotational momentum to the tiny planets?
- In the Sun and in the Universe, hydrogen and helium are abundant, but the heavier elements are almost non-existent. In the inner planets, Mercury, Venus, Earth and Mars, the heavier elements are as abundant as hydrogen and helium, which makes these planets curiosities in the Universe. It is difficult to see how they could have formed from the same cloud as the Sun allegedly did.[25]

As discussed earlier, spacecraft have provided powerful evidence that the solar system was deliberately designed to look most *unlike* planetary "evolution." Who would have guessed that Pluto is a "double

---

25. J. W. G. Johnson, *The Crumbling Theory of Evolution* (Perpetual Eucharistic Adoration, Inc., 1986), p. 104.

HOW OLD IS THE UNIVERSE?

planet"? How are these solar system riddles explained? Why do planets have markedly different characteristics, defying man's understanding of them? Surely it is naïve to assert that the wonders of the solar system came naturally from a supposed Big Bang explosion. And if specialist scientists do not even fully understand the solar system, how can we begin to understand truly the much greater Universe?

A most important argument, invariably raised by those who favor the long-ages belief, is that of the immense distances which exist in the Universe and the fact that light from distant galaxies is coming to Earth from enormous light-years distances. How can this be so, if the Universe is not billions of years old?

In reply it must be asked: Is enough known as yet about the very nature of the mysterious Universe, to be sure about how light reaches Earth? Since the idea of *uniformitarianism* has been shown to be invalid upon Earth and in the solar system, why should it be assumed as true in the gargantuan Universe, which dwarfs the solar system?

Various other questions also arise: Is the Theory of Relativity fully understood and agreed upon? Is the controversy about the speed of light definitely settled? Does the Universe exist in a kind of time-warp? Does light travel to Earth through curved space? Is the red shift/Doppler effect understood with certainty? How many other solar systems have definitely been identified? Have scientists drawn mistaken conclusions from the *appearance* of age in the Universe? If aspects such as these are not yet fully explored, let alone understood beyond doubt, then caution is required.

Can we ignore the possibility that the Creator of the Universe rapidly transformed matter created on Day 1 into galaxies on Day 4, and stretched out starlight *instantaneously* throughout the Universe? Instantaneous Creation is also mature Creation.

Big Bang Problems

There are weaknesses in the Big Bang Theory. In the formation of stars, galaxies and planets, by what process did matter in the form of gas coalesce into lumps the size of planets after the primeval explosion? The theory tends to assume naïvely that an explosion of stupendous proportions, from a central point, gave rise to the bewildering complexity observable in the Universe. But, as Chris Chui points out, how could the many rotating galaxies arise this way?

How does rotation of galaxies originate from a Big Bang, since the momentum is radial and not angular? The issue of rotation has been grossly left out either in the Big Bang or the Inflation Model. Virtually every known astronomical object either rotates or revolves around some other objects in space. Initial expansion does not account for any of these rotations. To say that it is a property of matter is not enough, because charged particles only deviate from straight line motions upon the application of magnetic force fields.[26]

Commenting on the highly publicized COBE Satellite (1992) findings (which recorded temperature differences in outer space of "six parts per million, detectable only after considerable computer processing of the generally cold temperature readings"—no more than a hundred-thousandth of a degree), Chui also points out that it takes many assumptions to arrive at the belief that these signals must be remnants from a Big Bang.[27]

The radiation in the Universe is not uniform in all directions, nor is it uniform in any direction.[28] This is inconsistent with the standard Big Bang concept, as is the fact that the radiation is nevertheless far too "smooth" for gravity to have generated the "lumpiness" of the real Universe—clusters of galaxies, clustered further into super-clusters and massive "galaxy walls" that have only recently been discovered.

Edward Boudreaux contends that the Sun could not have formed according to the Big Bang scenario (i.e., as a consequence of thermonuclear fusion of hydrogen to helium) because the destabilizing forces would have exceeded the stabilizing forces by a factor of sixty to one.[29]

Another important criticism, from an agnostic scientist, refers to the oscillating idea of the Big Bang Theory. According to this idea, the Universe may be involved in an ultimate reversal of gravity, which would draw all the Universe back into a giant concentration of matter, and then begin once again with another cycle of explosion, expansion, evolution, final collapse and so on, repeated over and over. The Universe would be self-perpetuating, except for a crucial factor:

> Unfortunately, the latest measurements indicate that the expansion of the Universe will continue forever, because the amount of matter

---

26. Chris Chui, *Did God Use Evolution to "Create"?* (Logos Publishers, 1993), p. 293.
27. *Ibid.*, p. 295.
28. Morris & Parker, *What Is Creation Science?*, p. 224.
29. See Edward A. Boudreaux, "Hydrogen-Helium Ionic Model of Solar Evolution" (Dept. of Chemistry, University of New Orleans, New Orleans, LA 70148, Sept. 14, 1990).

HOW OLD IS THE UNIVERSE? 157

in the Universe has turned out to be ten times too little to exert the gravitational pull that would be needed to halt the outward movement of the galaxies.[30]

This problem for the oscillating Big Bang Theory has also been noted by other scientists. After four years' work, a six-man team of British, Australian and Chinese astronomers concluded that the mean density of matter in the Universe is only 14% of that required for the Universe to collapse back on itself.[31] Paul Davies notes that the stars constitute only 1% of the required density, but nevertheless, he is confident that sufficient "dark or invisible matter" will be found.[32]

In *Age of the Cosmos*, Harold Slusher discusses a range of data concerning the masses of clusters of galaxies. He explains why the missing mass is a strong argument against an age of billions of years:

> When the total mass of all the galaxies in the cluster is determined, the gravitational force can be calculated and considered with the observed velocity dispersion. In other words, we can calculate the amount of gravity that must be present to keep the galaxies together. Knowing the amount of gravity in turn allows us to calculate the amount of mass it would take to hold the clusters together. The result has surprised and astonished evolutionist astronomers. In the Coma Cluster the mass is too small to counterbalance the velocity dispersion by a factor of seven. In other words, for every 7 kilograms of mass necessary to hold the cluster together, only one kilogram can be accounted for. This is not a trivial matter. *There is only fourteen percent of the matter in the cluster that should be there in order for the cluster to stay together. Astronomers have looked "high and low" for this "missing mass," but it is nowhere to be found.* Things get worse in this search when clusters other than the Coma Cluster are studied; from two to ten times the needed mass is "missing" for many. For the Virgo Cluster, it turns out that there should be fifty times more mass present than is observed. Ninety-eight percent of the mass expected is not found. This is called the "missing mass" problem.
>
> To sum up . . . There is a vast discrepancy found when the mass derived from consideration of the motions of galaxies in clusters is

---

30. Robert Jastrow, "Have Astronomers Found God?" *New York Times Magazine*, June 25, 1978.
31. *The Australian* newspaper, Nov. 11, 1983.
32. Paul Davies, *The Mind of God,* p. 51.

compared with the mass that we can observe, based on the assumption that these clusters are billions of years old.[33]

Slusher then discusses various possible explanations which could account for the "missing mass" and shows why they are unsatisfactory:

> *The obvious conclusion seems to be that the "missing mass" is not really missing, since probably it was not there to start with.* The Universe, thus, could be quite young, and other lines of evidence strongly indicate this. The break-up time for these clusters (the time for dispersion of the galaxies so that there are no clusters) is far, far less than the alleged evolutionary age of the Universe. This means that the clusters, since they have not been destroyed, are young, as well as the galaxies that form them. These galaxies contain stars that are alleged by the evolutionists to be the oldest objects in the Universe (nine to twenty billion years old in the evolutionary scheme of things). This rapid break-up of the clusters coupled with their presence in the universe would indicate that these allegedly old stars are not old at all. It is believed by most astronomers that the Coma Cluster could not be younger than the Milky Way. So, if the cluster is young, the galaxy is young and the objects within the galaxy are young. The break-up times of clusters are on the order of just a few millions of years at most. So the existence of clusters argues that the age of the Universe has not reached anywhere near millions of years, which is much less than the age demanded by the evolutionists. The observations clearly indicate all galaxies are members of clusters. It has been noted that the motions of the clusters look like those of bound systems which are not breaking up at all. If that is so, then the clusters would certainly be young, not having reached a stage where they are showing a looseness of organization indicative of a very old age.[34]

The fact that the universe is so immense by human standards does not mean that it *must* be billions of years old *in time as measured now on Earth.* On the contrary, God exists *beyond* time and has unlimited power. Since He created space, time and matter entirely of His own power, God is quite capable of creating the Universe to the size and order and time-frame that so please Him.

Astronomers using the Hubble Space Telescope now think that the Universe might be billions of years younger than previously thought.

---

33. Slusher, *Age of the Cosmos,* p. 12. Emphasis added.
34. *Ibid.,* p. 13. Emphasis added.

The correlation between a proposed model of Origins and the observed facts indicates the probability of the truth of the model.

Predictions can be made about the evidence that should be observed if a particular model is true.

Discoveries from field evidence and laboratory research are overwhelmingly in favor of the Special Creation model.

Consider the following range of evidences:

DESIGN—UNIQUENESS
Design Refutes Evolution/Affirms Creation

Cells, molecules	Neither matter nor life can create itself. Irreducible complexity within cells requires intelligent Designer.
DNA spiral helix/ coded information	DNA cannot of itself give rise to truly new, "higher" genetic information.
Mosquito head	A bewildering order is seen in the design of the head.
Fingerprints—odds Improbability—law/ science precedent	Belief in Evolution requires belief in impossibly high odds, much greater than those regarded as impossible in science of fingerprinting and law courts.
Eye cross-section	Many eyes are a complex system of systems which could not have arisen by slow incremental changes.
Wasps, spiders	Some display anatomical knowledge of their victims.
Platypus, kangaroos, pandas, koalas	Fascinating examples of unique creatures, with their own unique combinations of traits.
Marsupials vs. placental mammals	Many such skeletons are almost identical, but the pouch/placenta design differs greatly.
Birds' lungs	The avian lung differs radically from other lungs —birds could not have arisen from reptiles.
Birds' feathers	Delicate, exquisite feathers could not have arisen on reptiles with sharp scales.
Complementarity/ symbiotic relations	Cross fertilization of plants by birds and insects is an excellent example of interdependence in life forms.
Dogs, cats, fishes— great variety in kind	Great numbers of variations have arisen within each original "kind."
Woodpeckers, puffins, hummingbirds	Many types of birds have fascinating traits which defy explanation by Natural Evolution.
Migratory birds and butterflies	Some fly many thousands of miles and find the destination by instinct. Requires programming by a Designer-Creator
Migratory fish	Programmed to return home by instinct.
"Tree" phylogeny or "Vertical" ancestry?	Evolutionary descent theory is lacking in both field evidence and credible conceptual arguments.
Human race types	Wide variation could have occurred very quickly from the original perfect genes of Adam and Eve.
Human fetus	We are fully human from conception and are destined (each with a rational soul) for eternity.
Language riddle	Ancient languages were already highly complex. Languages tend to devolve, not evolve.
Computers—need for intelligence	Information Technology requires a human designer. Life forms require an unseen omnipotent Designer.

GEOLOGY—FOSSILS
Evidence Refutes Evolution/Affirms Creation

Geological column/ Evolution time scale	Uniformitarianism assumptions were accepted much too soon, before enough field evidence was evaluated.
Mountain strata— multi-folded	Such massive formations must have been deposited quickly, with great upheaval, while still soft.
Cross bedded strata	Cross deposits are best explained by cataclysmic Flood events and aftermath events.
Grand Canyon USA	Striking evidence of global Flood cataclysm.
Mount St. Helens (USA) volcano	Rapid strata/canyon formation processes were recorded in the early 1980s.
Fossilized jellyfish in Central Australia	Many hundreds now lie on the desert floor; these fossils could not have formed slowly.
Fish swallowing fish	Such fossils are found around the world; they must have been rapidly engulfed by sediment.
Polystrate trees standing through strata	Such fossilized trees were rapidly engulfed by sediment in less time than it takes for a tree to decay.
Frozen mammoths of Siberia	Smothered quickly in hailstones/muck/sediment— entombed since then in rock hard permafrost.
Fossilized giant life forms	Many giant creatures were rapidly entombed in sediment and became fossilized.
Fossil record—vast museum of death	The fossil record is massive world-wide evidence of a global cataclysmic Flood and violent death.
Cambrian explosion —much too quick for Evolution	Vast numbers of highly complex creatures found as fossils, in a time span which could not allow for evolution. Fossils cannot be used for dating strata.
Dinosaur fossils are found world-wide	Vast numbers are entombed in strata—some in Alaska are not yet mineralized into rock.
Opals now "grown" in bottles	Thousands of years are not needed to form opals; several months only are required.
Piltdown Man hoax	A fraud cited for 40 years as proof of Evolution.
Nebraska Man error	Based only on one tooth finding—a type of pig.
Skull/brain comparison	There are significant differences in brain sizes between australopithecines/primates/human beings.
Ayers Rock—central Australia	Massive deposition of "fresh," sharp-edged sand particles in vertical strata suggests "recent" origin.
Polonium radiohalos	Evidence of instant creation and "cooling" of Earth; findings in coalified strata suggest "recent" creation.
Carbon-14 dating method	Equilibrium has not yet been reached; this suggests the "recent" creation of the Universe.

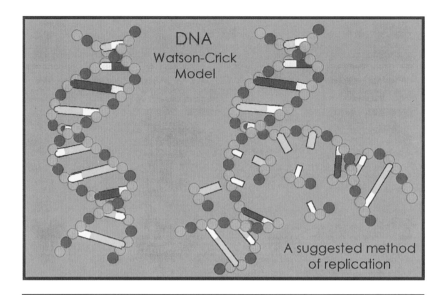

DNA Watson-Crick Model

A suggested method of replication

At the tiniest level of life forms there exist great complexity and interdependence, and DNA contains a bewildering amount of genetic information on the famous "double helix." Molecules are in fact incredibly complex "machines", which cannot work unless all parts function. Such irreducible complexity on a Lilliputian scale rules out the possibility of earlier intermediate stages. The evidence overwhelmingly suggests intelligent design.

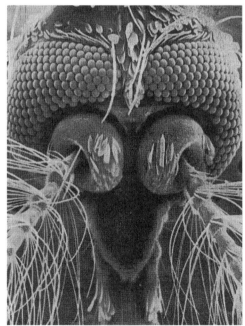

He may be small, but he's not simple!

Amazing complexity in design is seen in this greatly enlarged head-on view of a male mosquito's face.

The rows of "berries" in the upper part of the photo are lenses which form the mosquito's compound eyes.

The projections with threads attached are antennae, used to detect females.

Such masterful design surely points to a masterful Designer!

The small sparrow (circle) in this tranquil scene is almost lost amid a sea of beautifully colored flowers, while multitudes of bees go about their business of gathering pollen. Such cross fertilization shows how interdependence among plants, insects and birds has been designed into life forms.

Cleaner Fish: The Oriental Sweetlips lets the little Blue-Streak Wrasse come in and clean off loose flesh, parasites and residue from its teeth and lets it go without eating it. Such a complementary "cleaning symbiosis" defies an evolutionist explanation—both species have to be simultaneously programmed.
Photo Credit: Ocean Wide Images, Gary & Meri Bell.

Sixteen Specimens of Seven Species of Butterflies*

1. *Danaus chrysippus* (male)
2. *Danaus chrysippus* (female) f. *typica*
3. *Danaus chrysippus* (female) f. *dorippus*
4. *Hypolimnas misippus* (male)
5. *Hypolimnas misippus* (female) f. *typica*
6. *Hypolimnas misippus* (female) f. *inaria*
7. *Elymnias hypermnestra* (male)
8. *Elymnias hypermnestra* (female)
9. *Argynnis hyperbius* (male)
10. *Argynnis hyperbius* (female)
11. *Morpho rhetenor* (male)
12. *Morpho rhetenor* (female)
13. *Callicore mionina* (male)
14. *Callicore mionina* (female)
15. *Callicore astarte* (male)
16. *Callicore astarte* (female)

* Photo © Dr. Bernard D'Abrera, c/o Dept. Entomology British Museum (Nat. Hist.) London SW7 5BD.

Butterflies—Mimicry/Sexual Dimorphism and Polymorphism

Butterflies 1, 2 & 3 on the opposite page are poisonous if eaten, but 4, 5 & 6 are non-poisonous. Note that females 2, 5, 8 & 10 all bear a similar superficial resemblance. Some scientists believe non-poisonous species resort to mimicry of poisonous species in order to outwit predators. Butterflies 3 & 6 are also female—a supposedly mimic form known as form *dorippus*—and yet the pairs 7 & 8 and 9 & 10 have no form *dorippus*. Why not? If mimicry theory is valid, why do *all* butterfly species not look like poisonous ones?

Except for 1 & 2, the female forms shown on the photo do not resemble their males—11 & 12 are strikingly different—and this underlines the phenomenon of sexual dimorphism. In fact, many species have a wide range of variations of the female form, and this underlines the phenomenon of polymorphism. At what point in the supposed evolution of butterflies did the males and their respective females *both emerge concurrently in time,* so as to: (a) recognize each other as being of the same species, in spite of their different appearances, and (b) get together to mate successfully, to reproduce their own kind (remember that just one evolutionary stage of error in trial would guarantee its extinction)?

Rather than such phenomena among butterflies arising from Evolution by chance, the evidence overwhelmingly suggests intelligent design.

The neck of the giraffe could hardly have evolved simply by stretching higher—in fact, it even functions to prevent fainting when giraffes lift 550 pounds combined weight of head and neck. Little wonder that adults have hearts over two feet long which pump 20 gallons of blood each minute.

They also have four stomachs, eat about 75 pounds of food each day, drink about 10 gallons of water each time, and need a long neck to reach down to drink! In addition, each one has a unique pattern of spots.

The evidence is that of intelligent design.

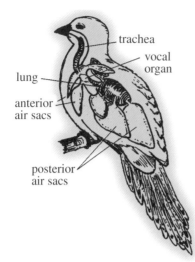

The lung and respiratory system of birds is unique and radically different from other lung systems. There is no evidence that the avian lung evolved from reptiles. On the contrary, there are incredibly complex conceptual problems facing such evolution.

Delicate birds like the Peacock and Brolga would be unlikely survivors if the Evolution scenario of "kill or be killed" were true. Male and female forms of birds are often radically dissimilar. Some, like the Parakeet, have magnificent colors, while others are plain. Some migratory birds perform amazing flying feats. Evidence found in birds suggests intelligent design.

Migratory Creatures Perform Amazing Feats of Navigation

Fairy Penguins inhabit various parts of southern Australia. Each day before dawn at Phillip Island, Victoria, they enter the sea and swim large distances, gathering food for their young, and return home unerringly each night at dusk to the same location. So punctual are they that their beach at Phillip Island regularly draws large numbers of overseas tourists, providing a major tourism asset for the State of Victoria. How do the penguins know to return home?

Short-tailed Shearwaters, commonly known as "mutton birds," also make Phillip Island their home for six months of the year, after flying an enormous distance from Siberia. After their young have been born, the adult birds fly out and migrate from southern Australia many thousands of miles back into the northern hemisphere. But the young birds fly out two weeks after the adults have left! How do they know where to go? Migratory birds must be able to fly across vast oceans on their first attempt—how did Evolution get this right the first time around?

Eels are born in the Coral Sea, between Australia and New Guinea, and swim thousands of miles to southern Australia, up Port Phillip Bay and into the Yarra River at Melbourne. They then swim about 50 miles through small creeks and over little waterfalls and live in this inland area for ten to twenty years, after which they return to the sea and swim back to the Coral Sea. There they reproduce and then die, and another generation of eels' migration begins once more. Why do they behave this way?

In North America, the tenacious return of the famous Salmon fish to their place of birth is well known. Some even find their way back up narrow drain pipes into the very pond where they were hatched! How do they know where to go? And what impels immense numbers of butterflies from northeast America to gather annually in one location in Mexico?

Such behavioral examples of migratory birds, fishes and butterflies can be cited many times around the world. There seems little doubt that they must have been programmed by an unseen Designer.

	AB	Ab	aB	ab
AB	AA BB	AA Bb	Aa BB	Aa Bb
Ab	AA Bb	AA bb	Aa Bb	Aa bb
aB	Aa BB	Aa Bb	aa BB	aa Bb
ab	Aa Bb	Aa bb	aa Bb	aa bb

Various Races Could Have Arisen Quickly From Adam and Eve*

Skin color is governed by at least two (possibly more) sets of genes, depicted above as "A" and "B," with correspondingly more "silent" genes depicted as "a" and "b." All human beings have the same basic skin color agent, melanin, but in different amounts. (The small letters in the punnet square depicted above code for a small amount of melanin.)

Human beings are not born with a genetically fixed amount of melanin, but rather with a genetically fixed *potential* to produce a certain amount in response to sunlight. Parents with genes that code only for very dark skin (AABB) will have only dark-skinned children. Similarly, parents with genes that code only for very light skin (aabb) will have only light-skinned children. And parents with genes that code only for middle brown colors will have only children with middle brown colors.

Since Adam and Eve were created in a state of high perfection and therefore possessed the full complement of genetic information, it is theoretically possible that a wide variety of human race types could have arisen very quickly. Not only would skin color have been affected, but many features of the body—color of eyes, color of hair, shape of nose, length of fingers, etc.

There are a fairly small number of main racial groups—such as Negroid, the Caucasoid, the Mongoloid and the Australoid, as well as different subgroups—but no race has anything which is, in essence, uniquely different from that possessed by another. For example, the Chinese eye, or almond eye, gets its appearance simply by having an extra fold of fat. Creationists believe that all main racial groups have arisen from Noah and his relatives who were aboard the Ark.

*Source: See Ken Ham *et al.*, *The Answers Book* (Creation Science Foundation Ltd., P.O. Box 6302, Acacia Ridge, DC Qld 4110, Australia, 1990), pp. 85-100.

"Under the circumstances it is impossible to offer decisive *proof* that no two fingers bear identical patterns, but the facts in hand demonstrate the soundness of the working principle that *prints from two different fingers never are identical.*" Prof. H. Cummins and Prof. C. Midlo, *Finger Prints, Palms and Soles* textbook (1943), p. 154.

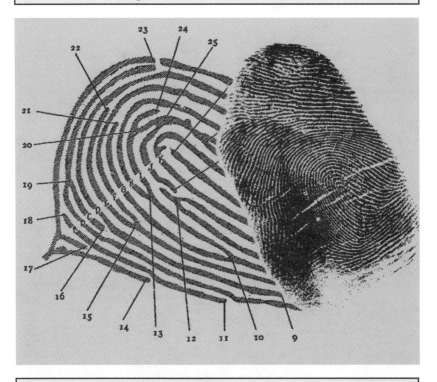

The odds against Evolution ever taking place are astronomical. Yet evolutionists say, "So what if the odds are very high? We choose to accept that, given enough time, it could have happened." But there is already a well-established precedent which ought to be deferred to. In the science of fingerprinting, the odds against two people having the same fingerprints are so astronomical, according to the head of fingerprint services at New Scotland Yard, F. E. Warboys, that they "put the question beyond the bounds of possibility." Over the past 100 years, hundreds of millions of fingerprints have been scrutinized and used for identification. Never has there been a single instance in which the uniqueness of any fingerprint has been put in any serious doubt. Evolutionists do not challenge fingerprint evidence that sends someone to prison—the odds are so high. Yet they accept greater odds by choosing to believe that Evolution happened by chance.

THE EVOLUTIONARY GEOLOGIC COLUMN

PERIOD		[MILLIONS] YEARS AGO
Pleistocene	Age of Mammals	2
Pliocene		7
Miocene	Tertiary	26
Oligocene		37
Eocene		54
Cretaceous	Age of Reptiles	136
Jurassic		190
Triassic		225
Permian	Age of Amphibians	280
Carboniferous		345
Devonian	Age of Fishes	395
Silurian		430
Ordovician	Age of Invertebrates	500
Cambrian		570
Pre Cambrian		

The standard geological column. Contrary to popular belief, all the rock levels are rarely, if ever, found together. Sometimes, supposedly "older" rock is found above "younger" rock. The evolutionist's concept of the geological column does not prove Evolution and was accepted too soon, before enough field evidence had been fully evaluated. It is simply an *idea*, which some well-qualified geologists reject in favor of alternative creationist explanations.

Were rock layers and canyon features of the Grand Canyon deposited and formed slowly over millions of years, as evolutionists believe, or quickly, during and soon after the cataclysmic Flood events? *Photo Credit: Ken Ham.*

Even in this older, black-and-white photograph, it is easy to see that the "swirling" effect of folded mountain strata must have taken shape while the sediment was still soft, and not formed slowly as solid rock. This explanation fits the creationist "catastrophe" model of Earth history.

This fossil jellyfish from Ediacara, South Australia, shows that such fossilization must take place rapidly. Such a soft creature would have to have been rapidly covered, for it would not have sunk into mud but would have rotted away if not covered quickly.

Fossil of a fish caught in the act of swallowing another fish. Such clear examples provide strong evidence of rapid catastrophic burial (in the Flood).

Beresovka Mammoth: Mammoths were "snap-frozen" in ice and permafrost across Siberia, Alaska and northern Europe. The evidence fits the creationist model of rapid events and permanent climate change across a large area following the global Flood. *Photo Credit: Dennis Swift.*

Fossilized tree trunks have been found projecting through two or more coal seams, such as this example (Swansea Heads, New South Wales, Australia). This fossil pine log's roots are broken off, so it did not grow there. The layers of coal and other strata surrounding the tree must have formed in less time than it takes for trees to rot—not over millions of years. *Photo Credit: Andrew Snelling.*

The Mount St. Helens volcano in Washington State, USA, erupted in May 1980. Much rapid erosion has taken place since the eruption. A flat plain of pumice was eroded to a depth of more than 30 meters (100 ft) by August 1984. Eight meters (25 ft) of layered rock formed in one day on June 12, 1980. . . . To all appearances, such canyons might seem to have been eroded very slowly over a long period of time. *Photo Credit: Ken Ham.*

Scientific Findings Indicative of a "Young" Earth*

- "Fingerprints" of creation (i.e., polonium halos found in some granites around the world) demonstrate that cooling may have occurred in less than three minutes, which is tantamount to creation of these rocks.
- The global Flood occurrence explains the vast accumulations of coal and oil deposits found around the world.
- The young age of coal and rapid formation of coal achieved in laboratory experiments fits only a young Earth scenario.
- Radioactive traces found in coalified wood defy the long-ages scenario.
- Dinosaurs found in mass graveyards and their tracks found in coal (displayed at Vernal, Utah) indicate they died in a global cataclysm.
- Excess helium found in deep granites shows that the Earth's crystal rocks must be "young."

Radiometric Age Dating is basically an Arithmetic Calculation*

Assumed Age = Constant x (Amount of lead ÷ Amount of Uranium).
A constant decay rate is assumed; isotopic ratios are measured in a laboratory. If the constant is really a variable, the assumed age cannot be trusted.

*Source: Robert V. Gentry, "The Young Age of the Earth,"
Alpha Productions video (1994).

U.S. physicist Robert V. Gentry is regarded as the world's leading authority on polonium halos—tiny spheres of discoloration formed in rocks by the radioactive metallic element polonium. In the Colorado Plateau coal deposits, Gentry found that the halos have been compressed and flattened, which indicates they developed when the coal was still somewhat soft in its formation process. Gentry found dual halos around a single center—one flattened and the other perfectly round. The flattened halos would have formed in 6-12 months, the round halos in 25-50 years. This coal therefore had to have hardened into its present shape within 25-50 years. (See Gentry's book *Creation's Tiny Mystery* for details.)

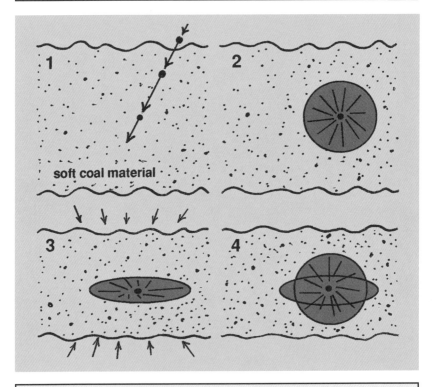

1. A radioactive bit migrates into the soft woody material and becomes lodged in a fixed position.
2. Within a few months a high-energy isotope present in the radioactive center produces a spherical halo.
3. The woody material is compressed and hardens into coal, flattening the initial radioactive halo.
4. A second isotope in the radioactive bit produces a second halo over several years. This second halo retains its spherical shape, and thus the dual halos, one flattened and one spherical, are now in place.

Source: Paul D. Ackerman, *It's a Young World After All* (Grand Rapids, Michigan: Baker Book House, 1986), p. 107.

One team has estimated 9 to 12 billion years, but another team reckon it might be 11 to 14 billion—much less than 15 to 20 billion years.

> But the difficult question of how old the oldest stars are then remains. "The ages [of the Universe] are coming out at the low end, and it's somewhat uncomfortable," said the Princeton University astronomer David Spergel.[35]

The Perplexing Universe

The Universe is very perplexing to human intelligence. Is it infinite, or does it have a boundary and, if so, what lies beyond the boundary? How can we comprehend the Universe existing within a boundary where time is in operation, and yet beyond the boundary there being no time? Questions such as these are baffling to contemplate, and, unfortunately, it seems impossible that human beings will ever be able to travel far enough to attempt to comprehend it fully.

Indeed, how can we begin to grasp adequately the sheer complexity and vast distances of outer space? Discussing the emergence of evidence that the Earth is being bombarded with thousands of huge ice comets, *adding about one million tons of water per day to the uppermost atmosphere,* Jerry Bergman captures some idea of the mind boggling Universe:

> It was once assumed that the Universe consisted of several thousand stars which were relatively closely spaced and somewhat randomly placed. *A century of intense astronomical research has now demonstrated that an incredible amount of unexplainable order, interdependence and complexity exists in the Universe.* An estimated over one-half of all stars are part of rotating binary or trinary star systems, and all known stars are parts of galaxy systems. Many of the galaxies in turn are grouped together in orbiting pairs. These pairs are organized in clusters, which tend to be consistently about 400 million light years apart. Our solar system is part of the Milky Way Galaxy, an orbiting family of stars about 100,000 light years across. Our galaxy is in turn part of a galaxy collection called the *local group* that is part of the Virgo Super-cluster. The recent discovery of the *Great-Wall galaxy cluster,* the largest object in the Universe—an estimated 200 million light years across and 500 mil-

---

35. "Astronomers Wrestle With Space Age Puzzle," *The Age* newspaper, Melbourne, May 11, 1996.

lion light years long, larger than the entire Virgo Supercluster—has forced a reassessment of the major evolutionary cosmological theories.[36]

Regarding the nature of light, it has been described variously by specialist scientists as a wave and as a particle. Some creationists have proposed that the speed of light has slowed down, and that this sheds light on the age of the Universe.[37] Measurements taken for about 300 years were claimed to show a "downhill" decreasing trend, indicating that the speed of light may have once been much faster than that of today.

A distinction was drawn between dynamic time and atomic time, and the point made that atomic clocks (those which measure time by the orbit of an electron) have been compared directly to dynamic clocks (those based on orbiting astronomical bodies) and the two are running at different rates, indicating that atomic time has slowed down.

This theory of the decaying speed of light is fascinating and has attracted a great deal of interest among Origins researchers because the major methods by which scientists have sought to demonstrate the vast age of the Universe tend to involve atomic time—e.g., the decay of radioactive minerals, the nuclear half-cycles of stars, and the time required to travel to distant objects in the Universe. If light was faster in the past, so was the rate at which radioactive elements decayed. The vast ages supposedly given by radiometric dating would need to be recalibrated significantly downwards to take this into account.

However, various creationist physicists have disagreed with the reducing-speed-of-light idea, citing especially doubts about the credibility of the data gathered many years ago. Russell Humphreys (physicist) proposes instead that the theory of General Relativity has validity, and that this makes explicable the idea of vast distances in the Universe within only about a 6,000-year time period (Earth time).[38] He suggests that the Universe was created initially in a type of time-warp, and this gives new insight on aspects such as the "red shift" phenomena which are regarded as indicative of the speed and direction of objects in outer space. ("Red-shift" refers to energy decrease in light and lengthening

---

36. Jerry Bergman, "Advances in Integrating Cosmology: The Case of Cometesimals," *Creation Ex Nihilo Technical Journal,* Vol. 10, Part 2, 1996. Emphasis added.
37. See Trevor Norman and Barry Setterfield, *The Atomic Constants, Light and Time* (Flinders University of South Australia, August, 1987).
38. See Russell Humphreys, *Starlight and Time: Solving the Puzzle of Distant Starlight in a Young Universe* (San Diego: Creation-Life Publishers, Inc., 1994).

of wavelength.) He also points out that an atomic clock at the Royal Observatory (Greenwich, UK) ticks five microseconds per year slower than an identical clock located about one mile higher in altitude at the National Bureau of Standards (Boulder, Colorado).

General Relativity in the Universe

In his 1994 book, *Starlight and Time,* Humphreys proposed a remarkable hypothesis which could explain the baffling problem of starlight traveling vast distances within a "young" Universe only about 6,000 years old. He points out that if the Universe has a substantial (water) boundary, this makes possible a radically different cosmology from the Big Bang concept, which has no center and no boundary. Gravity and black hole physics are of crucial importance.

> In particular, an experimentally measured general relativistic effect, called *gravitational time dilation* by some authors, causes clocks (and all physical processes) to tick at different rates in different parts of the Universe. (This is not the more familiar "velocity" time dilation of special relativity.) By this effect on time itself, God could have made the Universe in six ordinary days, as measured on Earth, while still allowing time for light to travel billions of light-years to reach us by natural means. The theory also appears to explain the two other major cosmological phenomena we see: the red shifts of light from distant galaxies and the cosmic microwave background radiation. Thus, this biblical foundation appears to lead to a young-Earth cosmology, which is consistent with Einstein's general theory of relativity and astronomical observations. As measured by clocks on Earth, the age of the Universe today could be as small as the face-value biblical age of about 6,000 years.[39]

Humphreys also points out that an "event horizon" is essential to both "black holes" and "white holes" cosmology, and that *strange things happen to time near an event horizon.*

The term "black hole" describes a collapsing star whose matter has all fallen within its Schwarzschild radius, and from which neither light nor matter can escape. The sphere defined by this radius is called an "event horizon" because one can never see events happening inside the collapsing star. (An "event horizon" is the point at which light rays try-

---

39. Humphreys, *Starlight and Time*, p. 54.

ing to escape from a black hole bend back in on themselves—*it is also where time is massively distorted*.) As more matter falls into a black hole, its mass increases and the event horizon is always moving outward. In contrast, a "white hole" is a term used by some astrophysicists to describe a black hole running in reverse. As matter is expelled and leaves the white hole, its mass decreases and the event horizon moves inward until it reaches radius zero and disappears, leaving behind a scattered collection of matter moving away from a central point.

Humphreys hypothesizes that the Universe was once within a huge white hole, that Earth is close to the center of the Universe, and that the Universe has a substantial boundary of water. Because of the curcial importance of explaining the vast distances/young Universe puzzle, Humphreys' scenario for the Creation Days is presented here at length:

> *Day One:* God creates a large 3-D space and within it a ball of liquid water, the "deep." The ball is greater than two light-years in diameter, large enough to contain all the mass of the universe. . . . Because of the great concentration of matter, this ball of water is deep within a black hole, whose event horizon is more than half a billion light years away. The Earth at this point is merely a formless, undefined region of water at the center of the deep, empty of inhabitant or feature. . . . the gravitational force on the deep is very strong, more than a million trillion "g"s. This force compresses the deep very rapidly toward the center, making it extremely hot and dense. The heat rips apart the water molocules, atoms, even the nuclei into elementary particles. Thermonuclear fusion reactions begin, forming heavier nuclei from lighter ones and liberating huge amounts of energy. As a consequence, an intense light illuminates the interior, breaking through to the surface and ending the darkness there. (This paragraph following is the most speculative of my reconstruction of events.) As the compression continues, gravity becomes so strong that light can no longer reach the surface, re-darkening it. . . . at this point the Spirit of God "moving [or 'hovering'] over the surface of the waters" (*Genesis* 1:2) becomes a light source, in the same way as He will again become a light source at a future time (*Rev.* 19:23, 20:5). . . .
>
> *Day Two:* By direct intervention God begins stretching out space, causing the ball of matter to expand rapidly, thus changing the black hole into a white hole. He marks off a large volume, the "expanse" within the deep, wherein material is allowed to pull apart into fragments and clusters as it expands, but He requires the "waters below" and the "waters above" the expanse to stay coherently together. Normal physical processes cause cooling to proceed as rapidly as the

expansion. Heat waves are stretched out to much longer wavelengths as a relativistic consequence of the stretching of space. Eventually, these stretched-out waves will become the cosmic microwave background radiation. Matter beneath the expanse expands until the surface reaches ordinary or present termperatures, becoming liquid water underneath an atmosphere. God collects various heavier atoms beneath the surface and constructs minerals of them, laying "the foundations of the Earth" (*Job* 38:4), i.e., its core and mantle. Gravity at the surface drops to normal or present values. Out in the expanse, matter is drawn apart, leaving irregular clusters of hydrogen, helium, and other atoms formed by the nuclear processes of the first day. The waters above the expanse stay together . . .

Day Three: Rapid radioactive decay occurs, possibly as a consequence of the rapid stretching of space. The resulting heating forms the Earth's crust and makes it buoyant relative to the mantle rock below it, causing the crust to rise above the waters, thus gathering the waters into ocean basins. I hypothesize that rapid volume cooling of molten rock deep within the Earth also occurs, again as a result of the rapid expansion of space, solidifying the rock. . . . The continuing expansion of space causes the waters above the heavens to reach the event horizon and pass beyond it. This causes the amount of matter within the event horizon to begin decreasing, which in turn causes the event horizon to begin rapidly shrinking toward the Earth. There are no stars yet, only clusters of hydrogen, helium, and other atoms left behind in the expanse by the rapid expansion.

Day Four: The shrinking event horizon reaches Earth early on the morning of the fourth day. During this ordinary day as measured on Earth, billions of years worth of physical processes take place in the distant cosmos. In particular, gravity has time to make distant clusters of hydrogen and helium atoms more compact. Early on the fourth morning, God coalesces the clusters of atoms into stars, and thermonuclear fusion ignites in them. The newly formed stars find themselves grouped together in galaxies and clusters of galaxies. As the fourth day proceeds on Earth, the more distant stars age billions of years, while their light also has the same billions of years to travel to the Earth. While the light is on its way, space continues to expand, relativistically stretching out the light waves and shifting the wavelengths toward the red side of the spectrum. Stars which are now farthest away have the greatest redshift, because the waves have been stretched the most. This progressive redshift is exactly what is observed.

Completion: God stops the expansion before the evening of the sixth day. Therefore, Adam and Eve, gazing up for the first time into the new night sky, can now see the Milky Way, the Andromeda galaxy, and all the other splendors in the heavens that declare the glory of God.[40]

The Missing Supernova Remnants

Jonathan Sarfati has highlighted an excellent argument against an age of billions of years (in time as measured now on Earth) for the Universe, and that is that the third stage remnants of supernovas (i.e., exploding stars) are missing.

When the nuclear fuel of a star is exhausted, gravitational collapse results in electrons and nuclei being crushed together to produce a core of neutrons. An explosion of gargantuan proportions then occurs, during which the "supernova" star will outshine all the billions of stars in its galaxy. (The Crab Nebula is the remnant of a supernova explosion which was seen in the year 1054 A.D.)

Supernova remnants (SNRs) are thought to pass through three stages: The first stage, that of hurtling debris, expands to about 23 light years distance over 300 years' time, after which a blast wave forms. The second stage expands to about 350 light years' distance, until it starts to lose heat energy over a time period of 120,000 years. The third stage expands to 1,500 light years over six million years' time.

Compared to 2,260 second stage and 5,000 third stage SNRs which should be observed from Earth if the Universe is billions of years old (Earth time), only 200 second-stage and *zero third-stage SNRs* have been observed. The fact that the third-stage supernova remnants are missing completely confounds the idea of the Universe's being billions of years old. If the Universe is billions of years old, we should expect to observe about 5,000 third-stage remnants. Why are there none? The observed evidence is far more consistent with a Universe which has been stretched out rapidly by God during the relatively "recent" Creation events.[41]

A Place in the Sun

Earth is ideally placed for life to exist—the right distance from the Sun, the right rotational speed, and with a tilt to give the four seasons

---

40. Humphreys, *Starlight and Time*, p. 33-38.
41. Jonathan Sarfati, "Exploding Stars Point to a Young Universe," *Creation* (Acacia Ridge, Qld, Australia), Vol. 19, No. 3 (June-Aug. 1997).

upon which so much of nature's interdependence relies. Any closer to the Sun would be much too hot; any further away would be much too cold. A slower rotation speed would make days and nights too long; a faster rotation speed would mean cyclonic conditions. Earth has water and gases which are harmonious to life and delicious food in the form of plants and fruits. It is the only known place in the Universe where biological life is possible.

Believers in Special Creation can assert positively that Earth was intended by God to be an abode for the crowning achievement of the physical Universe—human beings. The wonders of the Universe exist for man to stand in awe at the greatness of the Creator. From a naturalistic evolutionary viewpoint, however, a believer in chance processes must accept that he or she is incredibly lucky even to exist. To think that a gigantic explosion (The Big Bang) eventually gave rise to an earthly environment which is so well balanced that it could allow extremely complex life forms to flourish is truly bewildering. Therefore it is reasonable to conclude that truly gigantic odds must have been involved for all this to happen by chance.

If this conclusion is granted, however, the precedent which rejects the probability of immense odds, discussed in Chapter 2, also confirms that mankind's presence on Earth is definitely not a matter of luck. By the same reasoning, nor is the existence of the Universe a matter of lucky events occurring by chance over billions of years.

(Discussion of the question of "Age" resumes in Chapter 15.)

— Chapter 10 —

POINTERS TO THE CREATOR

Riddles for Naturalistic Evolution

As can be seen from the material already discussed, a number of riddles can be cited which pose great difficulties for the idea of chance Evolution. These riddles point strongly toward the existence of an unseen Designer. Consider the range of questions which cannot be addressed satisfactorily by Naturalistic Evolution—not even by appeal to the Anthropic Principle:

1. What can be said about the fundamental nature of matter and energy? Since matter itself is comprised of atoms and subatomic particles, did these in fact evolve from something else? Could pure chance alone account for the amazing complexity and incredible Lilliputian smallness of the world of atoms?

2. What constitutes the Universe; is it infinite, or does it have an end—and how does one account for the phenomenon of time? If the Universe did have a beginning in time, what can be said about the nature of space before the beginning?

3. If the Universe had a beginning in a supposed Big Bang explosion, before which all the matter in the Universe was compressed into a very small concentration, why did it explode and not collapse through its immense gravity? Indeed, what are the forces which could make it explode?

4. From where and how did the first protein originate, to facilitate the DNA code structure in each species? Proteins depend on DNA for their formation, and DNA cannot form without pre-existing protein.

5. How is it possible for marine plants suddenly to become adapted to growing on land, when radically different root systems and vascular systems are involved?

6. How could pure chance explain the balance in nature? What random process can account for the truly amazing interdependent wonders

of nature as seen, for example, in the behavior patterns of bees, ants and butterflies? How did symbiotic relationships between certain plants and creatures get started, when they depend on each other for survival?

7. What can explain the instinct which impels migratory birds to travel huge distances back and forward around Earth each year, some utilizing a number of different wind currents? What causes vast numbers of butterflies in the northeastern United States to assemble annually in one place in Mexico, after flying across the Gulf of Mexico? Some scientists speculate that migratory habits of certain birds, fishes and animals may be somehow guided by the Earth's magnetic field. Others suspect that birds are guided by stars.

8. Why do predatory wasps and spiders sometimes act as if they had an exact knowledge of the anatomy of their victims and know how best to paralyze them? How do spiders know to place a sticky substance on only central web fibers and not on outer support fibers?

9. How does one account for the fact that air-breathing creatures basically inhale oxygen and exhale carbon dioxide, and plants alternate between both, depending on darkness and daylight? What process of Evolution could "know" this and bring it about?

10. Why do creatures stop growing at maturity and not continue to grow indefinitely? How did the first immune system get started in time to avoid extinction, and why does a wound or an organ which has been damaged repair itself only to the old boundaries?

11. Could random beneficial mutations really explain the complexity of sense systems involving the brain, nerves, eyes, ears, nose, taste and touch? The human brain is composed of some 12 billion neurons, thus numbering about 120 trillion connections.

12. Where did "mind" come from—the self-awareness of human beings as distinct from the consciousness of animals? What is it that tells a person that something is either beautiful or ugly? What can be said about the relationship between the perception of colors and the conception of beauty? Man's memory, the possession of skills and the appreciation of the arts, delightful smell and of humor all seem unlikely to be simply accidental happenings.

13. Where did man's conscience and concept of right and wrong come from? Where did love and altruism come from, especially if the savage world of natural selection in its widest sense ("kill or be killed") is true?

Questions such as these can never be satisfactorily answered by naturalistic belief systems. *It requires a far greater act of faith to believe that the Universe and everything in it has come about by accident than to believe it is the work of an omnipotent Creator.*

Clues to the Creator

E. F. Schumacher, of *Small Is Beautiful* fame, had some outspoken views about the impact of Evolutionism on the consciousness of man. Even though he evidently believed that some form of biological Evolution had actually occurred, he nevertheless saw that

> Evolution as a process of the spontaneous, accidental emergence of the powers of life, consciousness and self-awareness, out of inanimate matter is utterly and totally incomprehensible. If the accidental emergence of the higher from the lower is possible, then anything and everything is possible, and there is no basis for human thought. Two plus two would not have to be four, but could just as well be five or anything else; nor would there be any necessity for us to believe that two minus two leaves nothing: why not believe it could accidentally make five?[1]

As to whether the existence of a Creator can be deduced from observation of the world around us, Schumacher proposed a very coherent process of thought in his "four levels of being," which point to life after death and thus help us scientifically to deduce the existence of a brilliant, unseen Intelligence, commonly referred to as God.

Noting that mankind has long been aware of four "kingdoms" in nature (mineral, plant, animal and human) he pointed out that rocks, soil, water and the like consist only of matter; that plants are also matter but have life as well, that animals are matter and have life but many also have consciousness; and that human beings are matter, but not only do they have life and consciousness, they also have self-awareness. Thus the mineral level could be expressed as "m," plants "m + x," animals "m + x + y," and human beings "m + x + y + z." This all points toward yet another level of existence:

> Aware of this, mankind has always used its imagination, or its intuitive powers, to complete the process, to extrapolate the observed curve to its completion. There was thus conceived a Being, wholly active, wholly sovereign and autonomous; a *Person* above all merely

---

1. Schumacher, *A Guide for the Perplexed*, p. 28.

human persons, in no way an object, above all circumstances and contingencies, entirely in control of everything: a *personal* God, the "Unmoved Mover." The four Levels of Being are thus seen as pointing to the invisible existence of a Level (or Levels) of Being above the human.[2]

As well as having self-awareness, human beings also possess the ability to discuss ideas and are vastly more endowed than all other creatures on Earth. This baffles some evolutionists.

> The mystery in all this is that human intellectual powers are presumably determined by biological evolution, and have absolutely no connection with doing science. Our brains have evolved in response to environmental pressures, such as the ability to hunt, avoid predators, dodge falling objects, etc. What has this got to do with discovering the laws of electromagnetism or the structure of the atom? John Barrow is also mystified . . . "Why should it be *us*? None of the sophisticated ideas involved appear to offer any selective advantage to be exploited during the preconscious period of our evolution. . . How fortuitous . . ."[3]

Other pointers can be cited. As the human body grows old and begins increasingly to degenerate, the mind within the person becomes more experienced and generally more wise. One could ask, for what purpose do people become wiser with age, if not for supernatural destiny—just as the fetus grows to a certain stage in the mother's womb with limbs and eyes and ears that are destined to be of use in a completely different environment.

Unfortunately, an attitude common to many people is the tendency to gloss over the difficult arguments raised against Evolution and, in effect, to assert that there are no substantial problems.

In a different vein, Robert Jastrow commented on the Big Bang Theory, which holds that the Universe began in a colossal explosion. He noted that such a theory would explain why the Universe appears to be expanding—thus invalidating the Steady-State theory of the Universe—and he believed therefore that it must have had a beginning. But some scientists are loath to concede the possibility of a First Cause at work in the Universe. Like everyone else, these scientists have human biases, which color their judgment:

---

2. *Ibid.*, p. 37.
3. Paul Davies, *The Mind of God*, p. 149.

Theologians generally are delighted with the proof that the Universe had a beginning, but astronomers are curiously upset. Their reactions provide an interesting demonstration of the response of the scientific mind—supposedly a very objective mind—when evidence uncovered by science itself leads to a conflict with the articles of faith in our profession. It turns out that the scientist behaves the way the rest of us do when our beliefs are in conflict with the evidence. We become irritated, we pretend the conflict does not exist, or we paper it over with meaningless phrases . . .

I think part of the answer is that scientists cannot bear the thought of a natural phenomenon that cannot be explained, even with unlimited time and money. There is a kind of religion in science; it is the religion of a person who believes there is order and harmony in the Universe, and every event can be explained in a rational way as the product of some previous event; every effect must have its cause.

Einstein wrote, "the scientist is possessed by the sense of universal causation." This religious faith of the scientist is violated by the discovery that the world had a beginning under conditions in which the known laws of physics are not valid, and as a product of forces or circumstances we cannot discover. When that happens, the scientist has lost control. If he really examined the implications, he would be traumatized. As usual, when faced with trauma, the mind reacts by ignoring the implications; in science this is known as "refusing to speculate"—or trivializing the origin of the world by calling it the big bang, as if the Universe were a firecracker . . .

Now we would like to pursue that enquiry farther back in time, but the barrier to progress seems insurmountable. *It is not a matter of another year, another decade of work, another measurement, or another theory. At this moment it seems as though science will never be able to raise the curtain on the mystery of Creation.* For the scientist who has lived by his faith in the power of reason, the story ends like a bad dream. He has scaled the mountains of ignorance; he is about to conquer the highest peak; as he pulls himself over the final rock, he is greeted by a band of theologians who have been sitting there for centuries.[4]

---

4. Jastrow, "Have Astronomers Found God?" Emphasis added.

—PART III—

CHRISTIAN INSIGHTS

—Chapter 11—
CHRISTIAN INSIGHTS

Granted that Scripture is truly free from error of any kind, empirical science will never discover any facts which will conclusively contradict Scripture; therefore, there is no need to fear its discoveries.

Some have argued that empirical science and theology do not mix, but on the contrary, notice ought to be taken of what empirical science has established about nature and about the Universe—not to make theology dependent on this form of science, but rather to see if modern discoveries shed further light on the true meaning intended to be conveyed by the sacred writers of Scripture.

Science has verified beyond doubt that a human being is conceived when a female ovum is impregnated by male sperm, and from the very beginning the tiniest human life has the full complement of genetic information. Thus the fetus does not develop into a human being, but is already fully human right from the moment of conception.

Because human life is a continuum from conception to the grave, each individual thus cannot be deemed to have only "potential" rights, but rather must be accorded full rights throughout all stages of life. Christian doctrine has always held to the intrinsic worth of each person (having a rational soul made in the image of God), but the empirical science aspects only recently have been clearly understood.

Within Christianity there rages a controversy concerning objective truth versus subjective concepts of reality, reflecting the conflict between traditional/conservative and liberal/revisionist beliefs on theology and exegesis. The debate is confused by the fact that there is no common agreement about how truth itself should be defined.

Traditional Christianity takes note of reality outside of man and regards objective truth as having an existence independent of man. Human beings are creatures made by God and subject to His laws, which have been made known by Divine Revelation. Since human beings consist of both body and soul, they are destined for eternal existence in the timeless world after death. Though man lives in a world beset by confusion, greed and sorrow, it is possible to rise above the difficulties of this life. We have been given a free will by God, and can become truly "set free" by choosing to live in conformity with God. Most importantly, human beings can discern objective truth.

On the other hand, the subjective viewpoint of reality concentrates on the inner realm of the individual. It tends to be concerned primarily with consciousness. One's thoughts, feelings, aspirations, search for identity and experiences all tend to take precedence over consideration of the world external to the self. The existence of a transcendent God may not seem so important in this approach to life.

These two respective viewpoints do not necessarily have to conflict, for they can be complementary. There does exist a world external to the individual, and it does not depend upon the consciousness of the person for its existence—but there is also an inner world of man's consciousness, wherefrom one is often forced to dredge up some meaning to life. When the respective truths are overlooked, narrow attitudes can result. At one extreme, individuals can become depersonalized, unduly authoritarian and lacking in compassion. At the other extreme, excessive preoccupation with the "self" can lead to breakdown of respect for moral standards and responsible authority.

Care must be taken to establish what is true and what is false in scientific theories. The Galileo case remains a constant reminder of this. The lingering impression is that some 17th century theologians allowed their subjective views to cloud an objective assessment of heliocentrism as proposed by Copernicus. The common belief now is that they would not look openly at the arguments and merely reacted out of prejudice. This belief is mistaken, but it nevertheless highlights an interesting challenge to modern scholars: One ought to expect that an objective assessment will be made of the detailed arguments now compiled by Origins researchers and that fair, detailed consideration will be given to them, rather than rejection out of hand.

Since the *Genesis* revelation is foundational to Christian beliefs, the arguments in favor of Special Creation cannot be dismissed as irrelevant. While *Genesis* contains only a partial account of the events of Creation, the quest for more complete understanding is worthwhile. We should expect to find that the revelation recorded in *Genesis* would tend to verify the model of Special Creation and to confound theological arguments used to support Evolution beliefs.

—Chapter 12—

PROBLEMS IN THEISTIC EVOLUTION

Theistic Evolution Now Pointless?

In examining the credibility of Theistic Evolution, it is necessary to recall that, long before science shed light on the existence of DNA, 19th century arguments being presented in favor of Evolution appeared to have a sound basis. Evolution Theory was seen increasingly as a credible science, but it contradicted the thrust of traditional Christian doctrine, which had generally accepted the concept of Special Creation.

In reaction to this, many Christian scholars proposed a synthesis to combine Creation and Evolution. Evolution would be accepted as the method of creation used by God, and implementation of the Creation events would be stretched out over eons of time.

Today, Evolution theories are seen as clearly lacking in convincing data and as being burdened with scores of conceptual problems. Ironically, *natural selection can no longer be cited in favor of Evolution.* Who would seriously deny that the mechanism of Evolution is still missing? It is known with virtual certainty that Evolution cannot occur, because DNA has been designed so that only variety within kind can occur. *The mechanism of Evolution cannot be found if it never existed.*

With Evolution effectively ruled out as a credible possibility, there is no point in Christian scholars continuing to devise all sorts of highly dubious "what-if" possibilities, trying to blend contrived divine intervention scenarios with fanciful concepts of "Evolution." There is no justification for flirting with Naturalistic philosophy and for seeking an evolutionary synthesis with Catholic doctrine. Any further speculation about "Theistic" Evolution is pointless and should be abandoned as a mistaken venture which has been very costly to the Church founded by Jesus Christ. Instead, greater focus should be placed on the rediscovery of the true story of Creation, with God the Creator at center stage.

Despite the claims of those Christians who still insist that Evolution occurred, the weight of modern evidence provides little support for the idea that God may have used Evolution as part of the means of Creation. One can confidently assert that the evidence is overwhelmingly consistent with the account given in *Genesis*.

Theistic Evolution writers almost invariably gloss over the hard details and fail to think through fully their own positions. For example, Fr. William Kramer is highly scornful of "young-age" creationists in his 1986 book, *Evolution & Creation—A Catholic Understanding*, but not surprisingly, he avoids defining the actual mechanism of Evolution. He insists that Evolution is *fact*, and presents many generalized, vague statements lacking in details. His treatment of Original Sin is meager.

In reality, the attempted synthesis known as Theistic Evolution stands exposed as contradictory to Catholic Tradition, especially on death, and this effectively renders it *impermissible*.

Human Death

If in favor of chance Evolution, one ignores the likelihood that God created Adam's body in adult form from inorganic elementary particles drawn from the ground, the question of death cannot be ignored.

According to theory, the evolution of human beings took place over many generations. Since every new level of Evolution was considered to be brought about by extremely rare beneficial mutations, many years must also have been involved. It is reasonable, therefore, to deduce that it would take a long time for early human beings to evolve slowly from cave man status to a high level of intelligence.

Theistic evolutionists are very vague about who exactly these evolving creatures were. As already discussed, the idea of so-called "hominid" evolution, especially of those possessing conceptual thought, is a myth anyway—they never existed! The very possession of a rational soul defines a human being, and Adam was the first human being. But if many generations were involved, leading up to the time when Adam would arrive and make his awesome decision, then many deaths *must* also have occurred among his ancestors, as they were slowly evolving up to the time when human beings would acquire sufficient intelligence with which to comprehend the result of disobedience to God. Further, if Evolution had taken place by random chance and life-death struggle, then death as a natural process was a deliberate part of God's means of making man the pinnacle of Evolution.

Scripture informs us otherwise. The death of human beings only entered into the world as a direct consequence of Original Sin; it only began to occur after the sin of Adam. St. Paul teaches that "as by one man [i.e., Adam] sin entered into this world, and by sin death." (*Romans* 5:12). The teaching in the Bible is quite clear—*death began with the sin of Adam*—"death reigned from Adam." (*Romans* 5:14).

The writer of the book of *Wisdom* indicates that death of human beings was not intended by God: "For God made not death, neither hath he pleasure in the destruction of the living. For he created all things that they might be. . . . But by the envy of the devil, death came into the world." (*Wisdom* 1:13-14, 2:24).

Vatican II reaffirmed this essential teaching in the Pastoral Constitution on the Church in the Modern World (*Gaudium et Spes*), and Pope Paul VI endorsed it in his *Credo of the People of God* (1968). Pope John Paul II also endorsed it in the 1992 *Catechism of the Catholic Church*:

> Finally, the consequence explicitly foretold for this disobedience will come true: man will "return to the ground," for out of it he was taken. *Death makes its entrance into human history.* (400).
>
> "God did not make death, and He does not delight in the death of the living." (*Wis.* 1:13). (413). Even though man's nature is mortal, God had destined him not to die. (1008).

Thus, this version of Theistic Evolution contradicts both Church doctrine and the Scriptural revelation that death was the direct result of Original Sin. In addition, the problem of death highlights another very difficult problem for Theistic Evolution: Evolution's general "upward" tendency away from imperfection toward perfection is at odds with the biblical account that God's perfect Creation degenerated "downward" toward corruption only as a direct consequence of the sin of Adam.

In contrast to the Creation deemed "good" by God, both Theistic Evolution and Progressive Creation (see Chapter 16) unavoidably require belief that conditions on Earth were grim right from the beginning and were intended so by the Creator. The disobedience of Adam was thus immaterial to this situation. Such belief stands in stark contrast to Catholic doctrine, reiterated again in the *Catechism of the Catholic Church*:

> Adam and Eve were constituted in an original "state of holiness and justice." . . . As long as he remained in the divine intimacy, man would not have had to suffer or die. The inner harmony of the human person, the harmony between man and woman, and finally the har-

mony between the first couple and all creation, comprised the state called "original justice." (375, 376). This entire harmony of original justice [was] lost by the sin of our first parents. (379).

But what about the death of life forms other than human beings? Were animals, birds and fishes, especially the predator types, affected by the Fall? Of creatures on Earth, only human beings possess rational souls and thus are destined for immortality. The Catholic Church is clear in the teaching on human death, but has not given a definitive teaching on death of other life forms. Much caution is therefore required here.

Non-Human Death

Regarding the death of plants, it is indicated in *Genesis* that God primarily intended for plants to be eaten by creatures on and above land, and so it seems fair to conclude that loss of plant life should not be equated with the death of creatures where bloodshed is involved.

The idea that Adam and Eve were created above a vast museum of death—a worldwide graveyard of dead creatures—while elsewhere animals were being torn apart alive by predators, is overdue for reconsideration. In this situation, only the Garden of Eden could have been a beautiful spot; outside of it would not have been very nice—not exactly a delightful place for human beings to enjoy paradise on Earth if Adam had chosen obedience to God.

One view is that carnivores were *intended* by God to devour herbivores to stop the herbivores from multiplying to such an extent that they would starve to death. Thus, predators would always have preyed on other live creatures, resulting in great bloodshed, with vultures feeding on the spoils and the spreading of diseases being quite rampant.

Can this death-riddled scenario be the tranquil scene deemed "very good" by God, or are we hampered by a distorted perception of the original Cosmos? Living in a fallen world, we have to contend with the consequences of Adam's disobedience, and we have little information about how life would have been if Adam had chosen obedience to God. But we do have information given to us in Scripture about the Creation events by the sole and absolutely trustworthy Eye-witness—God.

Since animals do not possess rational souls, there is no need to insist that they would not have died, even if the Fall had not occurred. But the actual mode of non-human death, peaceful or otherwise, in a non-fallen world is an important puzzle within the Origins debate.

The deliberate implementation of predatory savagery seems, to this writer, most inconsistent with the revelation about God—the picture coming through in Scripture is that death is abhorrent to God and was only brought on by sin. Therefore, was such savagery the deliberate choice of God, or was it brought into play by the tragic but inexorable consequences of the choice made by Adam in exercising his free will?

Mankind was given dominion over all other creatures on Earth (*Gen.* 1:28), and all land creatures were intended to live as vegetarians (*Gen.* 1:29-30), presumably in a state of tranquility. Many of the large beasts today (as well as most of the extinct dinosaurs) function as herbivores; is it unreasonable to conclude that the vicious nature of predators could have been overridden by grace? Snakes may not have felt the need to inflict their deadly venom, and an abundance of small sea herbs could have satisfied the appetites of what are now predatory sea creatures without the need for them to attack other sea creatures or human beings.

Were radical effects brought on by the Fall which would otherwise have remained overridden by grace? Consider the mysterious mention of "thorns and thistles" coming forth (*Gen.* 3:18)—what does it mean? Such things can come about as an adaptation to overgrazing, but this seems unlikely if the Earth had lush vegetation everywhere before the Flood catastrophe. Rather, the grim message in *Genesis*—"cursed is the earth in thy work; thorns and thistles shall it bring forth to thee; and thou shalt eat the herbs of the earth." (*Gen.* 3:17-18)—suggests that such things as "thorns and thistles" only arose because of Adam's disobedience. Perhaps they would have remained as harmless plants, if Adam had chosen obedience. They could have been created with soft, harmless "thorns" and "needles," which may have changed after Adam sinned.

Similarly, perhaps "thorns and thistles" (*Gen.* 3:18) include features such as viruses and diseases which otherwise would have remained dormant, never to be activated if Adam had chosen obedience. The truth is, of course, unclear. Microscopic creatures, devouring other life forms within larger creatures, seem an intrinsic part of the design of creatures.

It is not being argued here that the intrinsic nature of wild beasts was changed by the Fall, but rather that their *behavior* was affected. Adam was disobedient, and their docile behavior became disordered. The *Catechism of the Catholic Church* teaches that "human nature . . . is wounded in the natural powers proper to it" (405) due to the Fall—could not animals also have been wounded in their behavior by the Fall?

If the "natural powers" proper to non-human creatures (e.g., their reproduction rate and general behavior) had been kept intact by grace, this would provide for great harmony among all creatures. Perhaps even the digestive systems of animals and birds, and even fish, could have been such that they would have functioned only as vegetarians.

By way of analogy, we know that human beings, under the protection of grace during the preternatural period, would have kept their own desires under perfect control, and we would not have known physical death. The possibility for human death already designed into the human body would have remained overridden, and the aging process as happens today would not have affected human beings.

As is obvious, we are dealing with mysteries, and there is no neat set of answers to the problem of non-human death. If the Creation days were each 24 hours, it seems unlikely that any animal deaths occurred in the (probably) short time before the sin of Adam. But if the Fall had not occurred, what then? We do not even know for sure whether non-human deaths would have occurred before the human population reached the required number to repopulate Heaven—something known only to God.

This is *not* to suggest that non-human creatures would not have died if Adam had remained obedient. It must be admitted that, because such creatures do not possess rational souls, God may have intended that they die, even in a state of tranquility. But a further problem then arises concerning the mode of their deaths. With no enemies and no dread of others, what would be the actual cause of death? It can be deduced that entropy would have been in existence, even in a non-fallen world (e.g., unused fruit would decay)—perhaps a slow aging occurrence in animals would have been facilitated by the process of entropy not being fully replenished within such creatures. A very slow aging effect would result in the eventual peaceful death of animals.

This discussion of non-human death is not irrelevant.[1] It offers an alternative scenario to intentional evolutionary *"kill or be killed"* taking place upon a tranquil Earth, had Adam remained obedient. It is consistent with Scriptural references to future harmony between the lion and the lamb, and with the gracious disposition of Christ—the very same Creator of Adam (359)—who "groaned in the spirit and troubled himself . . . And . . . wept" (*John* 11:33, 35) on hearing of the distress caused by the death of Lazarus. The sheer repugnance of death, undesired by God, is a theme which deserves deeper contemplation.

---

1. For another discussion of such problems in Theistic Evolution, see *Création et Rédemption*, Fr. André Boulet (Chambray-les-tours Cedex, France: C.L.D., 1995 Edition). The book by Fr. Boulet is prefaced by the President of the Doctrinal Commission in France, Msgr. Henri Brincard, Bishop of Puyen-Velay.

If mankind is considered to be the product of chance processes, then further problems in Theistic Evolution must be addressed. What stage of body formation, brain development, consciousness and self-awareness did the evolving creatures reach before God first implanted a rational soul?

What Constitutes a Human Being?

The "soul" is the principle of life in any particular living thing, despite whether it be a human being, some other type of creature, or a plant. The important distinction, however, is that the soul of the human being is rational (i.e., capable of abstract reasoning), whereas other creatures are not rational and are dominated by instincts.

It is not possible for an animal to evolve to the stage of being fully human without also having a rational soul.

Scripture informs us that human beings are of an immensely higher order of Creation than the animals. Man was created "in the image of God" and given "dominion over the fish of the sea and over the fowls of the air . . . and every creeping creature that moveth upon the Earth." (*Gen.* 1:26-27). The implications of this revelation do not sit easily with the attempted synthesis known as Theistic Evolution.

(It is easy in the modern world to overlook that Adam and Eve probably possessed brilliant intelligence. When tempted that they could be "like God," this might have seemed very plausible to them.)

If Adam and Eve were the end result of a long evolutionary process, what can be said about the other evolving creatures on the brink of becoming fully developed human beings? What happened to them? Did they also reach the stage of becoming almost human (but without rational souls) and then all die out? (The question of death is a huge conceptual problem confronting the idea of polygenism [i.e., many "first parents"], but it is not addressed satisfactorily by theistic evolutionist writers. See Chapter 13.)

One possible answer is that these others either ceased evolving or all died out, and that only Adam and Eve and their descendants survived. However this is a hypothetical and very convenient explanation with no supporting evidence in the fossil record. Nevertheless, it highlights a dilemma for the concept of Theistic Evolution.

If the evolving creatures are deemed to have reached the stage of being almost human, the theistic evolutionist must account for both their lack of rational souls and for their extinction as a group. On the other hand, if these creatures only reached the stage of hominid, then theistic evolutionist theorists are forced to argue, implicitly, that God intervened not only to implant the rational soul, but also to change the *kind* of creature radically to that of a human being.

When carefully considered, the idea of Evolution of human beings by chance processes is untenable. To avoid the problem of the death or extinction of man's ancestors, one must argue that animal parents suddenly gave birth to a human being—a completely different kind of creature. In addition, they would have to produce at least one male and one female so that the new human species could propagate. The genetic problems in all this necessarily involve immense odds, and this means zero probability.

It is more reasonable for the theistic evolutionist to suggest that God intervened in the process. But this only raises further problems.

Divine Intervention

Despite any claims to the contrary, the logic of Theistic Evolution does necessitate the abandonment of Evolution Theory in favor of divine intervention. Because of the problem of the death of Adam's forebears, Christians who prefer to believe in Evolution are forced ultimately to abandon standard theories and to appeal implicitly to divine intervention on innumerable occasions over billions of years. Since there are no credible "missing links" found in the fossil record, it is necessary for Christian evolutionists to introduce an act of divine intervention each time a creature is deemed to "jump" to a new and more complex type.

Macro-scale Evolution, with new *"higher"* genetic information, is now widely regarded as impossible, and its ever-elusive mechanism remains missing. Further, the idea of *punctuated equilibria* to explain each "jump" is lacking in credibility because of the immensity of odds against so many rapid beneficial mutations.

Therefore, for changes to occur which result in the gaining of new *"higher"* genetic information in each species, genetic "jumps" are required and one must resort to claiming repeated *Theistic Intervention* for its accomplishment. But to label this as "Evolution" tends toward deception, since it is very misleading to those who assume that the commonly understood concept of Evolution is involved. The Naturalistic process of Evolution—belief in which ironically gave rise initially to the quest in search of an acceptable version of Theistic Evolution—is not involved at all.

It can be argued, of course, that God intervened and caused a series of beneficial mutations to occur concurrently in the germ cells of two ape-like progenitors as they conceived an offspring. In this way, the body of Adam could have been brought into existence. The animals would not then be truly regarded as his parents since he differed from them in kind and not in degree. But is this valid reasoning? Apart from the fact that this concept has nothing to do with standard Evolution theories, the following objections can be raised:

Firstly, if the body of Adam was in any way derived from animal "parents," then he would have arrived as a baby boy. There would be no one to nurture him, talk to him, hug him and so on—all of the things that we know are vital for infants to cultivate their intelligence. (The feral children, raised in India by wolves, were found culturally impoverished, so one could hardly argue that Adam would have been raised by ape-like creatures. They are not his *kind*.)

We know that Adam must have possessed a very high level of intelligence; after all, the destiny of mankind was at stake in the decision which rested upon him. Therefore, his body *must* have been brought into existence in adult form. Alternatively, there seems little point in arguing that God may have changed an adult ape into that of a human being. This also is unrelated to Evolution Theory; one may as well believe that Adam was specially created as an adult human being, which *Genesis* strongly indicates.

Secondly, if a similar divine intervention occurred in the creation of the body of Eve, with animal "parents," then how could she in any way have been derived from the body of Adam?

Any evolutionary concept requiring divine intervention effectively means that Evolution Theory is no longer the issue under consideration. Even if an interventionist scenario is not specifically advanced by theistic evolutionist writers, nevertheless their beliefs require it and so a completely different concept of Origins is at issue. Whatever else may be its merits, it does not relate to Evolution Theory as commonly understood by scientists. It would have to be accepted on faith, with only a tenuous basis in Scripture, if any at all.

The concept of Special Creation also incorporates divine intervention on some occasions, but only as can reasonably be inferred from Scripture (e.g., when God brought on the Flood events).

Distortion of Doctrine

Pope Leo XIII declared that the literal-as-given sense should be held until rigorously disproved, and the concept of Special Creation harmonizes neatly with the account revealed in *Genesis*. In contrast, Theistic Evolution has to be *read into* Scripture and imposed upon it, but the sacred writer(s) of *Genesis* (including God as the principal Author of Scripture) gave no clue that this was how the Universe was created.

The 1992 *Catechism of the Catholic Church* (in section 406) notes that the Church's teaching on the transmission of Original Sin was articulated more precisely in the 5th Century, following St. Augustine's opposition to the ideas of Pelagius, who had reduced the influence of Adam's fault to bad example. It was further addressed at the Second Council of Orange (529 A.D.) and the Council of Trent (1546 A.D.). Under Pope John Paul II, the *Catechism of the Catholic Church* reiterates the central importance of this foundational doctrine: *"The Church, which has the mind of Christ, knows very well that we cannot tamper with the revelation of original sin without undermining the mystery of Christ."* (389). (Emphasis added).

Theistic Evolution does tamper with doctrine. It distorts the teaching on human death, and it can lead unfortunately to a weakening of faith and facilitate a gradual loss of interest in religious belief. There are many fine Christians for whom Evolution Theory poses no problems for their faith, and who may be unaware of the inherent conceptual difficulties in Theistic Evolution. But what of those who may lack a sound formation in Christian teachings? Some Christians, swayed by Evolution theories, can tend soon to ignore the seemingly plausible arguments of Theistic Evolution and to accept the logic of "pure" Evolutionism. Attitudes of relativism can thus be accepted gradually, and the perception of absolute principles can become blurred.

It is quite easy for those who lack a sound formation in Christian beliefs to accept that stories of Original Sin and Adam and Eve are myths, true only in a vague religious sense. Confusion over man's Origins leads directly to confusion over man's destiny, and this has contributed greatly to the modern collapse of faith in Christian doctrine. Once the truth of man's Origins is distorted, belief in other vital doctrines becomes weakened. Many now believe that Christ was not divine, miracles are unbelievable, and that Satan and Hell are only fairytale hangovers from a superstitious past.

Evolution almost certainly never occurred and, from the genetic information we now have, cannot occur, so why waste time and effort in trying to legitimize the erroneous idea of Theistic Evolution?

—Chapter 13—
THE POSITION WITHIN CATHOLICISM

Origins Still Awaiting Clarification

Many Catholics overlook the fact that prior to the modern era there was little need for Magisterial declarations on Origins, and they come to the conclusion therefore that Evolution beliefs pose little danger for Catholicism.

Cardinal Ernesto Ruffini, in his 1940's analysis showing how human evolution is irreconcilable with Catholic doctrine, sought to establish from the Bible and from Catholic Tradition what God has revealed about the beginning of the world and the origin of mankind. He noted that *the science of genetics gives no support to Evolution* and that the ever-elusive mechanism of Evolution is missing. (If Evolution never happened, there is no mechanism to find.) He also succinctly defined the role of the Catholic Church as personally commissioned by Jesus Christ to teach doctrine until the end of time:

> Christ established an organization to continue, until the end of the world, the bringing of men to eternal happiness. This organization is the Catholic Church. He endowed His Church with the prerogative of infallibility so that it cannot err in teaching and interpreting His doctrine. God's revelation to man, which is called a Deposit, is found in the Bible, written under divine inspiration, and in Tradition. By Tradition is meant not something vague or legendary, but the actual living teaching of the Church itself which, under the abiding assistance of God promised by Christ, ever continues to transmit to men the doctrines received from Christ and His Apostles. Important witnesses of this Tradition are those ecclesiastical writers of the early centuries who are called Fathers of the Church. Divine revelation is so vast in extent and so profound in content, that sometimes its meaning can be determined only after study and discussion. Progress in this field consists in the deeper understanding by men of what is

contained in revelation. The knowledge given to us by God is as unchangeably true as are the truths of mathematics.[1]

The Origins debate is essentially a fairly modern controversy within Christianity. Apart from speculation by ancient Greeks some 600 years before Christ, the Evolution Theories of Naturalists (which were supposedly based on phenomenal data) only began to impact man's thought substantially from the early 19th century. The idea that Scripture contains true history (described "according to appearances"), as well as presented in the literal-as-given obvious sense, was first explained by Pope Leo XIII as recently as 1893. Thus, various Origins aspects have not been fully addressed in Tradition and still await clarification by the Magisterium.

Many Catholics incline toward generalized concepts of *Theistic Evolution* and argue that "Evolution" could have been the method of Creation used by God. (At issue, however, is what God actually *chose* to do when creating the Universe and all creatures and plants, and not what He *could have done*.) Many are loath to accept as literally true any passage in the *Genesis* Creation accounts, and appeals to Tradition tend to be dismissed as irrelevant. The Pentateuch tends to be seen as primarily applicable in a "salvation history" sense, as if it were known with certainty that the Divine Author did not intend to convey history. *At stake in the Origins debate is nothing less than the integrity of* Genesis, *so foundational to the Church founded by Jesus Christ.*

Catholic proponents of Evolution find little real comfort in *official* Magisterial teachings. But their openness to Theistic Evolution necessarily involves a departure from standard Evolution Theory, in favor implicitly of innumerable divine interventions. In addition, they now face another problem of credibility. Not only was *polygenism* (many "first parents") effectively prohibited in the 1992 *Catechism of the Catholic Church* (endorsed by Pope John Paul II), but various Creation aspects were reiterated in it. *In contrast, the word "Evolution" was not specifically mentioned even once in this catechism!*

Many paragraphs in the *Catechism of the Catholic Church* relate to Creation themes, including the following: The existence of God can be known by reason; the Bible is totally free from error; the great trustworthiness of God, who cannot deceive; the very point of creation of the Universe was to create human beings; God did not make death—

---

1. Ernesto Cardinal Ruffini, *The Theory of Evolution Judged by Reason and Faith* (New York: Joseph F. Wagner, Inc.), p. 64. (1959 English translation by Fr. Francis O'Hanlon, Melbourne, Australia). Originally published in Italian in the 1940's.

which only came into the world because of Adam's sin (*Romans* 5:12); human nature is thus wounded in its natural powers and the whole of Creation groans also in result; the importance of secondary causes; the great need to be mindful of Catholic Tradition; the Flood is mentioned in covenant context. A few sections which could vaguely be said to support evolutionary concepts are also explicable by Special Creation beliefs. For example,

> The question about the origins of the world and of man has been the object of many scientific studies which have splendidly enriched our knowledge of the age and dimensions of the cosmos, the development of life forms and the appearance of man. . . . (283).
>
> The great interest accorded to these studies is strongly stimulated by a question of another order, which goes beyond the proper domain of the natural sciences. It is not only a question of knowing when and how the universe arose physically, or when man appeared, but rather of discovering the meaning of such an origin: is the universe governed by chance, blind fate, anonymous necessity, or by a transcendent, intelligent and good Being called "God"? . . . (284).

The Catholic Church is unlikely to enter into and make pronouncements on matters which belong only in empirical science, but where empirical science and theology overlap, the Church is entitled to and has declared on important matters which affect the salvation of souls (e.g., with regard to abortion, in vitro fertilization and contraception). Thus, it is entirely appropriate for the Church to declare upon Origins matters, such as the possibility of Adam and Eve's being derived from previously living matter, or upon the historical reality of the Flood, or upon other important Origins aspects in which the domain of scientists, theologians, and exegetes can overlap.

In view of actual Magisterial teachings, *Genesis* cannot be written off as applicable only in fanciful "salvation history evolutionist scenarios." Nor can it be written off as a "story" understood in existentialist scenarios, as proposed by Rudolf Bultmann in his quest to demythologize Scripture. Despite current widespread disobedience to its doctrinal teachings, the Catholic Church still officially forbids the teaching of Evolution as though it were already proved.

The Church is bound to prohibit belief in Godless Origins and Godless Evolutionism, and the teaching on Original Sin can never change because 1) it is central to the Redemption paid by our great Creator/Redeemer, and 2) the Church cannot overturn even one doctrine already defined as true; otherwise, all credibility of Catholic

Tradition would be lost. As the *Catechism of the Catholic Church* reminds us, "*the Church, which has the mind of Christ, knows very well that we cannot tamper with the revelation of Original Sin without undermining the mystery of Christ.*" (389). (Emphasis added.)

In view of so many weird Origins views constantly being advocated by so many scholars around the world—contradictory of doctrine and harmful to the Church founded by Jesus Christ—one can only hope and pray that the Magisterium of the Church will recognize the need for an updated encyclical on Origins, to bring enlightenment where now there is great confusion. One must also hope that a wide range of views will be consulted, and not only those of the present *Pontifical Academy of Sciences* and the present *Pontifical Biblical Commission.*

The Doctrine of Original Sin

At the heart of the Origins controversy lies the doctrine of Original Sin, a teaching central to Christianity. As well as Pope Pius XII in *Humani Generis* (1950), Pope Paul VI endorsed its foundational importance in quite unmistakable terms in his *Credo of the People of God* (1968):

> We believe that in Adam all have sinned. By that we mean that the original sin he committed affected human nature itself. In what way? Through his sin, human nature, common to all men, fell into a state in which it incurs the consequences of his act. This new state, then, is not the one in which human nature first existed in our First Parents. They, in their origin, were set up by God in a state of holiness and righteousness. They had no experience of evil or of sin. But it is their fallen nature which has been passed on to all their descendants. These are, in consequence, destitute of the gift of grace that once adorned human nature. They are wounded even in their natural powers. They have incurred a liability to death which Adam and Eve passed on to all their descendants. All that is what we mean when we say "Man is born in sin." In accordance with the teaching of the Council of Trent, we likewise hold that original sin is transmitted along with human nature, and not acquired by imitation. We hold therefore that it is in each one of us as something proper to each person.[2]

The 1992 *Catechism* also reiterated the teaching on Original Sin in unmistakable terms:

---

2. Pope Paul VI, *Credo of the People of God*, Section 16.

> How did the sin of Adam become the sin of all his descendants? The whole human race is in Adam "as one body of one man." By this "unity of the human race" all men are implicated in Adam's sin, as all are implicated in Christ's Justice. Still, the transmission of original sin is a mystery that we cannot fully understand. But we do know by revelation that Adam had received original holiness and justice, not for himself alone, but for all human nature. By yielding to the tempter, Adam and Eve committed a *personal sin,* but this sin affected *the human nature* that they would then transmit *in a fallen state.* It is a sin which will be transmitted by propagation to all mankind, that is, by the transmission of a human nature deprived of original holiness and justice. And that is why original sin is called "sin" only in an analogical sense: it is a sin "contracted" and not "committed"—a state and not an act. (404).
>
> Although it is proper to each individual, original sin does not have the character of a personal fault in any of Adam's descendants. It is a deprivation of original holiness and justice, but *human nature has not been totally corrupted: it is wounded in the natural powers proper to it*; subject to ignorance, suffering, and the dominion of death; and inclined to sin—an inclination to evil that is called "concupiscence." Baptism, by imparting the life of Christ's grace, erases original sin and turns a man back toward God, but the consequences for nature, weakened and inclined to evil, persist in man and summon him to spiritual battle. (405). [Emphasis added.]

Clearly, Adam was the one who committed the Original Sin, and all mankind is wounded in consequence. The Catholic Church does not *officially* accept the speculation that "we all have chosen sin," a revisionist view which tries to accommodate polygenism by inferring that each one of us, being sinful, has rejected God as did early mankind. (In such a view, Adam and Eve would be only symbolic representations of mankind.)

In view of the massive dissent from Catholic belief which erupted openly in the 1960's (emerging when the Church still *appeared* doctrinally united, with the traditional Latin Rite of Mass still in use throughout the Western Church), in view of the ongoing dramatic decline in religious practice and the revisionist distortion of Original Sin, the concern raised by Cardinal Ruffini in the 1940's has turned out to be quite prophetic:

> *If, in the question of man's creation, the obvious meaning of the Bible is abandoned, a meaning which has been received and confirmed by constant Catholic Tradition, what attempt can be made to*

> *defend the account of the earthly Paradise, of the fall of Adam and its consequences? If it be admitted that the body of an animal became fit in the course of centuries to be informed by the human soul, will the unity of the human race remain sufficiently established against polygenism? And if this unity collapses, what will be the fate of the doctrine of original justice and original sin which constitutes the foundation of our sacred religion?*[3]

Prior to release of the 1992 *Catechism*, the major pronouncement by the Magisterium on Origins was in the encyclical *Humani Generis* (1950), in which important doctrinal teachings of past Church Councils were reiterated by Pope Pius XII.

Catholic Tradition

Despite the vague, highly tenuous Evolution scenarios proposed by modern Catholic evolutionists, how could the Catholic Church ever *officially* accept that the bodies of Adam and Eve were the product of natural Evolution? How could Adam and Eve have evolved naturally from previously living matter when DNA will not allow it to happen? The only recourse for the evolutionist is to argue that God intervened and miraculously transformed non-human life into human beings, and instantaneously created their rational souls—but *Genesis* gives no hint that this actually happened.

Why not believe the *Genesis* revelation that God miraculously transformed some non-living "dust" (understood now as elementary particles) from the ground into the living body of Adam? Cardinal Ruffini also argued strongly (Ruffini, pp. 124-136) that famous teachers as far separated in time as St. Irenaeus, St. John Chrysostom, St. Ambrose, St Jerome, St. Thomas Aquinas and Pope Pius XI all taught that Adam was directly created by God from the "dust" or "slime" of the earth. A strong case can thus be made that truth known from Tradition effectively renders *impermissible* the idea of "evolution" of the human body:

> To these authorities—Holy Scripture, the Holy Fathers, major and minor theologians—we must add the Christian sense (*sensus fidelium*, the faithful echo of the Church's teaching), so universal on this question and so certain that almost no member of the faithful would be free from surprise and scandal if he heard the teaching that

---

3. Ruffini, *The Theory of Evolution,* p. 164.

Adam was born of beasts, that the blood in his veins was the blood of animals, that the human race, as regards the flesh, is related to the brute beasts.[4]

If it is true that the body of woman was formed directly by God and thus does not come by way of evolution, who will be persuaded that man's body comes from the brute beast? *What an absurdity!* . . . If we wish to stand by Holy Scripture we must accept it in its entirety. . . . She gets the name *Virago* (*ishah*: woman) because she is taken from the *vir* (*ish*: man); likewise the man is called *Adam* (=homo) because, as Genesis says, he is taken from the *adamah* (=humus). *Whenever Holy Scripture speaks of the origin of the human body, it always names the Earth and only the Earth.*[5]

In a learned analysis of Church doctrine in relation to the question of Origins, Fr. Peter Damian Fehlner has shown that Tradition demonstrates the Church's constant opposition to Evolution Theory down the centuries as regards the origin of the world and of the many species, especially of the human species. As well as drawing on the teachings of the early Church Fathers and of various popes, Fr. Fehlner notes the declarations of the Vienne, Fifth Lateran, Trent, and the First Vatican Councils. He also observes that

> Evolutionary theories stress the continuity of development between the species from the lower to the higher, as well as a sufficient duration to permit the operation of natural or artificial causes according to the laws governing these. Catholic teaching stresses an essential discontinuity in the case of those essences, whose limits were fixed by the Creator and which cannot be modified by the intervention of natural or artificial agents of a finite power. [The Church] has never pretended in any instance of observable species, on the basis of revelation, to know what those limits are. But that there are such limits, even at the level of inanimate existence, sound science as well as philosophy has tended to confirm.[6]

Fr. Fehlner also points out why *Evolutionism* cannot, by definition, be accepted by the Magisterium:

---

4. *Ibid.*, p. 137.
5. *Ibid.*, p. 123. Emphasis added.
6. Fr. Peter Damian Fehlner, "In the Beginning" (Three-part series published in *Christ to the World*, Rome: Via di Propaganda, 1988), Volume XXXIII, Nos. 1-4, p. 243.

> Good arguments can actually be adduced in fact to show that evolution is simply not a scientific hypothesis. It is a dogma providing the context for all scientific endeavors. And it is just this assumption of evolutionism as the universal paradigm that directly conflicts with the teaching of the Church . . .
>
> The doctrine of creation, in general and in all its detail, is intimately bound up with the mystery of salvation. That is why no Catholic may call into question any aspect of the doctrine of creation which in fact the Church believes related to the mystery of salvation without also doubting that latter mystery.[7]

In the light of Tradition, there can be little doubt that Adam was directly created when God made his body using the "dust" or "slime" of the earth. Eve was also directly created by God, using a portion of Adam's body around which the rest of her body was created. Their rational souls were also directly created at the same time as their bodies. As Cardinal Ruffini strongly argued in his book (Ruffini, p. 120), it is known with *certainty* that Eve's body was formed directly by God and definitely was not the product of Evolution. Ruffini also argued that the Church Fathers *unanimously* interpreted the text relative to the formation of Eve in its proper literal sense (Ruffini, p. 121), even typifying the origin of the Church from the side of Christ crucified.

In addition, as brought to our attention by Fr. Brian Harrison, O.S.,[8] Pope Leo XIII taught clearly in 1880 in the encyclical letter *Arcanum Divinae Sapientiae* (Christian Marriage) that Adam and Eve are our first parents and that Eve was created from a portion of Adam's body:

> We record what is to all known, and cannot be doubted by any, that God, on the sixth day of creation, having made man from the slime of the earth, and having breathed into his face the breath of life, gave him a companion, whom He miraculously took from the side of Adam when he was locked in sleep. God thus, in His most far-reaching foresight, decreed that this husband and wife should be the natural beginning of the human race, from whom it might be propagated, and preserved by an unfailing fruitfulness throughout all futurity of time.[9]

---

7. *Ibid.*, pp. 246, 247.
8. Brian W. Harrison, O.S., "Did The Human Body Evolve Naturally? A Forgotten Papal Declaration." *Living Tradition*, Nos. 73-74. (Jan.-March 1998).
9. Pope Leo XIII, encyclical letter *Arcanum Divinae Sapientiae—The Pope and the People* (London: Catholic Truth Society, 1910), p. 178.

Also, Pope Pelagius I, in writing to King Childebert I in the year 557, declared his own strong conviction that neither Adam nor Eve was born of other parents but were instead created, and that Eve was created from a portion of Adam's body:

> For I confess that all men from Adam, even to the consummation of the world, having been born and having died with Adam himself and his wife, who were not born of other parents, but were created, the one from the earth, the other [*al.: altera*], however, from the rib of man [cf. *Gen.* 2:7, 22].[10]

As Fr. Harrison points out, Vatican II, in *Lumen Gentium*, para. 25, recognized four conditions which must be fulfilled in order for a doctrine to be proposed infallibly by the ordinary Magisterium. He argues powerfully that these conditions had been fulfilled by the year 1880 with respect to doctrines regarding the origin of Adam and Eve recalled by Leo XIII. The *effective* infallible teaching of the ordinary Magisterium in Catholic Tradition, by Popes and bishops, has been rejection of naturalistic evolution of Adam and Eve. (And once infallible, always infallible—by definition.) The four conditions cited as amply fulfilled by Fr. Harrison are as follows:

1. The Catholic bishops teaching the doctrine must be in communion among themselves and with Peter's successor.
2. The bishops must be teaching authentically in matters of faith and morals.
3. The teaching in question must be one that the popes and bishops agree upon. (General unity, not necessarily absolute unanimity.)
4. The teaching of the popes and bishops must be presented as one to be held definitively.

Nevertheless, many Catholics (both clergy and laity) still cling to evolutionary scenarios, hoping—against the voluminous evidence—that the mechanism of Evolution will emerge. But there is no justification in resorting to highly fanciful, implausible, theistic evolutionary explanations for the origin of Eve. The *Genesis* account is specific; she was directly created when God fashioned her body from a portion of

---

10. Pope Pelagius I, from *Fide Pelagii* in the letter *Humani Generis* to Childebert I, April, 557, *The Sources of Catholic Dogma*, Henry Denzinger's *Enchiridion Symbolorum* [228a] (30th Edition, 1957).

Adam's body, and this fact was affirmed by the Ordinary Magisterium by Pope Leo XIII in 1880.

Regarding the origin of Eve's body, the attempted synthesis of Evolution Theory with Catholic theology was bound to founder on Catholic Tradition. In fact, it was clearly doomed back in 1880 by Leo XIII, long before modern science could shed clear light on the behavior of DNA. Ironically, the "father" of the science of genetics—Fr. Gregor Mendel (d. 1884)—was still alive in 1880 and his published but unheralded scientific findings would not be rediscovered until the early 1900s.

With respect to the possibility of "special transformism," as distinct from "natural transformism," can serious credence be given to divine intervention scenarios in any so-called "evolutionary" origin of Adam and Eve? Divine intervention scenarios for the "evolutionary" origin of Adam and Eve are really highly speculative and unwarranted and ought to be rejected. Such scenarios have an aspect in common with the highly fanciful notion of "punctuated equilibrium"—that of being open to the charge of trying to prove something by the complete absence of proof.

Polygenism Proscribed

Polygenism is the idea that human beings arose from many "first parents" via Evolution,[11] and that Adam and Eve are symbolic representations of mankind. On November 30, 1941, in an address to the *Pontifical Academy of Sciences*, Pope Pius XII identified three elements that must be retained as certainly attested by the Sacred Author of *Genesis*, without any possibility of allegorical interpretation:

1. The essential superiority of man in relation to other animals, by reason of his spiritual soul.
2. Derivation in some way of the first woman from the first man.
3. The impossibility that the immediate father or progenitor of man could have been other than a human being, that is, the impossibility that the first man could have been the son of an animal, generated by the latter in the proper sense of the term.[12]

---

11. The idea of multiple pairs of first parents in one locality is known as monophyletism; in various places and times is known as polyphyletism.
12. See Fr. John A. Hardon S.J., *The Catholic Catechism* (London: A Geoffrey Chaptman book published by Cassell Ltd., 1977), p. 92.

As Fr. John Hardon, S.J. pointed out, in context this statement of Pope Pius XII reads, "Only from a man can another man descend, whom he can call father and progenitor." Even if the Pope had primarily intended in this address to stress the great gap in kind which exists between animals and human beings, nevertheless, as Cardinal Ruffini also pointed out, the possibility that human beings could have been born of animal parents is untenable.

Nine years later, in the encyclical *Humani Generis,* Pius XII not only expressly forbade the teaching of Evolution Theory as though it were already proved, but *he also instructed that data pointing both for and against Evolution Theory must be given due consideration.* Further, he also declared that

> The first eleven chapters of *Genesis,* although properly speaking not conforming to the historical method used by the best Greek and Latin writers or by competent authors of our time, *do nevertheless pertain to history in the true sense* which, however, must be further studied and determined by exegetes; the same chapters in simple and metaphorical languages, adapted to the mentality of a people but little cultured, both state principal truths which are fundamental for our salvation and also give a popular description of the origin of the human race and chosen people.[13]

Catholics are not free to believe in Naturalistic Philosophy, and Pope Pius XII did *not specifically* define Evolution as an "open question"—so how can it now truly be regarded as an open question, as though it does not matter whether human beings evolved from ape-like ancestors? After all, scientists, theologians and exegetes were permitted by the successor of Peter in 1950 *to hypothesize only* about the possibility of evolution of the human body:

> The teaching authority of the Church does not forbid that, in conformity with the present state of human sciences and sacred theology, research and discussions on the part of men experienced in both fields take place with regard to the doctrine of evolution insofar as it inquires into the origin of the human body as coming from pre-existent and living matter—for the Catholic Church obliges us to hold that souls are immediately created by God.
> However, this must be done in such a way that reasons for both

---

13. Pope Pius XII, *Humani Generis*, Denzinger p. 2329 (Wanderer Press translation). Emphasis added.

opinions, that is, those favorable and those unfavorable to evolution, be weighed and judged with the necessary seriousness, moderation and measure, and provided that all are prepared to submit to the judgment of the Church, to which Christ has given the mission of interpreting authentically the Sacred Scripture and defending dogmas of faith.

Some, however, highly transgress this liberty of discussion when they act as if the origin of the human body from pre-existing and living matter were already completely certain and proved by facts which have been discovered up to now, and by reasoning on those facts, and as if there were nothing in the sources of Divine revelation which demands the greatest moderation and caution in this question.

When, however, there is a question of another conjectural opinion, namely polygenism, children of the Church by no means enjoy such liberty. For the faithful cannot embrace that opinion which maintains either that after Adam there existed on this Earth true men who did not take their origin through natural generations from him as from the first parent of all, or that Adam represents a certain number of first parents.

Now it is in no way apparent how such an opinion can be reconciled with that which the sources of revealed truth and documents of the teaching authority of the Church propose with regard to Original Sin, which proceeds from sin actually committed by an individual Adam and which *through generation* is passed on to all and is in everyone as his own.[14]

In commenting on the Fall and in addition to referring to the Council of Trent and to Pope Paul VI, the *Catechism of the Catholic Church* referred to the above section in *Humani Generis* concerning polygenism:

> The account of the fall in *Genesis* 3 uses figurative language, but affirms a primeval event, a deed that took place *at the beginning of the history of man*. Revelation gives us the certainty of faith that the whole of human history is marked by the original fault freely committed by our first parents. (390).

Although Pope Pius XII expressed grave concern about polygenism, did he nevertheless leave an opening for it with the words, *"it is in no way apparent how such an opinion can be reconciled . . ."*? Various theistic evolutionists since 1950 have asserted that they can envisage

---

14. *Humani Generis,* Denzinger, pp. 2327-2328. Emphasis added.

possible (if highly implausible) ways in which many "first parents" can be reconciled with the doctrine of Original Sin. But J. Franklin Ewing S.J., professor of anthropology and *a theistic evolutionist*, was personally convinced that Pope Pius XII regarded polygenism as irreconcilable with Original Sin. Writing in 1956, Fr. Ewing candidly acknowledged this:

> Although the exact doctrine that Adam and Eve were the first parents of all men since their time has never been defined, still one is struck by the fact that all the ecclesiastical documents concerning them take this for granted. The Council of Carthage in 418; the Council of Orange in 529; and the Council of Trent in 1546—to mention outstanding and ecumenical examples, all speak of original sin, and in this connection of *one* Adam. The Biblical Commission, in 1909, mentions "the unity of the human race" as one of the fundamental doctrines reported in *Genesis*. All the Scriptural references dealing with our first parents plainly take it for granted that there was one man and one woman. Pope Pius XII, however, does not so much lean on the Scriptures in drawing up his condemnation of polygenism. *He emphasizes the evident irreconcilability of Catholic doctrine concerning original sin with polygenism.*[15]

Conceptual Weakness in Polygenism

As well as the Magisterial prohibition by Pope Pius XII, the idea of polygenism has a conceptual weakness anyway which poses problems for those who suggest that Adam and Eve were not the original parents, the first two human beings from whom all mankind have descended.

The problem relates to the test of obedience given to man by God (*Gen.* 2:16-17). Who was involved in the decision taken by man—was it only one person (i.e., Adam) as *Genesis* suggests, or were several, perhaps many, human beings involved? If God chose to give the choice to a *group* of men, He first would have had to address the following possibility: What if there were disagreement in the group?

The effects of the alternative choices are incompatible. By the choice of obedience to God, mankind would have been preserved from bodily death and would have lived in a state of paradise upon Earth before entering Heaven. However, by the choice of disobedience to God,

---

15. Fr. J. Franklin Ewing S.J., *Human Evolution-1956* (with Appendix "The Present Catholic Attitude Towards Evolution"), *Anthropological Quarterly*, p. 138, Volume 29, No. 4, October, 1956, The Catholic University of America Press, Washington DC. Emphasis added.

mankind was no longer preserved from bodily death, was subjected to inherent sinful tendencies, and has had to live with disease, violence and a harsh physical environment.

It is hard to see how the effects of the alternative choices could be implemented simultaneously. If there were many men, some might have chosen to resist the devil's temptation, while others might have chosen to eat the forbidden fruit, thereby becoming corrupt. How could perfect human couples co-exist with corrupted ones?

And what of the physical environment itself—how could it be a state of paradise for some, while a state of harsh climactic conditions for others at the same time? It seems more reasonable to hold that only one choice could be implemented. If this is granted, then a dilemma for the idea of polygenism must be faced: If more than one person were involved in the choice and there were disagreement in the group, only one party could have its choice implemented. The choice of the other party would have to be overridden. This would hardly be fair, and the members of the party concerned could feel that God misled them.

But we know that God can neither deceive nor be deceived, and it seems reasonable to conclude that "Adam" must refer to an individual person, as *Genesis* clearly records, and not to a group of people.

Against this, it can be argued that God selected one person among many human beings to make the choice on behalf of others. Or that a group of early human beings could have delegated authority to one leader. But this idea is totally dependent on every person in the group granting stupendous decision-making powers to this one person. Each could not afford the slightest doubt; total faith in the leader would be required. If only one person baulked at the idea, this concept is flawed. And God would have had to leave open the possibility of disagreement, a position considered above to be untenable.

One problem for such versions of polygenism is that Original Sin would not be transmitted *through generation* to the descendants of the sinless others in the group, but rather by the direct action of God. And there are other conceptual problems: What about Eve—was she "the mother of all the living" (*Gen.* 3:20), or only one of the group? Where did the group come from in the first place, since Evolution is effectively ruled out by modern findings in biochemistry and genetics? What Scripture passages can be cited in support of the concept? What clue or hint is there in *Genesis* or Tradition that the single couple—Adam and Eve—were in fact "many first parents"?

The most satisfactory solution to the above problem facing polygenism is that only Adam could have made the choice on behalf of mankind. Eve was involved, of course, in the shaping of Adam's choice, but the responsibility was his alone. The wording of *Genesis* points strongly in this direction, and the wording of *Humani Generis* confirms it (Pope Pius XII wrote about "an individual Adam").

The *Catechism of the Catholic Church* twice (pars. 28 and 360) quotes *Acts* 17:26: "From one ancestor [God] made all nations to inhabit the whole Earth." This quotation strongly suggests a reference to one person, but various Bibles are unclear on this point. That the *Catechism* refers to a single person is confirmed in footnote 226 of paragraph 360, which cites *Tobit* 8:8: *"Thou madest Adam of the slime of the earth and gave him Eve for a helper. From them the race of mankind has sprung . . . ."* The "one ancestor" could only be Adam. This teaching is also confirmed in section 359, which quotes St. John Chrysostom: "St. Paul tells us that the human race takes its origin from two men: Adam and Christ . . . The first man, Adam . . . was made by the last Adam." Therefore, polygenism should be regarded as *irreconcilable* with Catholic doctrine.

The Views of Pope Pius XII

Catholics believe that each pope, as the successor of St. Peter, is deserving of great respect and enjoys the privilege of infallibility. (See Denzinger Systematic Index IIIf.) However, loyal and informed Catholics also understand that each pope can otherwise be mistaken when expressing private, non *ex cathedra* opinions, and in matters concerning prudential judgement.

Pope Pius XII taught quite definitely in the encyclical *Humani Generis* (1950) that Adam and Eve were real human beings, the first parents from whom all of mankind has descended; they are not symbolic representations of mankind as a whole, and there were no other human races existing on the Earth from whom human beings could have descended. To hold otherwise, he declared, is to endanger the doctrine of Original Sin. He reiterated that the rational souls of Adam and Eve were divinely implanted by God in acts of Special Creation, and he reaffirmed the teaching of the Church on Original Sin: "Sin actually committed by an individual Adam and which through generation is passed on to all and is in everyone as his own."[16]

In view of truth known from Tradition and highlighted by Cardinal Ruffini, why did Pope Pius XII even allow any discussion about the

---

16. *Humani Generis*, Denzinger, p. 2328.

possible evolution of Adam's body, as though human Evolution *could* somehow be true? What need was there for further discussion—surely, enough was known already from Tradition? One can only speculate. Since "Evolution" is still today the subject of confusion and conflicting definitions and mechanisms, and the model of DNA was not even fashioned until three years later (1953), perhaps the Pope in 1950 opted for caution and in-depth clarification from scientists, not anticipating that his encyclical would soon be so blatantly distorted.

Although he may have inadvertently left an opening through which Modernist concepts could penetrate into Catholic consciousness, Pope Pius XII did not give the impression that he believed Evolution to be compatible with the Faith. His encyclical *Humani Generis* (dated August 12, 1950, and appropriately subtitled, *On Certain False Opinions Which Threaten to Undermine the Foundations of Catholic Doctrine*) was issued *in response* to the danger posed by evolutionists (such as Teilhard de Chardin) who were pushing pantheism. Perhaps the Pope did not anticipate the lengths to which Modernists would go in trying to overturn official Catholic doctrine, but he certainly had no illusions about the danger of *Evolutionism*:

> If anyone examines the state of affairs outside the Christian fold, he will easily discover the principal trends that not a few learned men are following. Some imprudently and indiscreetly hold that evolution—which has not been fully proved even in the domain of natural sciences—explains the origin of all things, and audaciously support the monistic and pantheistic opinion that the world is in continual evolution. Communists gladly subscribe to this opinion so that, when the souls of men have been deprived of every idea of God, they may the more efficaciously defend and propagate their dialectical materialism.[17]

Notwithstanding the fact that he allowed discussion about the possible evolution of Adam's body, how can it now be said that Pope Pius XII truly regarded Evolution as a serious hypothesis? After all, it was well known in 1950 that the "crucial mechanism of Evolution" was still truly missing. In a speech given only three years after *Humani Generis* was issued, long before exact details of discoveries in biochemistry and molecular biology would be available, Pope Pius XII in

---

17. *Ibid.*, p. 2305.

fact expressed very serious reservations about the scientific credibility of Evolution:

> In recent works on genetics one reads that the connection between living things cannot be explained better than by supposing a common genealogical tree. It is, however, necessary to remark that *what we have here is an image, a hypothesis, not a demonstrated fact.* . . . If most research workers speak of genealogical descent as a fact, they are premature in doing so. Other hypotheses are possible [in addition to that of evolution] . . . [Besides,] scientists of repute have pointed out that in their opinion one cannot as yet say what is the real and exact meaning of terms such as "evolution," "descent" and "transmission"; that we know of no natural process by which one being can beget another of a different kind; that the process by which one species begets another is altogether unintelligible, no matter how many intermediate stages be supposed; that no experimental method for producing one species from another has been found; and finally that we have no idea at what stage in the evolutionary process the hominoid suddenly crossed the threshold of humanity . . . [In conclusion] one is forced to say that the study of human origins is only at its beginnings: there is nothing definitive about present-day theory.[18]

A great deal of scientific research has taken place since the early 1950s, and the case against Evolution is overwhelming. Regarding the origin of Adam's body, the most probable explanation is this: "Yes" to very rapid transformation/creation from "dust" (i.e., inorganic non-living matter)[19] and "No" to Evolution from living matter. However, there has been no fully comprehensive discussion of Origins between specialists within the Catholic Church. Rather, it seems that views favorable to Evolution have tended to predominate, and commensurate attention has not been given to information unfavorable to Evolution.

Unfortunately, the papal permission of 1950—for specialists *to hypothesize only* about man's origin from living matter—was blatantly

---

18. Pope Pius XII, *Address to the First International Congress of Medical Genetics*, Sept. 7, 1953, quoted by Dr. Michael Sheehan, *Apologetics and Catholic Doctrine*, Part 11, 1962, p. 55. Emphasis added. Originally published in French in the *Acta Apostolicae Sedis*.
19. Special creation of Adam from pre-existent inorganic matter, in the form of elementary particles taken from the ground and instantly transformed into the body of an adult male human being.

exploited after the death of Pope Pius XII in 1958 by those who sought to accommodate evolutionary concepts to Catholic doctrine.

Humani Generis was misrepresented to convey the widespread impression that belief in Evolution is now accepted *officially* by the Magisterium. This distortion of truth about Origins has had a catastrophic effect among Catholics on their understanding of crucial Origins doctrines. The misrepresentation amounted to a debilitating assault on foundational doctrines such as Original Sin—an assault from which the Catholic Church is still reeling.

In spite of which Pope is reigning, Catholics loyal to the office of the papacy (many of whom have experienced the extremely distressing effects of loved ones leaving the Church in the unprecedented collapse of faith since the mid-1960s) are entitled to expect rigorous, fully informed comments from the successor of St. Peter on vital doctrinal matters which could affect the salvation of souls.

The Views of Pope John Paul II

Pope John Paul II appears to defer trustingly to his scientific advisers, especially to advice from the *Pontifical Academy of Sciences* (established in 1936 by Pope Pius XI), notwithstanding the possibility that some present members may be ambivalent toward Christian doctrine and despite the actual standing of the *Academy* within the Catholic Church:

> About this body I would say that it has no authority in matters of faith and doctrine and expresses only the views of its own members who belong to different religious beliefs.[20]

As of late 1996, the *Academy* had 86 members, over 20 of whom are Nobel prize winners, all apparently favorable to Evolution. It seems that there is not one creationist opponent of Evolution (Catholic or otherwise) in the *Academy* to give other views of modern science. Ironically, one member, the famous United Kingdom cosmologist Stephen Hawking, even promotes non-Christian views:

> . . . in 1981 my interest in questions about the origin and fate of the universe was reawakened when I attended a conference on cosmol-

---

20. Archbishop Luigi Barbarito (then Apostolic Pro-Nuncio, Australia), in correspondence to G. J. Keane (August 1, 1983).

ogy organized by the Jesuits in the Vatican. . . . At the end of the conference the participants were granted an audience with the Pope. He told us that it was all right to study the evolution of the universe after the Big Bang, but we should not inquire into the Big Bang itself because that was the moment of Creation and therefore the work of God. *I was glad then that he did not know the subject of the talk I had just given at the conference—the possibility that space-time was finite but had no boundary, which means that it had no beginning, no moment of Creation.*[21]

The reason for John Paul II's deference to the *Academy*, virtually as the exclusive scientific consultant, seems answered by himself:

> I had long been interested in *man as person*. Perhaps my interest was due to the fact that I had never had a particular predilection for the natural sciences. I was always more fascinated by man . . . when I discovered my priestly vocation, man became the *central theme of my pastoral work.*[22]

In a spirit of seeking the truth of Origins, the following comments are made with great respect and loyalty to the Pope as successor of St. Peter (who, despite his human frailty, was chosen personally by our Creator and Redeemer to lead the Church on Earth; see *Matt.* 16:18).

At the April 26, 1985 Vatican symposium on "Christian Faith and the Theory of Evolution," Pope John Paul II made the following comments, which no doubt reflect the collective view of the *Academy,* but which nevertheless appear inconsistent with the actual findings of modern science:

> Rightly comprehended, faith in creation or a correctly understood teaching of evolution does not create obstacles: Evolution in fact presupposes creation; creation situates itself in the light of evolution as an event which extends itself through time—as a continual creation—in which God becomes visible to the eyes of the believer as "creator of heaven and earth."[23]

---

21. Stephen W. Hawking, *A Brief History of Time: From the Big Bang to Black Holes* (London: Bantam Press, 1988). Emphasis added.
22. Pope John Paul II, *Crossing the Threshold of Hope* (London: Random House, 1994), p. 199.
23. Quoted by theistic evolutionist writer Fr. William Kramer, C.PP.S. in *Evolution & Creation: A Catholic Understanding* (Huntington, IN: Our Sunday Visitor, Inc., 1986), p. 114.

In view of the powerful case which can be made for Special Creation, has the Pope's trust in his present advisers been misplaced? Let us hope and pray that the Holy Father may see fit, in due course, also to consider the views of those highly credentialed Catholic scientists who firmly hold Evolution to be false science. After all, how can the truth of Origins emerge if all sides do not receive a fair hearing?

Unfortunately, the various private, non *ex cathedra* comments of Pope John Paul II on Origins have tended to be imprecise. What exactly did he mean by "a correctly understood teaching of Evolution"? The word "Evolution" means all sorts of things to all sorts of people, and a precise definition of terms is therefore desirable. Clearly, Darwin's general theory of Evolution has failed under searching scrutiny, and Natural Evolution is in the midst of a deep crisis of credibility.

The objective truth is that rigorous scientific disciplines—such as molecular biology, biochemistry and genetics—have revealed mind-boggling complexity within very tiny cells, and the reality of irreducible complexity within even such tiny cells clearly speaks only of Design. Numerous scientists have also shown that life forms are only capable of passing on the existing genetic information already possessed by them, and thus Natural Evolution *cannot* occur.

Modern research, therefore, suggests strongly that Natural Evolution *cannot* be historically true and that God chose to institute only *variety within kind*. This is not Evolution; since truly new "higher" genetic information is not being transmitted, life forms can only change within definite limits or boundaries. Nor can Theistic Evolution truly be defined as Evolution (as argued in Chapter 12); when fully considered, it reduces down only to innumerable instances of divine intervention— in contradiction to the observable facts of all reality, namely, that God works on all created things through *secondary causes*.

Instead of generalized private statements, one looks for precise terminology from the Pope, affecting vital related foundational Origins and Original Sin doctrine, which is not yet fully clarified. Addressing the *Academy* at the Rome conference on "Origins and Evolution of Life" (October 22, 1996), Pope John Paul II rightly pointed out that *"truth cannot contradict truth."* He also stated that, "the exegete and the theologian must keep informed about the results achieved by the natural sciences." But he did not specifically define what he means by "Evolution," nor did he fully elaborate how objective truth known from Tradition and theological reality can differ from the relative truth known from ever-changing scientific theories, especially since theology and empirical scientific investigation can sometimes overlap (e.g., on issues such as abortion, *in vitro* fertilization, contraception, euthanasia and the global Flood of Noah; all of these are proper subject matter for both theologians and scientists).

Has Pope John Paul II been inaccurately informed by his advisers, to the detriment of truth? Some of his other scientific comments made on October 22, 1996, seem most inconsistent with the actual findings of modern science: "Evolution is more than a hypothesis. . . . It is indeed remarkable that this theory has been progressively accepted by researchers, following a series of discoveries in various fields of knowledge. The convergence, neither sought nor fabricated, of the results of work that was conducted independently is in itself a significant argument in favor of this theory."

What discoveries? In light of modern scientific information, how can Natural Evolution now be regarded as anything other than failed attempted hypotheses? The highly complex information encoded on DNA suggests that only "like begets like." The missing mechanism of Evolution is doomed to remain ever-elusive because there is no mechanism to find, and evolutionary "convergence" never occurred because it cannot occur!

Further, has the Holy Father also been misinformed, in an effort to convince him that the biological aspects of Origins are somehow explicable by tying Naturalistic Evolution to divine intervention, as the following extracts from his October 22, 1996 speech imply: "ontological discontinuity," "consideration of the method used in the various branches of knowledge makes it possible to reconcile two points of view which would seem irreconcilable," "correlate them with the time line," "the moment of transition to the spiritual"?

In light of the pro-evolutionist scientific advice, is it accurate to say that, "In his encyclical *Humani Generis* [*On Certain False Opinions*, 1950], Pope Pius XII had already stated that there was no opposition between Evolution and the doctrine of the faith about man and his vocation, on condition that one did not lose sight of several indisputable points"?

Two years later (September, 1998), writing in the encyclical letter *Fides et Ratio* ("Faith and Reason"), Pope John Paul II lamented the crisis of meaning in the modern world and condemned the threat of *scientism*, which had accompanied the spectacular growth in modern scientific achievements. He described *scientism* as "the philosophical notion which refuses to admit the validity of forms of knowledge other than those of the positive sciences; and it relegates religious, theological, ethical and aesthetic knowledge to the realm of mere fantasy . . .

science would thus be poised to dominate all aspects of human life through technological progress." (Section 88).

John Paul II also wrote in praise of Pius XII, "In his encyclical letter *Humani Generis*, Pope Pius XII warned against mistaken interpretations linked to Evolutionism, existentialism and historicism. He made it clear that these theories had not been proposed and developed by theologians, but had their origins 'outside the sheepfold of Christ.' He added, however, that errors of this kind should not simply be rejected but should be examined critically." (Section 54).

One would hope for further Papal examination of many other crucial Origins aspects—such as polygenism, the origin of Eve's body, the impact of *Romans* 5:12 and conceptual problems in Theistic Evolution—and rigorous addressing of the historicity of *Genesis*.

Following the publication of the strongly pro-creationist *Catechism of the Catholic Church*, a comprehensive and rigorous encyclical on Origins—given high Magisterial status, if not fully *ex cathedra*—would be welcome news, as it would further help to clarify right from wrong and truth from error. It would also take its place forever in Catholic Tradition, for all future generations to consult.

—Chapter 14—

THE INERRANCY OF SCRIPTURE

One of the central beliefs of Christianity is that God has revealed some knowledge about Himself to mankind. Man can, of course, deduce the existence of the Creator by observation of the wonders of the Universe and through the use of reason. As well as this, however, we have been given by God a collection of sacred writings; the revelation contained in them is a gift to mankind.

Although written by an unknown number of human authors, the collection of books which constitute the Bible have God as the principal Author. While the human authors were the instruments used by God, they were not mere puppets. In a mysterious way, God, by supernatural power, so moved and assisted them that they correctly conceived with their minds the things which He had ordered, and then they faithfully determined to record them in the most suitable manner.

The Catholic Church teaches that all the books which it accepts as sacred and canonical[1] were written wholly and entirely at the inspiration of the Holy Spirit, and the human authors wrote all those things and only those things that God wanted written. Vatican Council I rejected the idea of Scripture's being merely a human production:

> The Church holds these books as sacred and canonical, not because, having been put together by human industry alone, they were then approved by its authority; nor because they contain revelation without error, but because, having been written by the inspiration of the Holy Spirit, they have God as their author.[2]

As greater insights are gained into this revelation, the Magisterium is empowered by Jesus Christ to define certain teachings as being truly revealed by God and therefore obligatory for Catholics to believe. In

---

1. The complete list is given in paragraph 120 of the *Catechism of the Catholic Church*.
2. Vatican I, Denzinger, p. 1787.

noting that the Real Presence is proclaimed only by the Catholic Church, through continuity of ordination going all the way back to Jesus Christ, Fr. William Most shows how the Catholic Church fairly lays claim to being instituted directly by Christ Himself and thus is qualified to proclaim with authority on Scripture:

> In summary then, we see a group that is commissioned to teach by a messenger from God [Jesus Christ], and promised God's protection for that teaching. These observations are made without treating the Scriptures as inspired. Now we not only intellectually may but are intellectually driven to believe what the body teaches. That body can then assure us that those ancient documents, used without knowing they are inspired, really have God as their author. And that body, the Church, can also tell us that the messenger sent by God is really God Himself. And, of course, it can guarantee countless other truths for us.[3]

Since human beings have only limited intellects and an incomplete understanding of the supernatural, some passages in Scripture (e.g., *The Apocalypse* of St. John) may still only be partly understood. Individuals are not empowered to decide by themselves, in isolation from the Magisterium, what meaning is intended by the Sacred Writer(s) in any particular passage of Scripture. This was reaffirmed at Vatican II:

> In order that the living Gospel might be forever preserved in the Church, the Apostles left as their successors, the bishops, "entrusting to them their own position as a teaching body." Hence Sacred Tradition and the Holy Scripture of each of the two Testaments resemble a mirror in which the Church, during her journeying on Earth, contemplates God, from whom she receives all things, until she may be brought to see Him, face to face, as He is.[4]
>
> The office of interpreting authentically the word of God, whether scriptural or traditional, has been entrusted exclusively to the living voice of the Church's Magisterium, whose authority is exercised in the name of Jesus Christ. This Magisterium is not superior to the word of God, but ministers to the same word by teaching only what has been handed on to it, insofar as, by divine command and with the assistance of the Holy Spirit, the Magisterium devoutly hears, reli-

---

3. Fr. William G. Most, *Free From All Error* (Prow Books/Franciscan Marytown Press, 1990), p. 13.
4. *Dei Verbum* (Dogmatic Constitution on Divine Revelation), par. 7.

giously keeps and faithfully explains the word, and from this one deposit of faith derives all those things which it proposes to us for acceptance as divinely revealed. It is clear, therefore, that Sacred Tradition, Holy Scripture and the Church's Magisterium are by God's most wise decree so closely connected and associated together that one does not subsist without the other two, and that all of them . . . under the impulse of the one Spirit of God, contribute efficaciously to the salvation of souls.[5]

Similarly, the *Catechism of the Catholic Church*, endorsed by Pope John Paul II, declares that we must preserve the true analogy of faith handed down (*Catechism,* 114) and that

> Sacred Tradition, Sacred Scripture, and the Magisterium of the Church are so connected and associated that one of them cannot stand without the others. Working together, each in its own way, under the action of the one Holy Spirit, they all contribute effectively to the salvation of souls. (95)

This catechism also discusses the various senses in Scripture:

> According to an ancient tradition, one can distinguish between two *senses* of Scripture: the *literal* and the *spiritual*, the latter being subdivided into the *allegorical, moral,* and *anagogical* senses. The profound concordance of the four senses guarantees all its richness to the living reading of Scripture in the Church. (115)

The *literal sense* is the meaning conveyed by the words of Scripture and discovered by exegesis, following the rules of sound interpretation. The *spiritual sense* enables the text of Scripture, and the realities and events about which it speaks, to be understood also in signs. (The *allegorical sense* considers events as signs of significance in Christ, the *moral sense* compels one to act justly, and the *anagogical sense* views events in terms of historical significance.)

In addressing the writings of St. Thomas Aquinas, Peter Kreeft notes that

> Modern hermeneutics (the science of interpretation) tends to create a great divide between the Modernist Demythologizers, who interpret as merely symbolic whatever passages are too miraculous or

---

5. *Ibid.,* par. 10.

supernaturalistic for their philosophy to stomach, and the Fundamentalists, who in reaction to the Modernists tend to be suspicious of all symbolism and confine themselves to literal interpretation of every passage (*except John* 6:48-56).

St. Thomas cuts across this either/or and maintains that a passage could rightly be interpreted both literally ("historically") and symbolically ("spiritually"), because God writes history as man writes words. That is, behind this hermeneutic is a metaphysic: the sacramental view of nature and history, according to which things and events as well as words can be *signs* as well as *things*, can be means by which other things are signified and known as well as being things known themselves.[6]

As the Origins controversy unfolded over the last two centuries, two Church Councils discussed the topic of errors in Scripture, and recent popes felt the need to address at length the topic of Biblical Studies: Pope Leo XIII in *Providentissimus Deus* (1893), Benedict XV in *Spiritus Paraclitus* (1920), and Pius XII in *Divino Afflante Spiritu* (1943) and *Humani Generis* (1950).

In contrast to the still-popular misconception that in matters of natural science and history the Bible may contain errors which do not impact on religious faith, the Church teaches that, *because God is the principal Author, there can be no errors whatsoever in the Bible.* There really can be no such thing as "limited inerrancy" of Scripture.

Pope Leo XIII's encyclical *Providentissimus Deus* (1893) established new criteria for understanding the various *senses* used in Scripture, in response to the gradually unfolding Origins controversy. Recognizing the very serious threat posed by Higher Criticism (which had arisen in German Protestant circles earlier in the 19th century), the Pope encouraged scholars to study diligently the languages in which the sacred books were written and to study the art of criticism. However, he strongly cautioned against extreme and unacceptable forms of Biblical criticism:

> There has arisen, to the great detriment of religion, an inept method, dignified by the name of "higher criticism," which pretends to judge of the origin, integrity, and authority of each book from internal indications alone. It is clear, on the other hand, that in historical questions, such as the origin and handing down of writings,

---

6. Peter Kreeft, *A Summa of the Summa* (San Francisco: Ignatius Press, 1990), p. 48.

the witness of history is of primary importance, and that historical investigation should be made with the utmost care; and that in this matter internal evidence is seldom of great value, except as confirmation. To look upon it in any other light will be to open the door to many evil consequences. It will make the enemies of religion much more bold and confident in attacking and mangling the sacred books; and this vaunted "higher criticism" will resolve itself into the reflection of the bias and the prejudice of the critics. It will not throw on the Scripture the light which is sought, or prove of any advantage to doctrine; it will only give rise to disagreement and dissension, those sure notes of error, which the critics in question so plentifully exhibit in their own persons; and seeing that most of them are tainted with false philosophy and rationalism, it must lead to the elimination from sacred writings of all prophecy and miracles, and of everything else that is outside the natural order.

In the second place, we have to contend against those who, making an evil use of physical science, minutely scrutinize the sacred book in order to detect the writers in a mistake, and to take occasion to vilify its contents. *Attacks of this kind, bearing as they do on matters of sensible experience, are peculiarly dangerous to the masses, and also to the young who are beginning their literary studies; for the young, if they lose their reverence for the Holy Scripture on one or more points, are easily led to give up believing in it altogether.*[7]

One century later, soon after the unprecedented collapse of belief within the Church, these papal words can be seen as starkly prophetic. The scholarly but Modernist agents of dissent did indeed wreak havoc upon the masses, both young and old, and the question ought be posed: Would the collapse have been so extensive if the relevance of Origins doctrines had been taught more rigorously in this century and less credence given to ideas such as "prehistory" and "Theistic" Evolution?

Historical Credibility of Scripture

In his address to the *Pontifical Academy of Sciences* (Oct. 22, 1996), John Paul II stated that, "in order to delineate the field of their own study, the exegete and the theologian must keep informed about the results achieved by the natural sciences." Contrary to the strongly pro-evolutionist stance of this *Academy*, the material presented elsewhere in this book shows that modern discoveries, by many highly credentialed

---

7. Pope Leo XIII, *Providentissimus Deus*, St Paul Editions. Emphasis added.

scientists, confirm that DNA was designed by God so that only variety within kind can occur. The objective truth now seems clear: Evolution did not occur, because it cannot occur.

Immediately preceding this statement by the Pope, he identified a most crucial aspect of the Origins debate, the perennial importance of which cannot be overstated:

> *It is necessary to determine the proper sense of Scripture, while avoiding any unwarranted interpretations that make it say what it does not intend to say.* [Emphasis added.]

Exactly! But what body actually determines the proper sense of Scripture and determines what is truly warranted or unwarranted in understanding *Genesis* 1-11? Should the present *Pontifical Biblical Commission* (which is no longer an organ of the teaching Magisterium) be given preferential consultation, when its 1993 document did not even recognize a Catholic concept of Special Creation, which is in no way Fundamentalist?

In the sacred books of Scripture, which have God as the principal Author, can we not expect to find some indication of what was meant to be conveyed? *Genesis* is obviously not a detailed scientific textbook, and Pope Leo XIII taught in *Providentissimus Deus* that the sacred writer(s) did not intend to teach men about the essential nature of things in the visible Universe. But who can deny without fear of contradiction that *Genesis* is primarily historical, and was intended to be understood as such? If one denies the weight of Tradition concerning the genuine historicity of *Genesis*, is there not an attendant danger of unwittingly imposing unwarranted and mistaken views of fallible modern scholars?

Indeed, how does one determine the historical credibility of ancient writings? C. S. Lewis made a most pertinent observation on the tendency of some modern scholars to discredit ancient writers. After seeing his own writings reviewed by people who read into them all sorts of erroneous ideas about what he had intended, Lewis noted the pitfalls involved in literary criticism. He observed how it was once fashionable to cut up Shakespeare's Henry VI between a dozen authors and to assign a share to each, and how this practice was itself then discredited. He criticized those modern scholars who are loath to concede that real history was recorded by ancient authors:

THE INERRANCY OF SCRIPTURE

> When a learned man is presented with any statement in an ancient author, the one question he never asks is whether it is true. He asks who influenced the ancient writer, and how far the statement is consistent with what he said in other books, and what phase in the writer's development, or in the general history of thought, it illustrates, and how it affected later writers, and how often it has been misunderstood (especially by the learned man's own colleagues) and what the general course of criticism on it has been for the last ten years, and what is the "present state of the question." To regard the ancient writer as a possible source of knowledge—to anticipate that what he said could possibly modify your thoughts or your behavior—this would be rejected as unutterably simple-minded.[8]

That something seems to have been seriously amiss for many decades concerning Origins beliefs, can be gauged from the tendency of various writers to introduce what appear to be unsubstantiated presuppositions upon the intentions of the sacred writer(s) of ancient texts of Scripture. Consider the views of Michael Sheehan, who proposed that a distinction be drawn when considering the Creation account. In making the following assertions, is he objectively correct, or rather, accepting and repeating without question unproven theories handed down from others?

> In the Bible account we must carefully distinguish what is principal from what is subordinate:
>
> *The principal element*: The Jews to whom the narrative was primarily addressed were surrounded by idolatrous peoples who believed in the existence of many gods, and paid divine worship to all kinds of creatures, to the sun and moon, plants and animals, and images of wood and stone. Hence, the sacred writer under the guidance of the Holy Spirit impresses on the Jewish race, with greatest emphasis, that there is but one God, and that He created the whole visible universe with everything in it, living and lifeless. The expression of this great truth is the chief element in his narrative. All else is secondary or subordinate.
>
> *The subordinate element*: The subordinate element is the popular dress in which the inspired message is clothed. Though for convenience sake we designate it "the subordinate element," it is nevertheless as truly a part of inspired Scripture as the principal element; it is the medium through which the Holy Spirit has chosen to speak to us.

---

8. C. S. Lewis, *The Screwtape Letters* (London: Collins Fontana Books, 1973), p. 139.

> The Church, while insisting that the account of the creation in *Genesis* is historical,[9] tells us at the same time that we may regard it as popular in form. Popular form implies popular expression and popular order [i.e., things and happenings not necessarily strictly chronological, described not in strict scientific language, but as they would appear outwardly to the senses]. Beyond these general directions and a general condemnation of all methods of interpretation which would impute real error to the sacred writings, the Church has decided nothing as to how the subordinate element in the Scriptural narrative is to be understood, except to indicate the importance of a study of primitive literary forms.[10]

Is all this true, or just conjecture? Are the views tabled by Sheehan in fact too restrictive to allow the genuine historicity to emerge? Who can prove for sure that both God and the human sacred writer really intended such principal and subordinate elements? Was not Scripture also intended for human beings in all later times? Why limit it to a local, ancient audience? Why not allow the Divine Author of Scripture to convey *both* a partial account of the true story of Creation and the idea of fallible human beings standing in awe at the marvelous Creator?

Another example: the famous account of the *Red Sea Crossing* has long been subjected to all sorts of opinions being imposed upon it, to reduce the events down to purely natural occurrences. In *Exodus* 14, we are informed that the Egyptians were destroyed in the Red Sea. If their destruction is seen basically as a combination of fortuitous natural occurrences which resulted in their drowning as the tide returned, then it is easy to overlook the fact that God instructed Moses: "But lift thou up thy rod, and stretch forth thy hand over the sea, and divide it: that the children of Israel may go through the midst of the sea on dry ground," and later, "Stretch forth thy hand over the sea, that the waters may come again upon the Egyptians . . ." (*Ex.* 14:16, 26).

It seems, according to the Bible, that there was an actual wall of water on either side of the Egyptians. The Hebrew word *chomab* can apparently mean "protection" as well as "wall." The reason why the Egyptians proceeded into the trap was that God confused them.

---

9. Footnote by Michael Sheehan: "Historical," i. e., not fiction. It is an account, in the popular form of the age, of things that really happened. There are many kinds of literary forms in the Bible, and the sacred writers, primarily interested in theological truth, did not always intend to treat of historical matters in the same way in which a modern historian would.
10. Michael Sheehan, *Apologetics and Catholic Doctrine, Part II* (1962), pp. 25, 26.

Perhaps, too, they were blinded by their own arrogance: "And I will harden the heart of the Egyptians to pursue you: and I will be glorified in Pharao, and in all his host, and in his chariots, and in his horsemen." (*Ex*. 14:17).

We are told that God sent "a strong and burning wind blowing all the night" (*Ex*. 14:21) to drive out the sea, and later caused the sea to return and trap the Egyptians. They were probably surprised by the speed and size of the returning water. The point is that it was an event with a supernatural dimension, for God intervened in the laws of nature, as He did on various other occasions.

Whether this activity by God is defined as a miracle or as the outworking of God's Providence may be subject to debate. (A miracle is a suspension of or alteration to natural laws; whereas Providence refers to God's use of natural laws to bring about His designs.) In a sense, every event in life has a supernatural dimension. And thus the history of all mankind is one which could be defined generally as "salvation history." After all, what *is not* "salvation history"?

We know that God is all-powerful and all-knowing, allows secondary causes to operate and sustains in existence all of Creation. He therefore could cause purely natural forces to coincide and achieve a particular purpose. However, we also know that He sometimes chooses to cause abnormal phenomena to occur. If the Red Sea account in *Exodus* is merely a symbolic "salvation history" story, immaterial as to whether it really happened, is not all of Scripture open for revision? Why hold some passages to be definitely believed, but treat other passages with ambivalence? Fortunately, concerning the discernment of history in *Genesis*, there is a precedent in Tradition which should be deferred to: In 1909 the Pontifical Biblical Commission (then an official organ of the Magisterium) declared its ruling on the historical character of the first three chapters of *Genesis*. The PBC ruled that the literal, historical sense cannot be excluded, that these chapters were not legend derived from mythologies belonging to older nations and purified of polytheistic error, but in some passages there is a figurative sense used by the writer. The question of whether the word *yom* could be interpreted as a day of 24 hours or a longer period of time was left open.

A decree issued by a Commission does not carry the same weight of authority as does a dogmatic decree or an encyclical issued by a pope; it remains binding authority only as determined by a pope. The decree in discussion here has still not been superseded. Fr. J. Franklin Ewing, S.J., Professor of Anthropology and a *theistic evolutionist*, acknowledged (in 1956) the binding effect of these PBC declarations:

> Although the decrees ... were handed down in 1909, they are not merely historical, since they are still in force. The Catholic is required to give internal prudential assent to these decrees, since they are approved by the Pope and represent a specialized form of the ordinary teaching function (Magisterium) of the Church.[11]

The present Pontifical Biblical Commission issued a document in Rome (April 23, 1993) entitled *The Interpretation of the Bible in the Church*, with a preface by Cardinal Ratzinger. The document includes a summary of current methods employed in Christianity for understanding Scriptural texts.

Many methodologies are discussed: the historical-critical method, patristic exegesis, the redaction method associated with the documentary theory, liberation theology, feminist theology, existentialist concepts, fundamentalist interpretation and so forth. The writers of the 1993 document sympathize strongly with the historical-critical method, with emphasis placed upon intensive analyses of literary genres, and they seem at pains to distance themselves from any form of literal sense.

On the surface, who would quarrel with the idea of studying literary patterns of writing to help discover what the sacred writer(s) *wished to assert* in Scripture passages? But what is very striking about the 1993 PBC document is that creationist exegesis seems equated mostly with Fundamentalism. And yet there exists a genuine Catholic version of Special Creation which is *not* Fundamentalist, which defers to Tradition and thus places great store on the collective "mind" of the Fathers, which recognizes that various senses are employed in Scripture, and places great importance upon rigorous exegesis. Why is there no mention of such an approach?

This PBC document displays very technical concepts and terminology, but one wonders whether something has gone wrong in striving for deeper understanding of *Genesis*. Perhaps the truth about *Genesis* has been staring us in the face all along, but it has simply not been believed. *Why not consider the possibility that the sacred writer (including God as the principal Author) primarily intended to convey history in Genesis?* How else could we obtain a clear overall grasp of Creation and of Original Sin, if a partial account of the true story of Creation

---

11. Ewing, "Human Evolution—1956," in *Anthropological Quarterly* (Washington, D.C.: Catholic University of America Press), October, 1956, p. 130.

THE INERRANCY OF SCRIPTURE

were not revealed to us? Is there not some danger in ignoring the only reliable Eyewitness who was there—God—and accepting instead the speculative theories of modern "experts" who were not there?

Fr. Stanley Jaki, O.S.B. devoted his book *Genesis 1 Through the Ages* to refuting the idea that there could be concordance between the *Genesis* 1 account and scientific theories of Origins, but does he thus ignore the obvious? In disputing this book, Msgr. John F. McCarthy, O.S. shows that Fr. Jaki fails to draw a clear distinction between mathematical science, philosophical science and theological science. He also charges that

> While the book, from its title, seems to promise a review of the insights into this chapter that the great commentators have expressed over the ages, its actual theme is that they have no insight.[12]

The absolutely reliable divine Eyewitness should be trusted concerning *Genesis* in preference to fallible modern human opinion. As Pope John Paul II confirms in his book, *Crossing the Threshold of Hope* (Pope John Paul II, p. 21), it is known that Jesus Christ is both Creator and Redeemer. Thus, the "Second Adam" was also intimately involved as the principal Author of Scripture. He obviously knew/knows everything about such things as the gargantuan Universe, the minute world of atoms and all the bewildering information impressed on DNA in all life forms.

A powerful case can be made that Christ accepted the Old Testament as real history. His comments on the fate of Capharnaum in comparison to the fate of Sodom and Gomorrha (*Matt.* 11:23-24) were obviously made in earnest. His teaching on marriage (*Mark* 10:6-9) clearly shows that He accepted the historical existence of the first human parents, as well as the mode of Eve's creation. Since Christ claimed to come upon Earth to redeem mankind from the sin of Adam, there can be no doubt that He was only too well aware of the historical existence of Adam and Eve. How could it be otherwise?

The sheer weight of evidence supports belief in His divinity. The fact that He actually lived on Earth about 2,000 years ago is amply attested to by historians, and the Resurrection is a proved fact. To assert otherwise is to deny the many witnesses who saw Christ after He rose from

---

12. See Msgr. John F. McCarthy O.S., "Not The Real Genesis 1," *Living Tradition* (Rome: Sedes Sapientiae Study Center), March, 1994.

the dead. It is to reject as being gullible fools the multitudes of early Christians, who willingly died for their Christian beliefs. Very few people will lay down their lives even for a good man, but how many would suffer the supreme sacrifice for someone who was a fake, for someone who promised to return from the dead and did not? Christ answered in the affirmative to the high priest's challenge (*Matt.* 26:63-64) as to whether he was God. St. Paul, in writing that Jesus Christ, "Who being in the form of God, thought it not robbery to be equal with God: But emptied himself, taking the form of a servant . . ." (*Phil.* 2:6-8), indicates his certainty that Christ was God.

Since strong evidence can be cited for the divinity of Christ and for His strongly stated belief in both the Biblical personalities and historical events recorded in the Old Testament, how can the true historicity of *Genesis* be ignored? Why doubt His words? (*John* 14:24).

While the Church declares that Scripture is totally free from error, this does not mean that every passage can be read in the literal-as-given sense. It is known that a variety of literary genres exist in Scripture—poetry, sacred psalm, legal material, prophecy, apocalypse—but not myth. Judgment has to be made as to whether the meaning intended to be conveyed by the sacred writer is given in one of several possible senses—e. g., literal-as-given, figurative or parable. Confusion arises, however, from the fact that the term "literal" has two distinct meanings: a) the apparent and obvious sense as written, and b) the deeper, proper, sense intended to be conveyed by the sacred writer. Strictly speaking, every Scripture passage has a literal meaning.

The sacred writer may be describing real history as it actually happened, or describing real history using figures of speech—in other words, using the language of appearance. (Some biblical figurative expressions, such as "the ends of the Earth" [e.g. *Ps.* 98:3], are still in everyday use.) He may be using poetry to praise some great work, perhaps relating a parable, or pronouncing a prophecy regarding the drift of a people from God, and so on. *The Catechism of the Council of Trent* noted in passing the use of figures of speech in Scripture:

> [In] the words "He sitteth at the right hand of the Father" . . . we observe a figure of speech; that is, a use of words in other than their literal sense, *as frequently happens in Scripture*, when, accommodating its language to human ideas . . .[13]

---

13. *The Catechism of the Council of Trent* (1923 edition; Rockford, IL: TAN Books & Publishers, Inc., 1982), p. 74. Emphasis added.

However, the primary rule of Biblical interpretation, made clear by St. Augustine and long recognized in Tradition, was reiterated by Pope Leo XIII in *Providentissimus Deus* (1893): "carefully observe the rule so wisely laid down by St. Augustine—*not to depart from the literal and obvious sense, except only where reason makes it untenable or necessity permits.*" (Pope Benedict XV in *Spiritus Paraclitus* (1920), and Pope Pius XII in *Divino Afflante Spiritu* (1943) and *Humani Generis* (1950) all later wrote in support of Pope Leo XIII.)

Thus, Scripture texts must each be taken as meant literally, unless reason or necessity demands that we understand it otherwise. Most importantly, *the onus of proof is therefore upon the person who contends that another sense is superior to the literal and obvious sense.*

Scripture cannot, of course, always be understood in the literal-as-given sense, and spiritual meanings may also be applicable. The idea of Adam and Eve hiding after they "heard the voice of the Lord God walking in Paradise" (*Gen.* 3:8) suggests a figurative sense, for God is spirit and does not "walk" in the way that human beings do; perhaps His presence was made known to them somehow by the use of matter.

The "serpent" in *Genesis* 3 is understood to mean Satan, but the manner in which he "talked" to Eve is mysterious. Did Satan speak, using human words spoken through a reptile or dragon-like creature, or did he plant thoughts in her mind to tempt her to doubt God?

While Scripture is divinely inspired and totally free from error, this is not to say that everything recorded in it actually happened. Christ's advice, "if thy hand scandalize thee, cut it off" (*Mark* 9:42) was very likely intended as an exaggeration for the sake of emphasis. The purpose of parables seems primarily to convey moral teaching and not necessarily to record history. *To insist upon complete inerrancy is therefore neither slavish literalism nor fundamentalism*:

> Defending the classical doctrine of biblical inerrancy does not require us to insist that the Bible's cosmological assertions be expressed in the kind of language which modern scientists would regard as precise and accurate. The tired old platitude that "the Bible is not intended to be a textbook of science" is true in that sense. (It is false when, as is often the case, it means that biblical assertions about the cosmos, when taken literally, can be just plain false or erroneous.) Leo XIII's observation in *Providentissimus Deus* to the effect that there is no true error involved when the biblical author affirms some physical phenomenon "according to the appearances" is well-known. What the author truly affirms in *Joshua* 10:13, for instance, is that a miracle took place in which the sun was *seen* by human eyes to

remain motionless for a whole day. Whether, scientifically speaking, it was the Earth that stopped, the sun that stopped, or neither that stopped (in which case there would have been a sun-miracle like that of Fatima, in which the sun and the Earth appeared to come closer together, but did not really do so) is immaterial to the point being made by the author.[14]

Nor does inerrancy mean that Scripture contains a complete and strictly chronological order of events. Only some things are recorded (see *John* 21:25), and only those things that suited the purpose of each sacred writer. For example, when did Jesus teach the Lord's Prayer to His disciples? St. Luke places it at the time of the journey to Jerusalem and to His Passion (*Luke* 11:2-4); whereas St. Matthew locates it at the Sermon on the Mount (*Matt.* 6:9-13). Perhaps it was taught twice.

The apparent chronological differences of events recorded in *Genesis* 1 and 2 does not mean they are in contradiction. If the primary purpose of *Genesis* 2 was to focus on the creation of the first parents, the other comments given perhaps may be regarded as secondary to information already addressed more precisely in *Genesis* 1.

If the notion of inerrancy of Scripture cannot be understood in isolation from knowledge of the intended purpose of the sacred writer, a problem arises: How can one know for sure the intended purpose of the sacred writer? This is a difficult question even for New Testament books, let alone Old Testament books.

Some scholars contend that inerrancy applies only in a supposed "salvation history" sense, and that the idea of limited inerrancy allows the existence of real errors in Scripture. Claims have been made that obvious errors can be seen in certain Old Testament passages, but how does one substantiate such claims? Consider the following examples of alleged errors cited in *The Jerome Biblical Commentary*:[15]

> The allegedly inerrant Bible contains statements that in any other document would be regarded as erroneous. Why then should the Bible be considered inerrant? For convenience and with no intention of offering an exhaustive classification, Scriptural statements that cause difficulty may be arranged under four headings:

---

14. Fr. Brian W. Harrison, O.S., "Bomb-Shelter Theology," *Living Tradition,* May, 1994, No. 52 (Rome: Sedes Sapientiae Study Center), p. 16.
15. Fr. Richard F. Smith, S.J., "Inspiration and Inerrancy," *The Jerome Biblical Commentary* (Prentice-Hall Inc.), Section 66:75, 1968, p. 512. Comments added.

1. Biblical self-contradictions, e.g., Noah's flood lasting 40 days and nights in *Gen.* 7:17, but 150 days in *Gen.* 7:24.
   [*Comment:* These incidents are not contradictory. The rain lasted 40 days and nights and later the waters prevailed upon the Earth for another 150 days.]
2. Errors in natural science, e.g., the Universe enwrapped in waters held back by a solid bell-shaped barrier called the firmament.
   [*Comment:* Where is the error? Scripture does not refer to a "bell-shaped barrier"—early man envisioned this concept. The use of the word "firmament" (more accurately "expanse") in *Genesis* appears to indicate space, either the atmosphere above the Earth or outer space in the Universe—or perhaps both.]
3. Errors in history, e.g., the inaccuracies of *Deuteronomy* 5.
   [*Comment:* Are there really errors here, or is this simply a repetition of the Law already given by Moses?]
4. Moral errors, e.g., *herem*—Total destruction of an enemy people or group, considered as carrying out the will of Yahweh (*Joshua* 11:14-15).
   [*Comment:* Right and wrong—therefore morality—is by definition a matter of what God decrees, and we understand that God can somehow bring about a greater good out of what seems like a catastrophe in man's eyes. Further, can it really be said of a divinely inspired writing that it actually contains moral errors?]

How can modern scholars be so sure that Scripture contains errors? How can it be stated with certainty that the writer of *Genesis* did not have an accurate understanding of the events of Creation?

Consider the views of Fr. Raymond E. Brown, S.S. An advocate for the existence of errors in Scripture, he proposes existential reinterpretations for the meaning of doctrine.[16] However, he faces a conceptual problem which confronts all those who oppose total inerrancy. *Why would the omnipotent God, who is incompatible with error, allow the human authors of the Bible to incorporate errors in their writings, and thus mislead mankind through all generations?* We could never in this life know the truth intended to be asserted by the sacred writers!

It is most interesting that, in outlining his views on what revised Scripture should mean for modern man, Fr. Brown seems strongly influenced by his belief in Evolution. One wonders about the accuracy

---

16. See Fr. Raymond E. Brown, *The Critical Meaning of the Bible* (London: Cassell Ltd., a Geoffrey Chapman book, 1982).

of his information, however, when within one paragraph he makes two conflicting statements:

> It is embarrassing to see the rightists [in the Catholic Church] lionizing the maverick scientist whose views can be used to challenge the overwhelming scholarly consensus in favor of human evolution. . . . To my mind the attitude toward science is crucial in Christianity's role in the modern world. When in the course of history Christianity has been at its best, it has been able to digest and profit from every new major body of knowledge it encountered.[17]

Contrary to this unquestioning acceptance of Evolution as fact, modern science has shown overwhelmingly that Evolution *cannot* occur! Brown's many arguments regarding revision of doctrine are poorly based, since they are influenced greatly by a mistaken grasp of modern scientific findings concerning Origins. His further assertion (Brown, p. 90) that "the author of *Genesis* . . . would scarcely have had access to an exact tradition coming down from the time of those events, thousands or millions of years ago" is gratuitous, and ignores the fact that God is the principal Author of Scripture. (The Creation account could have been handed down from Adam on clay tablets.)

Is there not something amiss within biblical scholarship in the modern Catholic Church, when eminent scholars such as Fr. Brown speculate as follows about Mary?

> We Catholics have not accepted biblical criticism if we cannot deal honestly with the unfavorable view of Jesus' natural family (including Mary) in Mark 3:21-35 and 6:1-6 . . . the sequence indicates that Mark *judged* that Jesus' mother was among "his own" and that she thought he was beside himself—scarcely a graceful picture of Mary.[18]

What is to be gained from such speculation, especially when we know that Mary was conceived free from Original Sin? In contrast is *The Catechism of the Council of Trent's* great teaching on typology:

> The Virgin Mother we may also compare to Eve, making the second Eve, that is, Mary, correspond to the first, as we have already shown that the second Adam, that is, Christ, corresponds to the first Adam. By believing the serpent, Eve brought malediction and death on mankind, and Mary, by believing the Angel, became the

---

17. *Ibid.*, p. 88.
18. *Ibid.*, p. 79.

Evolution was once thought to require eons of time, and belief that the Universe must be billions of years old came to be largely accepted as proven beyond doubt.

In turn, the historical credibility of **Genesis** *was increasingly doubted, and Modernism became popular.*

But the length of time is irrelevant— if the design of life forms does not allow Evolution, millions of years will not enable it to occur.

Discoveries in the Universe cannot be considered in isolation from the historicity of **Genesis**—*the reality intended to be conveyed by the sacred writer(s) cannot be ignored.*

SOLAR SYSTEM—UNIVERSE
Evidence Refutes "Evolution"/Affirms Creation

Solar system—great variety, complexity	Design defies uniformitarianism—how could the solar system arise from a Big Bang explosion?
Planets/moons–dual rotation behavior	Simultaneous opposite direction rotations suggest intentional Design rather than the results of explosion.
Janus & Epimetheus (satellites of Saturn) only 30 mi. apart	Two satellites swap places every four years instead of colliding. Were they programmed to behave this way by intelligent Designer, to thwart uniformitarianism?
"Shepherd satellites" accompany rings of giant gas planets	Such precise functioning is unlikely to result from a central Big Bang explosion. Evidence of intentional Design in the solar system.
Pluto—unique "double" planet	Pluto and its moon Charon spiral around a center of gravity which lies between them. How could this result from a central Big Bang explosion?
Short period comets	Should not exist if Universe is billions of years old. Where is proven evidence for the "Oort Cloud"?
Bombardment throughout the solar system	It seems likely that an immense, mysterious object (a planet?) exploded and impacted throughout the solar system. Did it also trigger the Flood events?
Earth's finely balanced distance from the Sun	So delicately balanced and precisely located to allow life on Earth—good fortune, the Anthropic Principle at work, or design by omnipotent Creator?
Spiral galaxy arms in the Universe	Why are spiral "arms" not wound up, if the Universe is many billions of years old?
Rotating galaxies in the Universe	Galaxies rotating in various localized directions would not result from a Big Bang explosion point, but are explicable in the Day 4 Creation scenario.
Gigantic gas pillars in Eagle Nebula	Discoveries by the Hubble Space Telescope show that the gargantuan Universe is barely understood by finite human beings. How can uniformitarianism be assumed about events in the Universe?
Entropy in Universe	"Time's arrow"—pointing downward to disorder and eventual total loss of energy. The Universe surely must have had a beginning.
Speed of light—vast distances in Universe	Were stars and light waves created in nanoseconds throughout the Universe? Is there a "time warp" in the Universe, making possible vast distances within a "young" Universe?
Origin of mysterious Universe—awesome beyond words	The mind-boggling, mysterious Universe could not have created itself. Surely an awesome Creator/Designer must have been involved!

TRADITION—CHRISTENDOM
Tradition Refutes "Evolution"/Affirms Creation

Satanic war against God—relevance of Origins debate	When crucial *Genesis* foundations are placed in doubt, belief in Catholic truths is greatly weakened, and Modernism flourishes.
Doctrine of Original Sin rejected/distorted	Crucial foundational beliefs are rejected by revisionist scholars. Objective truth becomes obscured.
Trustworthy Genesis account	Catholic Tradition holds that *Genesis* was intended to be understood primarily as historical revelation.
Ancient clay tablets	Clay tablets show use of colophons. Was the true Creation account recorded on clay tablets?
Noah's Ark— fact or fiction?	Noah's Ark is held in Catholic Tradition as a fact foreshadowing the Church founded by Christ.
Inerrant Scripture of crucial importance	The Bible is totally free from error—there is no such thing as limited inerrancy.
Pope Leo XIII (d. 1903)	Declared that the literal, obvious sense of Scripture holds ground unless another sense is shown superior.
Pope St. Pius X (d. 1914)	Condemned evolutionary-inspired Modernism 100 years ago; his body remained incorrupt after death.
Pope Pius XII (d. 1958)	Cast doubt on Evolution, dismissed polygenism. Declared that Evolution must not be taught as fact, and that its pros and cons must be fully addressed.
Myth/illusion/phantom of Evolution	Many atheists claim Evolution as "fact" and refuse to acknowledge possible existence of Designer.
Evolution beliefs can undermine doctrine	Evolution gives rise to belief that everything, including doctrine, is in a constant state of "becoming."
Naturalism vs. supernaturalism	Will moral standards/beliefs of Judeo-Christian civilization be further eroded by anti-God forces?
Wars/revolutions— result of Fall of Man	The many atrocities caused by some brutalized human beings were not intended by the Creator.
Social disharmony/ personal alienation	Distressing effects in society are inevitable if the Creator's moral absolutes are disregarded.
Pestilence/famine	Famine and starvation were not intended in the "good" Creation, but were brought on by Adam's sin.
Predatory behavior of animals/sharks	Were vicious predators part of the "good" Creation, or was their behavior also affected by Adam's sin?
Secondary causes	Secondary causes, not Evolution, instituted by God.
Jesus Christ often referred to Genesis	The trustworthy Creator/Redeemer, incapable of error or deception, took *Genesis* very seriously.
Restoration within Catholic Church	Rediscovery of Creation can bring personal liberating effect of truth and renewal in worship of God.

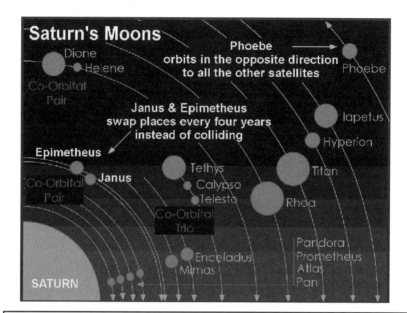

Janus and Epimetheus are co-orbital moons of similar size, 118 miles and 75 miles respectively, and their orbits are only 30 miles apart. Because they travel at different speeds, they should collide but, about every four years, they swap places instead of colliding! The inner, faster satellite moves to the outer orbit and slows down, and the outer, slower satellite moves to the inner orbit and speeds up. Such mysterious behavior would hardly result from a Big Bang explosion. Intentional design by the Creator of the Universe is a more convincing explanation.

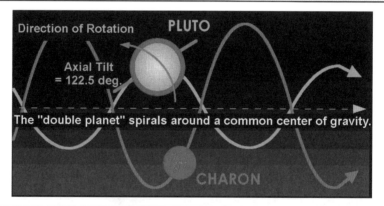

Recent discoveries are shedding light on many mysterious features in the solar system. (Six of Neptune's eight moons were only found in 1989.) Pluto, which is tipped over on its side, is known as the "double planet." Its moon Charon was only discovered in 1978. Curiously, both planet and moon rotate around a common center of gravity which lies in the space between them.

Above and Below: The four giant gas planets all have rings, but Saturn's are the most spectacular. Some of the rings are accompanied on each side by "shepherd satellites," to restrict the particles from drifting in or out. This suggests intelligent Design, rather than the result of a supposed Big Bang.

Below: Compared in size to Jupiter's Great Red Spot, Earth is very small indeed.

How can we fully comprehend the awesome Universe?

Photo Credit: NASA/STScI.

Clockwise from top left: Cat's Eye Nebula, Cartwheel Galaxy, Triple Rings Supernova, Spiral Galaxy M100, M16 Gaseous Pillars in the Eagle Nebula. Awesome features in the mysterious Universe are fully understood by the trustworthy Creator God. Where is the proof beyond doubt that the Universe is billions of years old? Why not creation of space, time and matter on Day 1, and instantaneous transformation of matter throughout the Universe on Day 4 by the Creator? *Photo Credit: Hubble Space Telescope NASA/STScI.*

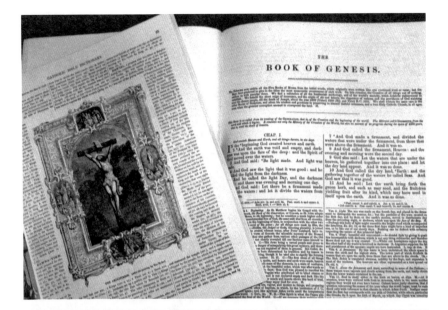

The Creator of the Universe, the Redeemer of fallen mankind and the Word made Flesh (*John* 1:1-14) is one Person—the Second Person of the Divine Trinity, Jesus Christ—recognized in the Church also as the "second Adam." His various strong references to *Genesis* indicate that He accepted *Genesis* as an historical revelation. Where is the literary justification in the *Genesis* text itself, and the weight of evidence of the Fathers in Tradition, to now conclude that the sacred writer(s) (including God as the principal Author of Scripture) definitely intended to convey only a symbolic interpretation of *Genesis* 1-11?

The Church has always recognized the sober reality of angelic forces (both good angels and bad angels) at work behind the scenes in the created world. This artwork is on display inside St. Peter's Basilica, Rome.

Above left: Mary, the wonderful Mother of God, has long been venerated in the Church as the "second Eve," who helped repair the harm done by the first Eve. She is depicted here crushing the serpent (Satan).
Above right: The ruins of the Roman Coliseum remain a poignant reminder of the many human beings who were slaughtered there by wild beasts and of how the pagan world was overcome by Christian truths.

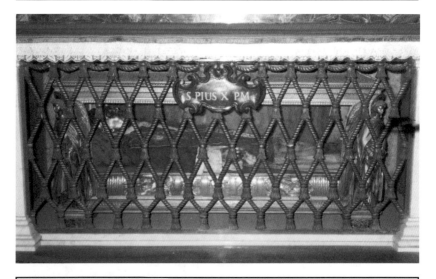

The body of Pope St. Pius X (d. 1914) remained incorrupt after his death, and it now lies in state within St. Peter's Basilica, Rome. He must have been very pleasing to God for him to be so honored. His condemnation of Modernism (which relies upon Evolution as fact) is still very relevant even since 1907.

instrument of the divine goodness in bringing life and benediction to the human race. From Eve we are born children of wrath; from Mary we have received Jesus Christ, and through Him are regenerated children of grace.[19]

Another conceptual problem facing revisionist Scripture scholars can be pointed out. If one is convinced that errors do exist in Scripture, then one must also ultimately concede that no parts of it are free from future revision of any kind. Unfortunately, that scholar in turn can soon be superseded by someone else, who claims greater insights. On what grounds would his subjective views be superior to those of others? And who is to decide—a new Magisterium comprised only of theologians who reject the pope? Revisionist theology is, in reality, self-defeating and can only lead to chaos.

Unfortunately, the ambiguous statement in the *Dogmatic Constitution on Divine Revelation* from Vatican II (Chapter 3) provides an opening for revisionist errors. To say that *"the books of Scripture teach certainly, faithfully and without error the truth that God for our salvation willed to be recorded in Holy Writ"*[20] leaves it wide open for anyone to assert that God only intended *Genesis* to be understood in a "salvation history" sense. It regards as irrelevant whether real history is contained in *Genesis*, and that it does not matter whether the writers of Scripture made any errors in their writings, since only the religious affirmations of *Genesis* are subject to inerrancy. That this statement was likely to be abused was already foreseen during the deliberations of the Council, as Fr. John McKee describes:

> The original text was not "for the sake of our salvation" but "pertaining to our salvation" and 184 Council Fathers asked that the phrase should be dropped for the precise reason that it might be taken as restricting inerrancy to faith and morals. The Theological Commission dragged its feet and, on October 8, 1965, a group delivered a memorandum to the Pope, claiming openly that the phrase had been deliberately inserted to restrict inerrancy in a way contrary to Catholic teaching. After an investigation, the Pope sent observations on this and other matters to the Theological Commission. He said that the matter involved "great responsibility for him towards the Church and towards his own conscience." The Commission was asked to drop the expression "truth pertaining to salvation" from the text. After dis-

---

19. *Catechism of the Council of Trent*, p. 46.
20. *Dei Verbum*, par. 11.

cussion and voting, the Commission adopted the text as we now have it. When one compares the two versions, one sees that a tightening has taken place in order to placate the Pope and the traditionalist section of the Council, but one detects still some foot-dragging on the part of the liberals. The text could have been sharper and met more fully the wish of the Vicar of Christ. If the Council had dropped the dangerous phrase as requested, instead of replacing it with an improved one, there could have been no misrepresentation.[21]

Fr. William G. Most also wrote about the controversy over inerrancy which took place at Vatican Council II:

> The most decisive proof of what Vatican II really meant in the sentence under consideration is this: Pope Pius XII in his great encyclical *Divino Afflante Spiritu* had quoted the words of Vatican I saying that God Himself is the chief author of Scripture, and commented that the words of Vatican I were "a solemn definition." Of course Vatican II would not contradict a solemn definition! That would be heresy. And no matter on what level Vatican II was teaching at this point, it could not possibly teach heresy. Really, Vatican II even added four footnotes to this very passage in which it refers to the text mentioned of Vatican I and to several pages of Leo XIII's encyclical, and to two passages of Pius XII—all of which insist more than once with great care that there is no error of any kind at all in Scripture.[22]

The truth of the inerrancy question *must* be that there are no errors whatsoever in the Bible and that modern scholars still have much to discover about the truth being asserted in Scripture. Who knows what valid insights are yet to be discerned from Scripture?

A most important Origins controversy, affecting the credibility of Scripture generally, is the possibility that the first five books of Moses (commonly known as the Pentateuch) may be the work of many editors ("redactors"), who revised the documents until they reached their final form. It is contended that the work was finally developed long after the time of Moses and has been incorrectly ascribed back to him. Involved also is the question of whether the belief in only one God (monotheism) was itself developed from earlier beliefs in many gods.

---

21. Fr. John McKee, *The Enemy Within the Gate* (Houston: Lumen Christi Press, 1974), p. 264.
22. Fr. William G. Most, *Free From All Error*, p. 38.

The Documentary Theory of the Pentateuch

The Documentary Theory (known also as the JEDP Theory) has its origins in antiquity, but the modern movement gained great momentum from the evolutionary inspired religious preconceptions of liberal scholars such as K. H. Graf (d. 1869) and Julius Wellhausen (d. 1918). The theory rejects the belief that Moses was substantially the writer or editor of the first five books of the Old Testament. What is proposed instead is that four "traditions"—Yahwist, Elohistic, Deuteronomic, and Priestly—are spread throughout these books.

Scholars who support the JEDP Theory simply assert that the first *Genesis* Creation account is chronologically later than the second account, and that *Genesis* is explicable as mythology. However, creationists offer another possible explanation which does not resort to calling *Genesis* "mythology." The initial account given in *Genesis* 1 may be entirely the supernatural inspiration of God given to the sacred writer, since no human beings could have witnessed all the Creation events. The remainder of *Genesis* may be the work of Moses, under divine inspiration certainly, but also drawing as editor on documents or clay tablets handed down by earlier patriarchs (in the line from Adam to Noah and Abraham and finally to Moses), and perhaps drawing on oral transmission of the earliest events handed down to Moses. It was then recorded in the form of expression popular at the time of writing.

Whereas *Genesis* 1 is very likely the actual order in which Creation took place, the remainder may only be a record of the most significant events which took place. Thus, only some information is recorded. It would not necessarily have to be in strict chronological order.

If the Earth is no more than about 6,000 years old, it would easily be possible for an accurate account of the earliest events to have been handed down to Moses. Man's knowledge of these events could have been transmitted by both oral tradition and written methods, such as clay tablets. However, with respect to the age of the Earth, God was also involved in the production of the Old Testament; it did not depend solely on human beings.

The Documentary Theory has long been rejected by many scholars as being both incorrect and obsolete. Critics of it point out that the so-called "doublets" (e.g., repetition of various incidents), *alleged* to be found in the Pentateuch, do not mean that multiple authorship was involved. Also, the Graf-Wellhausen system predates the recovery of vast libraries of the ancient world, and the information discovered therein of the methods and styles used by ancient scribes.

About 20,000 clay tablet fragments, ranging in thickness from one inch to two feet and dating back to 2300 BC, were recovered at Ebla, Syria, in the 1960's. As Clifford Wilson notes, these discoveries show that the ancient Canaanite culture was contemporary with ancient Sumeria and Egypt, and all three civilizations date back only to about 5,000 years ago. The cuneiform inscriptions on clay tablets dealt with trade, literature, dictionaries and religious texts; and colophons [publishers inscriptions] were used when indicating a series of tablets. Among the Ebla discoveries was an account which attributed Creation to one great being, known as "Lugal, the Great One." This shows that written clay records about the Creation were in existence about 1,000 years before the time of Moses.[23]

Archaeologists have discovered that in ancient Mesopotamia, a title phrase known as *colophon* was often placed at the end of each tablet in a series of tablets. Charles V. Taylor supports the theory of P. J. Wiseman[24] and argues that *Genesis* is likely to have been edited from nine original volumes. He shows how the colophon was used:

> Understanding it this way involves accepting that the earliest written records, presumably on baked clay tablets, did not place the titles at the beginning, though they sometimes referred to a record by its first words. What we could call a title is both in our day and theirs a phrase summarizing contents or some striking feature. This was placed at the end of the last tablet, or as we would say, the last page of the record. In that position it's known as a *colophon*.
>
> However the colophon also functioned as a run-through line, similar to the idea of modern printers who put chapter headings at the top of each page today. Thus the initial words or some summary words would be found at the foot of each tablet. By this means a set of tablets would be kept in correct order. Thus we might find at the foot of a tablet:
>
> Tablet 5 of IN THE BEGINNING.
>
> This is why we find these rather strange duplications, or sometimes abrupt breaks, wherever there's a reference to 'generations' in *Genesis*, i.e., in these verses:
>
> 2:4, 5:1, 6:9, 10:1, 11:10, 11:27, 25:19, 37:2.
>
> These will then represent places near the links between the nine

---

23. See Clifford Wilson, *Visual Highlights of the Bible* (Victoria, Australia: Pacific Christian Ministries, 1993), Vol. I.
24. P. J. Wiseman, "New Discoveries in Babylonia about *Genesis*," republished by D. J. Wiseman in *Clues to Creation in Genesis* (Marshall, Morgan and Scott, London, a member of the Pentos Group, 1977).

volumes which Moses put together into one new book, Genesis, called in Hebrew *Bereshith*, or *"At the Beginning (of Things)."*[25]

Another point of interest here is that statements which begin with *"these are the generations of"* are likely to refer to information possessed by the owner or writer of the particular tablet, rather than to the history about the person named. For example, the term *"these are the generations of Noah"* would refer to the record of history known or possessed by Noah, rather than be an account of the history of Noah. Any genealogies listed refer to the ancestors of the particular person and not to his descendants.

The Documentary Theory has itself undergone many revisions. It was rejected in 1906 by the *Pontifical Biblical Commission* (then an organ of the Magisterium), which declared that Moses was the human author of the Pentateuch, but that he may have utilized and edited earlier documents and oral traditions, and that there may have been some modifications or additions or even some faulty readings on the part of the copyists after the time of Moses.[26] Although still discussed as though credible among scholars, it seems impossible that the Catholic Church can ever *officially* accept the Documentary Theory because JEDP speculation conflicts with crucial non-evolutionary, Original Sin doctrine already declared in Tradition (*which is thus unchangeable*), including *Genesis* being primarily historical and devoid of mythology.

The standing of the PBC early this century was recognized by Pope Pius X in his *Motu proprio, Praestantia Scripturae* (Nov. 18, 1907): "We do now declare and expressly order, that all are bound by the duty of conscience to submit to the decisions of the Biblical Pontifical Commission . . . just as the decrees of the Sacred Congregations which pertain to doctrine and have been approved by the Pontiff."

The Protestant scholar, Oswald T. Allis, noted a crucial weakness in the Documentary Theory. According to this theory, the entire Mosaic law is supposed to have been practically non-existent until centuries after the time of Moses. Thus, the "law of Moses" supposedly did not

---

25. Charles V. Taylor, *The Oldest Science Book in the World* (Qld, Australia: Assembly Press Pty Ltd., Slacks Creek, 1984), p. 21.
26. For a comprehensive analysis of the controversy concerning the origin of the Pentateuch and of the official teaching of the Catholic Church in relation to it, see Msgr. John E. Steinmueller, S.T.D., *A companion to Scripture Studies* (Texas: Lumen Christi Press, 1969), Vol. 11. See also Br. Thomas Mary Sennott, *The Six Days of Creation* (Cambridge: Ravengate Press, 1984) and Fr. William Most, *Free From All Error* (Libertyville, IL: Prow Books/ Franciscan Marytown Press, 1990), Chap. 16.

precede by many centuries, but was itself later than the golden age of prophecy. We should therefore speak of "the prophets and the law," rather than "the law and the prophets," as referred to in *Matt.* 22:40.[27]

On another aspect, it is noteworthy that the mid-1980s computer study conducted by Jewish scholars found that

> All these reservations notwithstanding, and with all due respect to the illustrious Documentarians past and present, there is massive evidence that the pre-Biblical triplicity of *Genesis*, which their line of thought postulates to have been worked over by a late and gifted editor into a trinity, is actually a unity.[28]

Christ's own words strongly support the Mosaic authorship: "If you did believe Moses, you would perhaps believe me also, for he wrote of me. But if you do not believe his writings, how will you believe my words?" (*John* 5:46-47), "And beginning with Moses and all the prophets, he [Jesus] expounded to them in all the Scriptures the things that were concerning him." (*Luke* 24:27). If we believe in Christ the Creator about His promised Second Coming, with the destruction of Earth by fire, why not also believe Him in his assertions about Origins?

The Galileo Case

Since the time of Galileo, the Catholic Church often has been accused of making empirical science subservient to theology, and even of being anti-science. The idea that Galileo's trial was a showdown between blind religious faith and enlightened scientific reason is a myth which still lingers strongly today, and the position of the Church in this affair continues to be misrepresented.

Early in the 17th century, as far as was then known, both the old Ptolemaic geocentric theory (Earth-centered) and the newly developing heliocentric theories (Sun-centered) appeared to provide plausible explanations of how the solar system actually works. Since the implications for Christian beliefs of a Sun-centered Universe were thought to be quite profound for theology, Church authorities at the

---

27. For Protestant sources supporting the Mosaic authorship, see Oswald T. Allis, *The Five Books of Moses* (Phillipsburg, NJ: Presbyterian and Reformed Publishing Co.). Also William Henry Green, *The Higher Criticism of the Pentateuch* (Baker Book House).
28. Yehuda T. Radday and Haim Shore, *Genesis: An Authorship Study in Computer-Assisted Statistical Linguistics* (Rome: Rome Biblical Institute Press, 1985), p. 190.

time preferred to regard the heliocentric theories as hypotheses awaiting verification.

The Ptolemaic theory increasingly was seen to contain problems, for it required a series of epicycles (a process rather like looping the loop) in the behavior of planets so that they would each have perfectly circular orbits. Ironically, the actual Copernican theory, even though Sun-centered, required even more epicycles than the Ptolemaic model, for it too was based on perfectly circular orbits by planets. It seems that Copernicus himself came to realize that his model was heavily flawed; hence his reluctance to publish the manuscript. His *Book of the Revolutions of the Heavenly Spheres* was only published in 1543, when Copernicus was on his deathbed, about seventy years before the Galileo case.

Galileo, who had greatly improved the recently invented telescope and discovered four moons of Jupiter, promoted the perfectly circular orbit theory of Copernicus and insisted that the Church fall into line with his views. However, his belligerence provoked a strong counter-reaction against him. In time, both the Ptolemaic and Copernican systems were overturned, following the work of scholars such as Johannes Kepler, who showed that Mars travels around the Sun in an elliptical orbit, and Isaac Newton, who proposed that the movements of the planets and their moons are determined by the interaction of the force of gravity and centrifugal forces. Arthur Koestler, a non-Christian scholar, wrote about the case in his book, *The Sleepwalkers*:

> It must be remembered that the system which Galileo advocated was the orthodox Copernican system, designed by the Canon himself. . . . Incapable of acknowledging that any of his contemporaries had a share in the progress of astronomy, Galileo blindly and indeed suicidally ignored Kepler's work to the end, persisting in the futile attempt to bludgeon the world into accepting a Ferris wheel with forty-eight epicycles as "rigorously demonstrated" physical reality . . . When, carried away by sudden fame, he had at last committed himself, it became at once a matter of prestige to him. He had said that Copernicus was right, and whosoever said otherwise was belittling his authority as the foremost scholar of his time.[29]

---

29. Koestler, *The Sleepwalkers* (London: Penguin Books, 1986), p. 444. Emphasis added. See also *Planets: A Smithsonian Guide*, p. 20 for a photograph from the book by Copernicus, showing the diagram of six planets tracing perfect circles around the Sun. Galileo was quite mistaken in insisting upon adherence to the strict Copernican theory.

The conflicting heliocentric theories of Copernicus and Kepler had been open to public discussion for about 70 years, but it seems clear that Galileo's arrogant insistence on acceptance of the strict Copernican theory, as though already proved, provoked the Holy Office of the Inquisition into a too-hasty reaction. Of great concern was the following problem: If the Sun were indeed immobile, does this mean that the Bible contained an error in *Joshua* 10?

In 1616, Copernicanism was proscribed by the Holy Office. But what actually was proscribed? And were *ex cathedra* papal pronouncements involved which could not be reversed by future popes?

The first proposition—that the Sun is in the center of the Universe and immobile—was declared by the Congregation to be philosophically absurd and formally heretical, since it contradicts Holy Scripture in many passages, according to the proper meaning of the language used and the sense in which they have been expounded and understood by the Holy Fathers. The second proposition—that Earth is not in the center of the Universe, and moves as a whole with diurnal (rotation) motion—was declared by the Congregation to deserve the same censure philosophically and, theologically considered, to be erroneous in faith.

However, as Koestler emphasizes, the findings remained a judicial opinion, without signed papal endorsement, and thus the immobility of the Earth never became an article of faith, nor the immobility of the Sun a heresy.[30] Perhaps the lack of open papal endorsement was providential, thus enabling future popes to declare more fully on the various senses which are used in *Genesis*.

While the situation was inflamed by Galileo's belligerent attitude, it seems that some of his many opponents were not free from blame. A document was written, apparently mischievously, to the effect that Galileo was prohibited from teaching the Copernican Theory in any way whatsoever. This dubious document was later to be instrumental in Galileo's trial of 1633, when he was threatened and forced to recant. Brother Thomas Mary Sennott, in his book *The Six Days of Creation*, comments on the trial of 1633:

> With St. Robert Bellarmine now dead and a sympathetic Pope, Urban VIII, now reigning, Galileo hurried down to Rome to get permission to publish on the Copernican system. (He had been forbidden to publish by St. Robert.) He suggested to Urban that he present

---

30. *Ibid.*, p. 463.

the problem in the form of a dialogue in which one speaker presented the Ptolemaic system and another the Copernican. Urban agreed, but told him not to endorse either system, but to let the dialogue end in a draw, since there was still the unresolved problem of stellar parallax. Galileo agreed, but unfortunately did not live up to his agreement. He had his Copernican spokesman utterly demolish the Ptolemaic speaker, his final climactic argument being his [Galileo's] theory of the tides. Urban was understandably furious, and after learning from the Jesuit scientists at the University of Rome that *Galileo's theory of the tides was just bad physics*, he ordered Galileo to be tried by the Inquisition. Galileo was condemned and sentenced to confinement at his estate near Florence for the rest of his life.

Had he lived up to his agreement with Urban and treated each of the world systems in an equitable manner, there would have been no Galileo Case. Evidently he did not inform the censors of the substance of his agreement with the Pope, and thus was guilty of subterfuge in obtaining the *Imprimatur* ("Let it be printed"). Also possibly if he had been able to demonstrate the Copernican system convincingly, there would have been no Case, but his false theory of the tides was fatal.[31]

At the trial Galileo was questioned several times about his book, *The Dialogue on the Great World Systems,* in which the Ptolemaic spokesman, named Simplicius, was made to look a fool. Galileo (who was now seventy years old) answered each time that he did not hold to the Copernican theory. However, as Koestler describes,

> Both the judges and the defendant knew that he was lying; both the judges and he knew that the threat of torture (*territio verbalis*—as opposed to *territio realis* where the instruments of torture are shown to the accused) was merely a ritual formula which could not be carried out [old or sick people, and Galileo was both, could not be tortured]; and that the hearing was a pure formality. The "formal prison" took the form of a sojourn at the Grand Duke's villa at Trinita del Monte, followed by a sojourn in the palace of the Archbishop Piccolomini in Siena where, according to a French visitor, Galileo worked "in an apartment covered in silk and most richly furnished." He returned to his farm . . . and later to his house in Florence, where he spent the remaining years of his life.
>
> From the purely legal point of view, the sentence was certainly a

---

31. Sennott, *The Six Days of Creation,* p. 184. Emphasis added.

miscarriage of justice. On the other hand, the judgment hushes up the incriminating contents of the book, stating that Galileo had represented the Copernican theory as merely "probable"—which is a whale of an understatement. The gist of the matter is that Galileo could not legally be convicted without completely destroying him—which was not the intention of the Pope or the Holy Office. Instead, they resorted to a legally shaky concoction. The intention was, clearly, to treat the famous scholar with consideration and leniency, but at the same time to hurt his pride, to prove that not even a Galileo was allowed to mock Jesuits, Dominicans, Pope, and Holy Office; and lastly to prove that, in spite of his pose as a fearless crusader, he was not the stuff of which martyrs are made.[32]

Pope Urban VIII was not directly involved in Galileo's trial, and only seven of the ten judges signed the sentence; the others abstained. Papal infallibility cannot be delegated, and thus there was no papal declaration *ex cathedra* [i.e., by the pope "from the Chair" of Peter, about "faith" or "morals," and therefore infallible]. Subsequently, in 1757, Pope Benedict XIV suspended the earlier decree. If the Galileo case had truly involved infallibility, this would not have escaped the notice of many opponents of the Catholic Church, who ever since would have fully exploited the consequences of such as disaster—far exceeding the misrepresentation which already exists. Fr. Peter Damian Fehlner has shown that, as far as the case touched matters of belief, two points stand out:

1. The astronomical theories of Galileo touched points also mentioned in Scripture. His views, propounded as proven fact, would seem to render Scriptural references to the Earth either false or meaningless: the decision to place the works of Galileo on the *Index of Forbidden Books*, and to forbid him to publish anything more on the subject, was not a condemnation of scientific theorizing as such; it was an insistence that his particular theory be held merely as a hypothesis, *until such time as the Church should have resolved the exegetical questions*; to publicize the same in circumstances where it might easily be taken as proven fact by the uninformed would act to the detriment of their faith. One may discuss whether this was the best manner to handle the pastoral problem; but it hardly constitutes intellectual tyranny. . . . Galileo's celestial mechanics was not condemned, neither was Aristotle's canonized.

---

32. Koestler, *The Sleepwalkers*, pp. 499-501.

2. The immediate concern of the Church was not the justification of astronomical theory, but the guardianship of the deposit of faith and its correct interpretation. Revelation does contain references to what seems to be the immobility of the Earth. The Fathers of the Church, as St. Robert Bellarmine noted, also seem to attest to this as a fact. If the heliocentric theory is true then, as St. Robert observed, our understanding of these passages must be reexamined to discover the faulty interpretation, but it is not permitted in the meantime to say God has stated something false or engaged in pious deception. If the theory is merely possible, this is not a sufficient basis as yet for doubting the literal sense of Scripture attested by the Fathers.[33]

Thomas Huxley, an eloquent 19th century English evolutionist who was famous as Darwin's "bulldog," investigated the Galileo case and decided not to pursue it further. It seems that he concluded the facts of the case gave no grounds for criticism of the Catholic Church.

As a non-Christian, Koestler can hardly be accused of bias in favor of the Catholic Church, and he paid tribute to Galileo as being among the men who shaped human destiny. Galileo, in his seventies, went on to rediscover his real vocation as the founder of the science of dynamics, after having spent 25 years on his propaganda crusade for the heavily flawed Copernican system. But Koestler also noted that many achievements have been wrongly attributed to Galileo:

> Contrary to statements in even recent outlines of science, Galileo did not invent the telescope; nor the microscope; nor the thermometer; nor the pendulum clock. He did not discover the law of inertia; nor the parallelogram of forces or motions; nor the Sun spots. He made no contribution to theoretical astronomy; he did not throw down weights from the leaning tower of Pisa, and did not prove the truth of the Copernican system. He was not tortured by the Inquisition, did not languish in its dungeons, did not say *"eppur si muove"* ["But still it moves"]; and he was not a martyr of science.[34]

Koestler also laments the fact that Galileo's ill-conceived crusade precipitated the divorce of science from faith. Unfortunately, the misrepresentation made against the Catholic Church continues unabated:

---

33. Fr. Peter Damian Fehlner, "In The Beginning," *Christ to the World*, Volume XXXIII, 1988, Rome, p. 160. Emphasis in original.
34. Koestler, *The Sleepwalkers*, p. 358.

> *The Galileo affair was an isolated, and in fact quite untypical, episode in the history of the relations between science and theology. . . . But its dramatic circumstances, magnified out of all proportion, created a popular belief that science stood for freedom, the Church for oppression of thought.*[35]

In consequence, ever since the Galileo affair, the Catholic Church has often been cast as intransigent to change. Of particular harm is the fact that many Catholics now cite the Galileo case as justification for 1) avoiding literal senses being granted in *Genesis*, and 2) consigning *Genesis* to unbelievable mythology on the flawed ground that never again will there be another clash between science and theology.

This is ironic when one reflects on the experience of F. Sherwood Taylor, who set out to document the Galileo case from a position hostile to Catholicism, only to convert to the Catholic Church. His book *Galileo and the Freedom of Thought* was published in 1938.

Geocentrism/Heliocentrism

In the early 17th century, the emerging possibility of the heliocentric system must have seemed very daunting to theologians. If geocentrism was invalid, what did this hold for religious beliefs? Such a profound problem for religious beliefs surely deserved long and careful consideration. If Galileo had not been so belligerent in trying to force belief in the strict Copernican model (including *circular* planetary orbits, rather than *elliptical* orbits) and in provoking a counterreaction to his insensitive attitude, theologians of the time most likely would have reflected at length over many years in trying to resolve the matter.

Unfortunately, the lingering idea of a split between science and theology—supposedly resulting from the Galileo case—still remains a major problem today, affecting the mindset of many learned scholars within the Catholic Church. Some argue that theology must never again enter into matters of science, but this attitude overlooks the fact that the Church has every right to be heard when empirical science overlaps upon matters which rightly also belong in the domain of theology and exegesis. In fact, by way of example, recent popes have not hesitated to declare views on behalf of the Catholic Church in certain matters which involve both empirical science and moral theology (e.g., induced abortion, *in vitro* fertilization, contraception).

---

35. *Ibid.*, p. 533. Emphasis added.

As to the scientific truth of the geocentrism/heliocentrism issue, it may be too soon to know if all the astronomical data is available and if all scientific aspects are fully understood. Some researchers argue that a valid scientific case can still be made for geocentrism, and various experiments have been conducted with respect to solar eclipses. Unfortunately, direct observation from outer space is impossible; to observe the solar system fully, in order to judge relative motion, would require traveling an impossibly long distance away from Earth. Nevertheless, things are certainly not black and white about the physical form of the Universe, as physicist Wolfgang Smith notes:

> In the light of 20th century physics one can no longer say that Copernicus was right and Ptolemy wrong. All that one can legitimately claim is that the Copernican coordinate system leads to simpler laws of planetary motion (which, incidentally, is precisely what Copernicus himself did claim).[36]

In arguing the same point about modern physics, Gerald Aardsma (physicist) quotes well-known astronomer Fred Hoyle and points out,

> The generally overlooked lesson here is that scientific theories do not provide a very secure basis from which to interpret Scripture. In the course of the last 500 years, the weight of scientific consensus has rested in turn with each of three different theories about the form of the Universe: first geocentricity, then heliocentricity, and now relativity.[37]

For the layman, in addition to trying to cope with baffling concepts such as relativity, the issue is further complicated by difficulties in definition of terms. What exactly *is* geocentrism? Is Earth located at the center of the Universe? Is Earth regarded as stationary, or does it rotate on its axis, and does it move through space? Does geocentrism function according to the old Ptolemaic theory, with the Sun and all the planets rotating around the Earth in circular orbits, which would involve the old problem of epicycles?

Could the Sun really travel about 585 million miles *daily* around the Earth (over 24 million miles per hour), with no sign of a fiery trail? And

---

36. Wolfgang Smith, *Teilhardism and the New Religion* (Rockford, IL: TAN Books and Publishers, Inc., 1988), p. 134.
37. Gerald E. Aardsma, "Geocentricity and Creation," *Impact* (El Cajon, CA: Institute for Creation Research), No. 253, July, 1994.

how could the incredibly distant galaxies also travel *daily* around the Earth, at speeds far exceeding the speed of light?

Further, does the Bible actually *assert* geocentrism? A strong case can be made that the Earth is the cultural center of the Universe, as this is the only known place where life exists or has existed. So far there has been no completely confirmed evidence that other solar systems even exist. And it must not be overlooked that the whole point of the creation of Earth was to create human beings; creation of the physical Universe must surely be secondary to this primary purpose. The Earth was simply an abode within space, time and matter for human beings who possess rational souls "made in the image and likeness of God"; with the Universe created as a backdrop for man to reflect in awe at the greatness of the Creator.

It seems most unlikely that life forms with rational souls could exist anywhere else in the Universe. The Church teaches that the whole Universe was adversely affected by Adam's sin of disobedience, and it is hard to see how other creatures, all possessing rational souls and living in various galaxies throughout the Universe, would be bound by the disobedience of Adam. If they existed, they probably would be put to tests of obedience and God would have to anticipate, ahead of their creation, the consequences of their respective choices.

If they all chose obedience, how could they then be expected to live in a Universe which the *Catechism of the Catholic Church* says "groans in travail" as a result of man's disobedience on Earth? But if various ones chose disobedience to God, what then? Would Christ also be required repeatedly to become the Redeemer, another "Second Adam," perhaps many times? Would there be need for other Marys—many "Second Eves"—many Mothers of God? Thus, it is hard to believe there could be multiple Mothers of God in the Universe. (Pope Pius XII declared in 1950 as certain, obligatory to be believed by Catholics, that there is only one Mother of God.)

The idea of other life forms beyond Earth *in possession of rational souls,* therefore, seems theologically unlikely, and so it would seem reasonable to conclude that life does not exist anywhere else in the Universe. However, the Earth being the cultural center of the Universe is one thing, but what about its actual physical location? Are we at the very center of the Universe, or somewhere else, and where does Scripture really teach that Earth is physically at the center of the Universe?

Catholic Tradition has been virtually silent about geocentrism, and this silence is not favorable to its credibility. Recent popes have taught definitively that figurative senses—conveyed "according to appearances"—are also employed in the Bible, and we know that Scripture must be understood in the sense intended by the sacred writers.

Various passages (e.g., in *Psalms* 93, 96 and 104) often cited as proof that geocentrism is asserted in Scripture would require acceptance only in the literal-as-given sense, but the same passages are open to understanding in a figurative sense. Thus they are not necessarily supportive of geocentrism. In commenting on the Galileo Case, Fr. Brian W. Harrison, O.S. discusses geocentrism:

> But even assuming that Galileo's inquisitors were scientifically wrong (and there are now—since the 1970's—some Catholic and Protestant scholars with PhD's in physics and astronomy who maintain that they were scientifically right, i.e., that geocentrism is the truth), their error was *not* in supposing that if the Bible makes assertions about physical reality, these must be accepted as revealed truth (a supposition which they did indeed make—and very rightly). Rather, their error lay in faulty exegesis: in supposing that the Bible does in fact assert a particular physical proposition (geocentrism) which it does not really assert. We have to say that *that* was the error which led them to trespass unwittingly into the autonomous domain of science.[38]

The account in the book of *Joshua* of the Sun standing still (*Josh.* 10:13) cannot be cited as strong evidence in favor of geocentrism. The Sun was recorded as standing still for about a whole day, *and the moon also*: "And the sun and the moon stood still, till the people revenged themselves of their enemies." (*Josh.* 10:13). Whether this happened in fact, or only appeared to happen, gives no support to geocentrism *per se*, but may in fact do the opposite, since it strongly implies a (real or apparent) cessation of Earth's rotation as the cause of the Sun's "standing still." Joshua needed the Sun to stand still, but the moon's motion was otherwise irrelevant; the fact that the moon also stood still suggests that this event recorded in *Joshua* was an abnormal occurrence.

Similar phenomena occurred at Fatima on October 13, 1917, when the Sun was seen by over 70,000 people to spin like a wheel of fire in multi-colored rays and then come rushing toward Earth. How is this explained? If the Sun had in fact come rushing towards Earth, it seems reasonable to suppose that a massive blast of heat would have hit Earth

---

38. Fr. Brian W. Harrison, O.S., *Living Tradition*, p. 3.

and been felt around the world. It was not. The many eyewitnesses who recorded the phenomenon may have experienced a genuine miracle (intended to bring mankind back to a sensible fear of God) that was carried out by God "according to appearances."

Proponents of geocentrism have pointed to other aspects which are of great importance: Deep-seated cultural problems exist in the modern era, and many scientists seem to regard physics as somehow more enlightening than theological reality. Some scholars even give the impression that theology, to be relevant, must ever-hasten to harmonize its concepts with the supposed findings of secular scientists regarding the form of the Universe and its *apparent* age of "billions of years."

Addressing the impact of Einstein's General Theory of Relativity upon 20th century society, the historian Paul Johnson articulates deep concerns felt by many about the continuing, amoral, cultural drift:

> For most people, to whom Newtonian physics, with their straight lines and right angles, were perfectly comprehensible, relativity never became more than a vague source of unease. It was grasped that absolute time and absolute length had been dethroned; that motion was curvilinear. All at once, nothing seemed certain in the movements of the spheres. . . . At the beginning of the 1920's the belief began to circulate, for the first time at a popular level, that there were no longer any absolutes: of time and space, of good and evil, of knowledge, above all of value. *Mistakenly but perhaps inevitably, relativity became confused with relativism.* No one was more distressed than Einstein by this public misapprehension. . . . He lived to see moral relativism, to him a disease, become a social pandemic, just as he had lived to see his fatal equation bring into existence nuclear warfare.[39]

Johnson argues that Einstein was a classic example of the dual impact made by great scientific innovators. They can change prevailing ideas. He contends that a) Galileo's empiricism created the ferment of natural philosophy in the 17th century which ushered in Scientific and Industrial Revolutions, b) Newtonian physics formed the framework of the 18th century Enlightenment and helped facilitate modern national-

---

39. Paul Johnson, *Modern Times* (London: Orion Books Limited, 1992), p. 4. Emphasis added.

ism and revolutionary politics, c) Darwin's notion of the survival of the fittest became central to the Marxist concept of Class Struggle and to the racist philosophies which shaped Nazism, and d):

> *The public response to relativity was one of the principal formative influences on the course of 20th century history. It formed a knife, inadvertently wielded by its author, to help cut society adrift from its traditional moorings in the faith and morals of Judeo-Christian culture.*[40]

In discussing the culture-shock upon Christian culture which came with the advent of heliocentrism, Wolfgang Smith also touches upon deep issues in the geocentrism/heliocentrism issue—important in an era which is plagued with confusion about Creation:

> With the disappearance of the Ptolemaic world-view, Western man lost his sense of verticality, his sense of transcendence. Or rather *these finer perceptions had now become confined to the purely religious sphere, which thus became isolated and estranged from the rest of the culture.* So far as cosmology—Weltanschauung in the literal sense—was concerned, European civilization became de-Christianized. . . .
>
> Man is indeed a "microcosm," a universe in miniature; and that is the reason why, symbolically speaking, man is situated at the very center of the cosmos. . . . No doubt the reason for this centrality is that man, having been made "in the image of God," carries within himself the center from which all things have sprung. . . .
>
> Along with the Ptolemaic theory, the ancient anthropology fell likewise into oblivion. . . . Meanwhile all the ideal aspects of human culture, including all values and norms, have become relegated to the subjective sphere, and truth itself has become in effect subsumed under the category of utility. Transcendence and symbolism out of the way, there remains only the useful and the useless, the pleasurable and the disagreeable. There are no more absolutes and no more certainties; only a positivistic knowledge and feelings, a veritable glut of feelings.[41]

It is beyond the competence of this writer to know the truth about rel-

---

40. *Ibid.*, p. 5. Emphasis added.
41. Wolfgang Smith, *Cosmos & Transcendence: Breaking Through the Barrier of Scientific Belief* (Peru, IL: Sherwood Sugden & Co., 1990), pp. 142-144. Emphasis added.

ativity. Clearly we are faced with a most mysterious Universe, and its physical form ought to be investigated in a spirit of searching for truth. Since it is so vast and puzzling, perhaps it cannot be discerned without considering the clues given by the totally trustworthy Creator, who alone knows the full truth about the form and age of the Universe.

— Chapter 15 —

THE QUESTION OF AGE

Relevance to Crisis of Faith

As already outlined in Chapter 9 regarding the age of the Universe and when the Creation events took place, there are several distinct sets of beliefs now competing for acceptance within Christianity:

- God created the Universe billions of years ago and used Evolution and/or divine intervention in the creation of life forms and subsequent changes *beyond* kind.
- God created space, time and matter only about six thousand years ago and specially created each kind of life form in fairly rapid succession, and set genetic variation into operation to allow subsequent changes *within* kind.
- God created the Universe billions of years ago and much later specially created each kind of life form, with repeated divine intervention at great intervals of time, and set genetic variation into operation to allow subsequent changes *within* kind.

Evolution or Special Creation? The Universe billions of years old, or only about six thousand years old (in time as measured now on Earth)? What does it matter, anyway? Such questions are not irrelevant to the recent massive collapse of faith which has occurred within Catholicism, because belief in Evolution Theory and/or belief that the Universe *must* be billions of years old was used to discredit the integrity of *Genesis* and open the way over the last 150 years for Modernist theology subtly to gain widespread acceptance.

The Origins debate is not fully comprehensible in isolation from consideration of the age of the Universe, and belief in an age of billions of years is perhaps *the* key factor holding back the full dissemination of Origins doctrine in Catholic institutions. Unfortunately, many "conservative" Catholics now regard *Genesis* 1-11 as only explicable in a vague "religious" sense; hence, the widespread unwillingness to admit the possibility of six literal, 24-hour Creation days or a global Flood.

(To many human beings, the more remote in time the alleged origin of the Universe, the more remote becomes the idea of a Creator and the more incredible seems the integrity of the revelation contained in *Genesis*. God is perceived to be much more "distant" and remote from the daily cares of modern life than if Creation were accepted as having occurred only some 6,000 years ago.)

At issue is not simply the age *per se* of the Universe, but also the problem of ascribing a "non-literal interpretation" to *Genesis*, purely on the basis of extra-biblical, scientific or philosophical theories, when there is no literary justification in the text itself for suspecting that this is what the Divine Author intended.

Once the foundational beliefs were placed in doubt, all of the Bible was thrown open to assault, on the flawed ground of making it relevant to modern man. Not surprisingly, the New Testament became a target also for consignment to mythology (via concepts such as Higher Criticism), accused of being incredible to the modern world. Because *Genesis* is so foundational to doctrine, once its integrity *appears* destroyed, the rest of Scripture and central doctrines of the Catholic Church lose an indispensable base. Surely, no small matter!

Contrary to widespread assumptions, therefore, the age of the Universe is not a minor side issue within the Origins debate which can conveniently be left aside. The controversy over the question of "Age" has been around for a long time already and cannot be left out of the debate—if objective truth is worth striving for. *Put simply, the question of "Age" is most important because it impacts directly on the very integrity of* Genesis *and upon the credibility granted to the senses of Scripture, the "revision" of which logically ends in the Rudolf Bultmann position of demythologizing* Genesis *of its true historicity and shedding the remainder of Scripture of its true meaning.*

In the opinion of this writer, because of many conceptual problems, *there is no credible third position between a young Universe without Evolution and an extremely old Universe with Evolution.* When fully considered, the opposing set of beliefs could hardly be more different and do not really allow for another position:

- The Creation Days were really each 24 hours; the genealogies in Scripture trace back only about 6,000 years; Earth was created along with the rest of the Universe only some 6,000 years ago (Earth time); the Flood of Noah is a fact of history; and, whereas

Genesis is a partial account, it is nevertheless truly historical.
- The Universe is somewhere between 10 and 20 billion years old and the Earth about 4.5 billion years old. *Genesis* is essentially a "religious" account virtually devoid of real history and must be "reinterpreted" away from Origins beliefs clearly accepted by Jesus Christ Himself. The Flood of Noah is denied and revised as a series of localized catastrophes which occurred over eons of time.

Many Scripture scholars seem loath to concede any place for the literal and obvious meaning in *Genesis*. They describe it as symbolic mythology, and we are supposed to accept that God wanted it understood this way. But are these scholars correct, or simply accepting without question mistaken ideas handed down from others? Anyone who now challenges such views runs the risk, automatically, of being considered tendentious [biased]. But truth must emerge sooner or later. Since Evolution Theory is now so discredited, those scholars who believe in Evolution have been shown to be wrong in their grasp of empirical science concerning Origins. Are they prepared to reconsider their theological Origins beliefs concerning the historicity of *Genesis*?

The Antiquity of Mankind

Many Church scholars who write on the subject of Scripture interpretation tend to assume without question that the Earth is extremely old and that mankind dates back many hundreds of thousands of years. Few seem disposed to doubt the presupposition of vast ages of time.

One wonders how many Scripture scholars have investigated, outside of their own discipline, the many assumptions inherent in the various dating methods which ensure that only vast ages are considered credible. After all, the long age timescale was developed as a consequence of the erroneous assumptions of *uniformitarianism,* which had become increasingly popular in the early 19th century. It was thought, also erroneously, that evolutionary descent needed millions of years, and dating methods therefore had to accommodate that amount of time.

It is now known, virtually beyond doubt, that there has not been enough time, nor could there ever be enough, to "save" Evolution. The amount of time available is irrelevant to "supposed mechanisms of Evolution," because creatures or plants cannot pass on truly new, "higher," information not already possessed, and eons of time cannot change that reality. The specific information encoded and impressed upon DNA within each *kind* determines the limits to change.

The fact that many scholars were convinced the Earth is extremely old can be seen in the following extracts, taken from a wide range of viewpoints concerning Scripture interpretation:

> We now know that the age of man on Earth is far vaster than *Genesis* could possibly have imagined. The age of man is to be reckoned not in thousands of years, but in hundreds of thousands; the oldest fossil remains of man may go back as far as a million years. It obviously surpasses belief that merely human tradition could span this vast age with any accurate information.[1]
>
> We now know that the distance between the dust, and the living being, man, is a period of milliards of years. The creative act of God, by which he created man, took milliards of years.[2]
>
> The origin of planets, including the Earth, also has a variety of hypothetical explanations, but with one factor in common: The planets are derivatives from the stars. It is fairly agreed that the Earth and other planets are about four and a half billion years old.[3]
>
> If we recall the centuries, the millennia, the hundreds of thousands of years, that separated the beginnings of the human race from the composition of these first eleven chapters of the first book of the Bible, we realize how extremely unlikely it is that an (even substantially) accurate account of certain events should have been preserved.[4]
>
> Our inspired story of Creation was not written until perhaps one hundred thousand years, which is a very conservative figure, after the appearance of the first man.[5]

In fact, there is no proof beyond doubt that the Universe is billions of years old (in time as now measured on Earth), and assumptions taking for granted that vast eons of time are proved, now held about the Old Testament, should be re-examined. The attributing of myths and

---

1. Dr. Bruce Vawter, C.M., *A New Catholic Commentary on Holy Scripture* (London: Thomas Nelson and Sons Ltd., 1975), par. 146a.
2. Dr. A. Hulsbosch, O.S.A., *God's Creation: Creation, Sin and Redemption in an Evolving World* (London: Sheed and Ward, 1965), p. 30.
3. Fr. John A. Hardon, S.J., *The Catholic Catechism* (London: A Geoffrey Chapman Book published by Cassell Ltd., 1977), p. 95.
4. Fr. Thomas Barrosse, C.S.C., *God Speaks to Men: Understanding the Bible* (Notre Dame, IN: Fides Publishers, Inc., 1964), p. 39.
5. Fr. Alexander Jones, S.T.L., L.S.S., *Unless Some Man Show Me* (London: Sheed and Ward, 1961), p. 78.

legends to *Genesis* was prohibited by the *Pontifical Biblical Commission* in 1909, and *Genesis* should be regarded primarily as containing descriptions of true events, despite the way the sacred writer(s) conveyed the account.

The Cultural History of Mankind

From a Christian perspective, the age of the Universe, in the opinion of this writer, should for various reasons be regarded as being closely related to the length of time that man has lived upon Earth. Christians who support the idea that the age of the Universe is billions of years have yet to address fully the argument relating to the relatively short cultural history of mankind. Even though this only dates back theoretically to the "Neolithic Period," i.e., deemed to be up to about 10,000 years B.C. (see Chapter 9), let us assume for the sake of this discussion that human beings have lived upon Earth for one million years. Some difficult questions must then be addressed.

Bearing in mind that the existence of polonium radiohalos in the basement rocks suggests strongly that these rocks were created fully formed *within three minutes*, that the various creatures and plants upon the Earth belong to separate distinct *kinds*, that transitional forms are conceptually untenable, and that Evolution by means of random beneficial mutations defies the discoveries of modern science and is plagued by the problem of immense odds, an obvious conceptual problem arises:

If the Earth is 4.5 billion years old, and assuming that mankind has been around for one million years, then, *since Evolution did not take place, what was happening upon the Earth during the earlier 4,499,000,000 years?* It is a colossal period of time and well beyond human comprehension. *If Evolution was not taking place, then nothing much was happening—the Earth was virtually idle!* The violent deaths of animals during all that time seem ruled out, since Earth was clearly intended by God to be in a tranquil state and would have remained so if Adam had chosen to be obedient to God—and anyway, *Genesis* informs us that death did not arise until the sin of Adam. Thus, the Earth would have been dormant—virtually idle—for all that mind-boggling length of time!

The idea of the Earth being 4.5 billion years old, therefore, makes little sense if Evolution Theory is not historically true. But in addition, *the sacred writer(s) of Genesis (including God as the principal Author) give no indication whatsoever that Genesis was meant to be understood other than historically, and this is of crucial importance.* In contrast to the possibility of billions of years, ample evidence is given that God deliberately chose six days of 24 hours when creating time and implementing the Creation events.

Does Scripture give any indication that God intended for about 4.5 billion years to pass after the creation of Earth, until human beings would at last be created? The revelation in *Isaiah* 45:18 indicates it is improbable that God would have left the Earth empty of life for billions of years: "*For thus saith the Lord that created the heavens, God himself that formed the earth, and made it,* the very maker thereof: he did not create it in vain: he formed it to be inhabited. I am the Lord, and there is no other." (*Is.* 45:18).

Mankind was intended to be the crowning achievement of this Creation, and Earth was to be home for human beings, so why attribute to God a period of billions of years before creating Adam and Eve?

> If we acknowledge a supernatural Creator at all, why not allow Him to do the work of creating and organizing the cosmos all at once, getting right to the implementation of His purposes for creating it in the first place? Why force Him to drag it out over tortuous eons, merely in order to accommodate evolutionary speculations for which there is not one iota of either scientific or Scriptural evidence? If the Creator actually employed unknown billions of years of universal decay, to allow a primordial ten minutes of miraculous integration to eventually produce man "in His own image," then He certainly selected the most wasteful, inefficient, cruel process that could be conceived to accomplish His goal.[6]

The Genealogy of Christ

The fact that there are two Scriptural accounts of Christ's own genealogy (*Matt.* 1:1-17, *Luke* 3:23-28) lays emphasis upon its historical reliability. Matthew, addressing the Jews, presents Christ as the son of David and traces his lineage back some 2,000 years to Abraham; each person is listed as the father of the next person. *Luke, addressing non-Jews and therefore perhaps intent on stressing the idea of Creation,* traces Christ's lineage to the first man, Adam; each person is listed as the son of the next person. Both genealogies link Christ to the Old Testament with its promises and prophecies of Redemption.

---

6. *Thermodynamics and the Development of Order*, Ed. E. Emmett Williams, p. 127.

If there were thousands of years between one person and the next, it would seem very odd to include the names of some men who receive no other mention in Scripture. Why pick them out for special attention? *The Roman Martyrology* places the birth of Christ in the 5,199th year of the Creation of Heaven and earth,[7] and the following genealogy time period given in Scripture—from Adam to Jacob—is only about 2,250 years. (With allowance for uncertain/disputed aspects, even if this period were off by several thousand years, it is still a "young" history.)

Adam	became the father of Seth when aged	130 years
Seth	became the father of Enosh when	105 years
Enosh	became the father of Cainan when	90 years
Cainan	became the father of Mahalalel when	70 years
Mahalalel	became the father of Jared when	65 years
Jared	became the father of Enoch when	162 years
Enoch	became the father of Methuselah when	65 years
Methuselah	became the father of Lamech when	187 years
Lamech	became the father of Noah when	182 years
Noah	became the father of Shem when	500 years+ 2?
Shem	became the father of Arphaxad when	100 years+10?
Arphaxad	became the father of Shelah when	35 years
Cainan	(listed only by Luke in the Greek Septuagint as the father of Shelah)	
Shelah	became the father of Eber when	30 years
Eber	became the father of Peleg when	34 years
Peleg	became the father of Reu when	30 years
Reu	became the father of Serug when	32 years
Serug	became the father of Nahor when	30 years
Nahor	became the father of Terah when	29 years
Terah	became the father of Abraham when	70 years
Abraham	became the father of Isaac when	100 years
Isaac	became the father of Jacob when	60 years

It may be objected that "became the father of" or "begat" should be understood as "became the ancestor of." However, there seems little doubt that Shem was Noah's literal son; why should the later persons

---

7. *The Roman Martyrology*, edited by Canon J. P. O'Connell, Typical Edition approved by Pope Benedict XV (Maryland: The Newman Press, 1962, English Edition), p. 279.

not be literal fathers and sons, also? It seems rather forced to try and squeeze great time periods in between these later patriarchs named in *Genesis* 5.

Jacob's son Joseph—later to become famous as second only to Pharaoh in all of Egypt—was born when Jacob was 91 years old, and few would doubt that Joseph lived less than 2,000 years before Christ. Thus, it seems highly likely that Adam lived 6,000 to 7,000 years ago. Bearing in mind that modern science has shown overwhelmingly that Evolution cannot occur, it is reasonable to conclude that human beings have lived on Earth for only 6,000 to 7,000 years.

The Flood of Noah—Fact or Fiction

Whether or not the Flood of Noah was a factual event is of great importance to the Origins debate and is not unrelated to the question of "Age." Evidence in support of an historical global flood has been presented in Chapter 4 of this book, and Scripture contains numerous references indicating that a flood definitely did occur.

In contrasting the Second Coming of Christ with the Flood of Noah, Christ Himself had this to say: "As in the days of Noe, so shall also the coming of the Son of man be. For as in the days before the flood, they were eating and drinking, marrying and giving in marriage, even till that day in which Noe entered into the ark. And they knew not till the flood came, and took them all away; so also shall the coming of the Son of man be." (*Matt.* 24:37-39). "As it came to pass in the days of Noe, so shall it be also in the days of the Son of man. They did eat and drink, they married wives, and were given in marriage, until the day that Noe entered into the ark: and the flood came and destroyed them all." (*Luke* 17:26-27).

Why not accept that Christ—the omniscient Creator/Redeemer—believed the Flood had occurred? The possibility that perhaps He contrasted a "story" about a non-existent flood with the glorious Second Coming is not convincing.

If we take seriously Christ's admonitions about His Second Coming, which will see the world destroyed in fire, why not take seriously His belief in the Flood, which brought global destruction? Unless one maintains that either He was in error (an impossibility) or that in these passages He is using a vague theme of "salvation history," then one must accept that a flood, of whatever proportion, did occur.

The Catholic Church teaches that *Genesis* is not fictional, but definitely contains true history. The 1992 *Catechism of the Catholic*

Church (pars. 56-58) discusses Noah and the Flood in a covenant context and gives no indication that the Flood was anything other than a true event in history. The 1566 *Catechism of the Council of Trent*, in discussing the crucial importance of *typology* in Scripture, recognized the historical fact of the Flood:

> The figures of the Old Testament have great power to stimulate the minds of the faithful and to remind them of these most beautiful truths. It was for this reason chiefly that the Apostles made use of these figures. . . . *Among these figures the Ark of Noah holds a conspicuous place.* It was built by the command of God, in order that there might be no doubt that it was a symbol of the Church, which God has so constituted that all who enter therein through Baptism, may be safe from danger of eternal death, while such as are outside the Church, like those who were not in the Ark, are overwhelmed by their own crimes.[8]

The *Genesis* account indicates that the Flood was of massive proportions, and the Ark was the size of a large modern ship: "Now the earth was corrupt in God's sight, and the earth was filled with violence. And God saw the earth, and behold, it was corrupt; for all flesh had corrupted their way upon the earth. And God said to Noah,

> The end of all flesh is come before me, the earth is filled with iniquity through them, and I will destroy them with the earth. Make thee an ark of timber planks: thou shalt make little rooms in the ark, and thou shalt pitch it within and without. And thus shalt thou make it: The length of the ark shall be three hundred cubits[9]: the breadth of it fifty cubits, and the height of it thirty cubits. Thou shalt make a window in the ark, and in a cubit shalt thou finish the top of it: and the door of the ark thou shalt set in the side: with lower, middle chambers, and third stories shalt thou make it. Behold I will bring the waters of a great flood upon the earth, to destroy all flesh, wherein is the breath of life, under heaven. All things that are in the earth shall be consumed. (*Gen.* 6:13-17).

St. Peter was quite specific as to how great the Flood was and how many persons survived it: ". . . when they waited for the patience of God in the days of Noe, when the ark was a building: wherein a few,

---

8. *The Catechism of the Council of Trent*, p. 107. Emphasis added.
9. The size of a cubit varied between about 18 and 22 inches.

that is, eight souls, were saved by water." (*1 Peter* 3:20). God "spared not the original world, but preserved Noe, the eighth person, the preacher of justice, bringing in the flood upon the world of the ungodly." (*2 Peter* 2:5). "Whereby the world that then was, being overflowed with water, perished." (*2 Peter* 3:6).

And St. Paul in the letter to the Hebrews touches upon the Flood, in reference to the faith of Noah: "By faith Noe, having received an answer concerning those things which as yet were not seen, moved with fear, framed the ark for the saving of his house, by the which he condemned the world; and was instituted heir of the justice which is by faith." (*Heb.* 11:7).

The crucial question is, of course, was the Flood global in extent, or merely local?

Proponents of the idea of eons of time are loath to grant credence to the idea of a global Flood and almost invariably will only recognize the possibility of a series of local floods. The reason is not hard to see. *To admit that the Flood was global in extent would destroy the basis for the long-ages interpretation of the actual strata evidence*, and so they are driven to deny there was a global Flood. But this raises further conceptual problems, for the *Genesis* account places the Flood as occurring long *after* the creation of human beings, whereas the long-ages view is ultimately driven to acknowledge a series of catastrophes long *before* Adam was created. *Thus, support for the long-ages scenario implicitly involves doubt being cast on Christ's own words.*

If a supposed theme of "salvation history," including mythology, is to predominate—instead of granting recognition to the historicity of *Genesis*—then one may as well argue that the Ark itself never existed and that the Flood account is only symbolic, a sort of "religious story" which is not historically true.

Why did the sacred writer give so much detail in describing the dimensions of the Ark and discussing the creatures being led on and off of it, if the account were not historically true? In what way do such details, even down to the exact days, months and years on which they entered and later left the Ark, serve to lead man to salvation: because *the Ark is important in typology, because it is a symbol of the Church founded by Jesus Christ, because it is not symbolic mythology!*

If the *Genesis* account of God's instructing Noah to build the Ark is only seen as "salvation history" and not true history, why not also reject the historicity of *Genesis* where it refers to the instructions of God which were given to Abraham? (For example, why not believe that the account of Abraham's being told by God to sacrifice his son Isaac

(*Gen.* 22) is only fiction? Indeed, on what grounds would any passages in the Old Testament be believed as true events?

It is hard to see how the Flood can be held to be merely a local event. Apart from the vast evidence found upon the Earth of phenomena consistent with such a massive catastrophe (e.g., burial, suffocation and instant snap freezing of mammoths down to minus 175° F), there are also sound theological arguments to consider.

If the Flood of Noah were only local in extent, and thus affecting only a small part of the world, there would have been no need to build an Ark, let alone one of massive proportions. God could easily have led Noah and his family to safer ground. There would be no need to worry about the fate of animals, since they could have survived elsewhere.

Genesis records that "the sons of Noe who came out of the ark, were Sem, Cham, and Japheth: and Cham is the father of Chanaan. These three are the sons of Noe: and from these was all mankind spread over the whole earth."(*Gen.* 9:18-19). Therefore, unless one holds that Scripture is in error or that *Genesis* is only symbolic, then it must be accepted that all of mankind outside the Ark perished in the Flood. If one denies this, a problem must be faced: God's promise to destroy mankind no more by a flood would appear to have been broken many times. For example, the tidal wave which struck Bangladesh during the night in 1971 claimed about 300,000 human lives.

The actual wording of God's "rainbow" covenant with Noah is worthy of reflection:

> Thus also said God to Noe, and to his sons with him, "Behold, I will establish my covenant with you, and with your seed after you: And with every living soul that is with you, as well in all birds as in cattle and beasts of the earth, that are come forth out of the ark, and in all the beasts of the earth. I will establish my covenant with you, and all flesh shall be no more destroyed with the waters of a flood, neither shall there be from henceforth a flood to waste the earth. And God said: This is the sign of the covenant which I give between me and you, and to every living soul that is with you, for perpetual generations. I will set my bow in the clouds, and it shall be the sign of a covenant between me, and between the earth. And when I shall cover the sky with clouds, my bow shall appear in the clouds: And I will remember my covenant with you, and with every living soul that beareth flesh: and there shall no more be waters of a flood to destroy all flesh. And the bow shall be in the clouds, and I shall see it, and shall remember the everlasting covenant, that was made between God and every living soul of all flesh which is upon the earth. And God

said to Noe: This shall be the sign of the covenant which I have established between me and all flesh upon the earth." (*Gen.* 9:8-17).

Note that God says, "There shall no more be waters of a flood to destroy all flesh." (*Gen.* 9:15). This indicates clearly that not only was all of mankind destroyed in the Flood, but all living creatures as well, except the passengers aboard the Ark and the inhabitants of the seas.

If the Flood were not global in extent, what other explanation can account for the immense sedimentary strata which cover about three-quarters of the Earth? What would account for the fossil graveyards full of thousands, and in some places millions, of creatures trapped in sedimentary death pits? The standard theory of slow uniformitarian processes, with periodic volcanic eruptions and the like, has conceptual problems, because various strata are greatly out of assumed order, nor can it account for phenomena such as fossilized jellyfish and polystrate tree trunk fossils, which *must* have been entombed rapidly.

If the *Genesis* account of the Flood is compelling belief, why not also trust the divine revelation given in *Genesis* about Creation Days?

Days of 24 Hours—Or a Million Years?

A most contentious issue directly affecting the question of "Age" is the meaning given to the seven days of Creation in *Genesis* 1. Were these days literally 24 hours, or do they each represent a long period of time, the length of which is unknown? Is there another possible explanation, perhaps even less than 24 hours? The crucial question is this: *What meaning did the sacred writer of* Genesis *wish to assert; what meaning was intended to be understood by the reader?*

The Hebrew word *yom* can apparently be used in various contexts to mean either a day of 24 hours, the daylight period of a day of 24 hours, or a period of time longer than 24 hours. Experts disagree as to its meaning in *Genesis,* but there are strong grounds on which to claim that a meaning of 24 hours harmonizes best with Hebrew grammar in these *Genesis* passages. Other Hebrew words would have been more appropriate if the sacred writer were referring to long periods of time.

It seems implausible that "day" would be used symbolically the first time it is mentioned in Scripture. More likely, it was deliberately defined by the sacred writer to include light and dark in each rotational 24-hour day.

Another usage of "day" is recorded later in *Gen.* 2:4: "in the day that the Lord God made the heaven and the Earth." The more likely context here is that of "era" (defined in Hamlyn's *Encyclopedic World Dictionary* as "a period of time marked by distinctive character, events, etc."), describing a series of events and given in past tense. David M. Fouts has noted the crucial distinction in the different uses of *yom*:

> As to יוֹם referring to a literal 24-hour day in *Genesis* 1 . . . with ordinal numbers (second, third, fourth, etc.), the word יוֹם always refers to a 24-hour day in Scripture, with only one exception. The one exception is in *Hosea* 6:2, a poetical format in which the numbers 2 and 3 are paralleled in the common Semitic x//x+1 formula . . .
>
> The use of *yôm* with the *bet* (בּ) preposition (*b*ᵉ*yôm*—בְּיוֹם) is in construct with the infinitive *'ăśôt* (עֲשׂוֹת), a syntactical construction which often is simply translated idiomatically as "when." So *Genesis* 2:4 may be safely translated, "when the Lord God made . . ." without any mention of "day" at all. Similar constructions are found in *Genesis* 2:17, 3:5, 5:1, 5:2, 21:8, 35:3 and *Exodus* 10:28, to name a few. English versions will vary between "in the day" and "when" in these instances. To negate the meaning of *yôm* as a 24-hour day in chapter 1 by using *b*ᵉ*yôm* in *Genesis* 2:4 is at best an imprecise argument.[10]

God is "outside" of time and eternity, is in a state of *timelessness*, rather than endless time. Human beings, of course, find it baffling to contemplate the idea of "time" and "no time"—at the same time—yet God has revealed it, and Tradition teaches it. How can we fallible human beings fully grasp the idea of the ever-present?

This reality about the timeless nature of God accounts for the passages, "one day with the Lord is as a thousand years, and a thousand years as one day" (*2 Peter* 3:8) and "for a thousand years in thy sight are as yesterday which is past" (*Ps.* 89:4)—often mistakenly thought applicable to the "Age" question and the amount of time taken in creating the Universe. "Day" must first have a literal meaning, before it can be contrasted with a period of one thousand years.

It must not be overlooked that the fossil record has revealed no trace of Evolution phylogeny whatsoever of any creatures. Not only are all the "missing links" still missing, but the crucial mechanism of Evolution also remains missing! Nor should it be overlooked that the

---

10. David M. Fouts, "How Short an Evening and Morning?" *Ex Nihilo Technical Journal,* Vol. II, Part 3, 1997, pp. 303, 307.

catalyst for trying to impose a long-ages timescale upon Scripture arose from 19th century uniformitarian/evolutionist assumptions that were mistaken.

The misguided attempt to weld a synthesis of Evolution and theology has proved futile. There is no need to attempt accommodation of eons of time with divine Creation events, especially in such stark contrast to a straightforward obvious reading of *Genesis*.

Has not the truth all along been staring us in the face in *Genesis*? The one reliable Eye-witness (i.e., God), who was there at the Creation events, has been ignored in favor of modern human experts who were not there. Since God is truly omnipotent, He could surely create anything in any shape, manner and time-span that so pleases Him. Out of all the possibilities open to Him, God may have desired to create the Universe in six days of 24 hours each, to harmonize best with the intended highest achievement of that Creation—human beings. (The concept of a seven-day week comes *only* from *Genesis*. Days, months, and years can all be derived from consideration of astronomy, but not weeks.)

Being well aware of man's bodily needs, it would be fitting for God to place human beings in an harmonious setting. Mankind has found it necessary generally to have one rest day in seven to maintain proper balance within the body, certainly within fallen man at any rate. Therefore the period taken to unfold the Creation may well have been tailored to suit the bodily design and well-being of human beings.

Other arguments affecting the question of "Age" must also be considered. *Some scholars want the days of Creation extended to millions of years, but at the same time, they implicitly want the years of the Patriarchs shortened to much less time!*

On the one hand—insofar as they grant any historical meaning at all to *Genesis*—they want the days of Creation to be much *longer* periods of time, generally to accommodate an evolutionary scenario and/or to allow the Big Bang to develop. On the other hand, they want the ages of the Patriarchs to be much *shorter,* because they view such long life-spans as incredible. Little effort is made to harmonize such inconsistencies in their reasoning. It is not simply enough to assert belief in the long-ages idea as though self-evidently true: *The onus is upon those who reject a meaning of 24 hours to prove their case.* The onus of proof is *not* upon those who believe in a "young" Universe.

After all, Scripture informs us about true history and records that the Flood began *"in the six hundredth year of the life of Noe, in the second month, in the seventeenth day of the month."* (Gen. 7:11). Surely this is a clear-cut, deliberate statement about a true historical event; such specific details are hardly consistent with vague mythology.

It smacks of selective reasoning to hold that the 24-hour days of Creation were really millions of years, while the number of years, months and days attributed to Noah's life were really only of a much shorter time, so that his lifetime was like that of humans today. How can one simultaneously "interpret" the days of Creation as an immensely *longer* period of time than 24 hours, and the lifespan years of Adam, Noah and other Patriarchs as a much *shorter* period of time than the hundreds of years clearly stated in *Genesis*? What Scriptural criteria will justify this? Simply to dismiss the texts as symbolic or mythical is to avoid the issue. (It must be remembered that Adam lived for the rest of Day 6, throughout the entire Day 7 and well beyond, until he died at the age of 930 years.)

Further, as given in Chapter 4, the fascinating argument articulated by Scott Hahn—that Shem and Melchisedech were the same person—adds weight to the *Genesis* revelation that Shem in fact lived for 600 years. Perhaps the sacred writer(s) wrote *Genesis* to be specifically *unlike* the long-ages scenario and to thwart future revisionist theology!

The most common reasons offered for disbelief in literal 24-hour Creation days is that the *Genesis* account states that the Sun did not appear until Day 4, and that the gargantuan Universe requires vast amounts of time for light to reach Earth. Certainly the nature of the mysterious light given as created on Day 1 is unknown, but Earth could have turned in rotation against directional light coming from a cloud of heated gas before the Sun, moon and stars were brought into the scene on Day 4. (See Chapter 9.)

In support of a literal meaning for *yom* ("day"), one can cite the use by Jesus Christ, Creator and Saviour, of a teaching-by-example method of communicating fidelity to obedience. As Fr. David Becker has pointed out that a) the account of the washing of the feet and b) the Sabbath day of rest following six days of activity both have a common theme of "follow My own example." Of course, God exists in a timeless state and needs no "rest," but perhaps He chose Creation events of 24 hours each, since these are time periods to which human beings can best relate:

> It is my belief that Christ our Divine Pedagogue, in creating this universe for the purpose of hosting and fostering the drama of human life and history, used a didactic method of creation in such wise that the origin of the universe contains a timeless pedagogy ever instruct-

ing human beings about how they are to structure their time for work, rest and worship. And it is my belief that Moses accurately transmitted to us the divine command to imitate God's example of work and rest which God *in reality* set for us during the creation week. Holy Mother Church in her outreach to the secularized masses must provide a sufficient *motive* for people to restructure the way they live out each week, so that they will place a limit on work, and take time for rest and worship. In the last analysis the only credible reason why people should live this counter-cultural lifestyle, the only adequate *motive* for doing so, is the biblical reason of God's creation week example and God's command that we follow it: *In six days the Lord made heaven and earth, and the sea and all things that are in them; and rested on the seventh day."* (*Ex.* 20:11).[11]

That the Universe was *not* brought into being over billions of years seems clearly indicated in Scripture. The quoted words of Jesus Christ (who as the Second Person of the Divine Trinity *must* have been involved as both the Creator of the Universe and the principal Author of Scripture) indicate that He knew the Creation was of "recent" vintage. In discussing marriage and thus focusing on the history of human beings, He said to the Pharisees: "But *from the beginning of the creation,* 'God made them male and female.'" (*Mark* 10:6, emphasis added).[12]

If this text is understood as "from the *beginning of human beings*, God made them male and female," it reads like a tautology; what else can they be but human beings? Alternatively, to understand it as "from the *beginning of Creation*, He made them male and female human beings" still does not convey much except perhaps to deny an evolutionary origin. So why else would Our Lord include *"from the beginning of Creation"* in this revelation?

A coherent explanation is possible. Christ, the omniscient Creator/Redeemer who cannot deceive, knew that Scriptural passages would be grossly distorted and disbelieved many centuries later in an era plagued with doubts about Christian doctrine. As well as asserting that fidelity in marriage is most important, He may have included the reference to

---

11. Fr. David Becker, "The Divine Pedagogy," *Watchmaker* (Shade Gap, PA: Morning Star Newsletter, Catholic Origins Society, April, 1995).
12. The RSV, Douay-Rheims, Navarre and Jerusalem Bibles all have this identical wording, which in Latin reads *"Ab initio autem creaturae masculum et feminam fecit eos Deus."* The Knox Version similarly states, "God, *from the first days of creation,* made them man and woman."

Creation to confirm to 20th century Christians that human beings were indeed created very soon after the creation of the Universe. Jesus would have known that all this was true and would simply be speaking in the context of actual history.

The words attributed to Christ are now disbelieved by many Christians, as has occurred with His references to the Flood of Noah and to Moses, but it is surely no light matter to place in doubt the very words of the Redeemer Himself—and possibly mislead others as a result.

Most importantly, as Cardinal Ernesto Ruffini pointed out (despite his belief that the Universe is billions of years old), many of the Church Fathers (not all of them) held that God created the world successively in six natural days, each of 24 hours, just as the first and obvious sense of the sacred text sets forth:

> This opinion, held by St. Ephrem, St. Basil, St. Gregory of Nyssa, St. John Chrysostom, St. Ambrose, St. Jerome, St. Gregory the Great and, in general, by the majority of the ancient interpreters, was almost universal in the schools up to the 18th Century. It had, as its foundation, the following principle: *"In the interpretation of Holy Scripture, it is not lawful to depart from the obvious literal sense, unless reason prohibits it or some necessity forces us to leave it."* (This is the wise advice of St. Augustine taught by Pope Leo XIII in his admirable encyclical *Providentissimus Deus* [On Sacred Scripture]). Now, it was said, *such a prohibition or necessity does not arise in the present case.*[13]

Despite this precedent within Tradition, the enchanting appeal of uniformitarian/evolutionary concepts made a profound impact upon many Church scholars in the last two centuries, with the result that the literal-as-given, obvious sense came to be considered virtually unbelievable. Perhaps some scholars were also influenced by the views of St. Thomas Aquinas, who preferred the theory of St. Augustine. Cardinal Ruffini also quoted St. Thomas (in II Sent. dist. XII q.1, art.2):

> Augustine's view is that at first in creation certain things were given existence in their own nature, according to their distinct

---

13. Ruffini, *The Theory of Evolution Judged by Reason and Faith,* p. 69. Emphasis added.

species, for example, the elements, the heavenly bodies and spiritual substances (angels); but that certain others—such as plants, animals and men—existed only in their germinal principles (*rationes seminales*)....

Ambrose and other saints hold that there was an order of time by which things were distinguished. *This opinion is indeed more generally held, and seems to accord better with the apparent literal sense of Scripture.* Still, the previous theory of Augustine is the more reasonable, and ensures a better defense of Holy Scripture against the derision of unbelievers. To this, insists Augustine, must the fullest heed be given: "the Scriptures are so to be explained that they will not incur the ridicule of unbelievers" and his theory is the one that appeals to me.[14]

With all due respect to the great Aquinas, his reason for not accepting the literal view, which he admits is more generally held in Tradition, seems quite astonishing. As it happens, we know that various saints sometimes held conflicting views, and some Fathers of the Church and saints sometimes held views later seen as erroneous:

> In the 12th century, St. Bernard of Clairvaux, who was so fond of devotion to Mary, clearly denied Mary's Immaculate Conception. Beginning with St. Bernard, then, most of the major theologians of the Middle Ages, including even St. Thomas Aquinas, denied the Immaculate Conception.[15]

Regarding the Origins views of St. Augustine, it is most interesting that Cardinal Ruffini saw fit to record the following note:

> The saint confesses (*De Gen. ad litteram 8, 2*) that, shortly after his conversion, in his work *De Gen. contra Manichaeos*, "*non praeiudicando meliori diligentiorique tractatui*," he had explained many things figuratively, which afterwards, "*diligentius intuens atque considerans*," he understood should be taken literally.[16]

(Modern scholars often give the impression that we are dealing only with the human author of Scripture, but the fact that God is the principal Author of the Bible cannot be ignored; the human writers of

---

14. *Ibid.*, p. 72. Emphasis added.
15. Fr. William G. Most, *Free From All Error*, p. 3.
16. Ruffini, *The Theory of Evolution Judged by Reason and Faith*, p. 122.

Scripture—mysteriously—wrote only what God wanted them to write.)

Cardinal Ruffini also made very relevant observations about the literal meaning of *yom*. The strong support in Tradition for a natural day of 24 hours cannot be ignored in the modern Origins debate:

> We may pass over the opinion that the term *yom* (day) can, *per se*, be taken also in the sense of period or epoch, although from biblical texts produced to support this interpretation, we get the meaning of a day of uncertain date rather than that of an epoch of uncertain duration. However, *we are not persuaded that in our present case day is equal to period. If the author really wished to speak of ordinary days, as we believe, what expressions should he have used that are clearer than those he adopted? Almighty God in verse 5 expressly calls yom the light of day, in verses 14-16 He created the luminaries of heaven to divide the yom from the night, and entrusted to the sun the office of presiding over the yom, and to the moon that of presiding over the night. Therefore, what reason could there be for holding that the word yom was used in the same context and, indeed, in the same verse, in two completely distinct senses, namely, that of a common day and that of a period?* . . . It is an evident principle that the best commentator, the most authentic interpreter of a book or a phrase is its author. Now in *Exodus* the Hexaemeron is taken to be an ordinary week. . . . Holy Scripture limits the account to the space of a week: for six days the Lord works in creating, on the seventh He rests. This is all. It goes no further; and precisely because it wishes to stop, it does not make the usual evening and morning follow the sabbath. *Therefore, the sacred text evidently speaks of civil, ordinary days and not of periods; and the unanimous interpretation of the Fathers bears witness to this.* If the periodists, in order not to go against [false] science, are constrained to take *yom*, with its evening and its morning, as a period or epoch, how can it be maintained that the rest of the narration should be taken in the proper literal sense?[17]

Other scholars concede this is what the sacred writer intended to assert. While accepting the idea of a very old Earth, Fr. Thomas Barrosse nevertheless saw no problem in a literal meaning of 24 hours:

> A few decades ago a so-called "concordist" explanation of these chapters was popular. According to this explanation the details of the

---

17. *Ibid.*, pp.77-79. Emphasis added.

creation account of *Genesis* 1, for example, are factual and can be harmonized or shown to be in accord (or concordance) with (geological) science. Thus, the six days of creation would be six periods or eras, not six 24-hour days. After all, the word "day" often means an era in our modern languages: we speak of the "day" of Washington or of our own "day and age." The futility of this opinion can readily be seen by a close examination of the six-day creation account. That the author definitely conceives of the days of which he speaks as 24-hour days appears from the fact that he assigns to each a morning and an evening.[18]

St. Bonaventure, an outstanding 13th century Franciscan who wrote on the subject of the Bible, also had no problem in accepting a literal 24-hour day, as Fr. Peter Fehlner explains:

St. Bonaventure acknowledges that the Church has never condemned St. Augustine's view, creation of all as if it were in a day. But what St. Bonaventure notes in opting for the literal interpretation of day in the first chapter of *Genesis* has been commonly overlooked in modern times. The *ratio seminalis* of St. Augustine is the equivalent of essence, not embryo. It is the same when the world began to operate on its own as it is now. Only God can make it, change it, annihilate it. And thus how long it actually took God to make these species, only God can answer, because no one else was there to observe. It might have taken a day, or 200 days, etc., says St. Bonaventure, but the only evidence we have is what God has told us. For Bonaventure, the philosophical and epistemological points Augustine wishes to defend can be made just as well or better by holding for six days of twenty-four hours; and for Bonaventure there is no other convincing evidence pointing to a merely figurative meaning.[19]

The crucial question concerning the Creation days is this: *What meaning did the sacred writer(s) of Genesis wish to assert, what meaning was intended to be understood by the reader?* According to James Barr (Regius Professor of Hebrew at Oxford University):

. . . probably, so far as I know, there is no professor of Hebrew or Old Testament at any world-class university who does not believe that

---

18. Barrosse, *God Speaks to Men*, p. 38.
19. Fehlner, *In the Beginning*, p. 70.

the writer(s) of *Gen.* 1-11 intended to convey to their readers the ideas that,
(a) creation took place in a series of six days which were the same as the days of 24 hours we now experience.
(b) the figures contained in the *Genesis* genealogies provided by simple addition a chronology from the beginning of the world up to later stages in the biblical story.
(c) Noah's flood was understood to be worldwide and extinguish all human and animal life except for those in the Ark.

Or, to put it negatively, the apologetic arguments which suppose the "days" of creation to be long eras of time, the figures of years not to be chronological, and the flood to be a merely local Mesopotamian flood, are not taken seriously by any such professors, as far as I know. The only thing I would say to qualify this is that most such professors may avoid much involvement in that sort of argument and so may not say much explicitly about it one way or the other. But I think what I say would represent their position correctly.[20]

(This is not to say these experts believe the account, but simply that they believe the sacred writer intended to convey the literal meaning.)

St. Thomas Aquinas on Creation

St. Thomas Aquinas (13th century) held that the text *"In the beginning God created heaven and earth"* was intended to refute errors by asserting the following truths:[21]

- Time was not eternal; it had a beginning.
- There were not two principles of Creation, one good and one evil; Creation was appropriated to Jesus Christ by reason of Wisdom.
- Creation of matter was not done by medium of an earlier spiritual Creation. There was one Creation, involving both spirit and matter.
- It is now known that Natural Evolution requires the gaining of new, "higher" genetic information not already possessed by parents. But, seven centuries ago, St. Thomas Aquinas seemed necessarily opposed to Evolution *per se*: "Some have maintained that

---

20. Prof. James Barr, in a personal letter dated April 23, 1984 to David C. Watson. Quoted by Dr. Clifford Wilson in *Visual Highlights of the Bible*, Volume 1 (*From Creation To Abraham*) (Australia: Pacific Christian Ministries, 1993), Supplement, p. 11.
21. St. Thomas Aquinas, *Summa Theologica*, Vol. 1, Questions 46, 65-70. (Westminster, MD: Christian Classics).

creatures proceeded from God by degrees, in such a way that the first creature proceeded from Him immediately, and in its turn produced another, and so on until the production of corporeal creatures. But this position is untenable . . . No secondary cause can produce anything, unless there is presupposed in the thing produced something that is caused by a higher cause."

St. Thomas held the Creation to be threefold works carried out by God over six days: *Creation* (on Day 1), *distinction* (on Days 2 and 3), and *adornment* (on Days 4, 5 and 6). It is hard to see how his understanding of these days can be "interpreted" as anything other than literal 24 hour days.

- "There can be no day without light. Therefore light must have been made on the first day. . . . Matter must be held to have been created at the beginning with substantial forms, afterwards receiving those that are accidental, among which light holds the first place. . . . The light [on Days 1-4] was the sun's light, formless as yet, being already the solar substance, and possessing illuminative power in a general way, to which was afterwards added the special, determinative power required to produce determinative effects."
- "Although to the senses there appears but one firmament; if we admit a higher and a lower firmament, the lower will be that which was made on the second day, and on the fourth the stars were fixed in the higher firmament."

Creation Days Scenario

Genesis 1 informs us that "In the beginning God created (or when God began to create) heaven and the earth" . . . "The earth was void and empty, and darkness was upon the face of the deep" . . . "And God said, 'Be light made'" (all of this before Day 1 was over!) and Scripture repeatedly states that He "*stretched out the heavens*" (*Job* 9:8, *Psalms* 103:2, *Isaias* 40:22, 42:5, 45:12, 51:13, *Zech.* 12:1, *Jer.* 10:12, 51:15). What does all this mean; what happened?

The following tentative Origins scenario is derived from a straightforward reading of *Genesis*, without imposing "unwarranted interpretations that try to make Scripture say what it does not say":

At the instant of Creation, God created many orders of angels in heavenly realms, the space of the entire physical Universe (i.e., the firmament of the heavens), time itself and all matter, and Earth in rudimentary form only—completely covered by water. Shortly after, He formed all atomic energy instantaneously and, in a flash of brilliant light, ignited clouds of superheated gas and dust to become "protostars" throughout the Universe.

On Days 2 and 3, God stretched out the physical Universe and all lightwaves within, created Earth's atmosphere and firmament, and brought forth dry land above the Sea. He also created all kinds of plant life upon Earth. (Earth could thus have rotated on its axis and received light from a heated cloud of gas before God transformed this cloud into our Sun on Day 4).

On Day 4, from the immensity of atomic energy spread as nebulae throughout the Universe, God instantaneously transformed/created all the stars, the many spiralling galaxies, our Sun and the whole solar system, as well as such things as comets. He directly created Pluto as a "double planet," installed the rings around the giant gas planets together with their "shepherd" satellites, set up the forces which allow Janus and Epimetheus to swap places instead of colliding every four years in their orbits around Saturn, set planets rotating in odd directions and some moons rotating in opposite directions to others, and He installed all the other baffling features of the solar system.

During the rapid Creation events, God thus ensured that light waves from stars were already reaching Earth: "Let there be lights made in the firmament of heaven to divide the day from the night; and let them be for signs and for seasons and for days and years . . . to give light upon the earth. *And it was so.*" (Gen. 1:14-15). *Since we know from Scripture that Earth was not intended to lie idle for billions of years without light from stars, then both stars and starlight must have been created rapidly for them to be immediately seen from Earth "for signs and for seasons."*

Perhaps God actually "stretched" out the stars and their associated light in *nanoseconds* as galaxies formed in the Universe—and effectively placed the lights in the firmament of the heavens like we now hang out lights throughout a large tent—after which the speed of light immediately reduced to that now measured in modern times. (By way of analogy, the cataclysmic process rates of the global Flood of Noah would have later stabilized down to slower rates as known today.)

Who can fully explain the functioning of the Universe? Perhaps human beings will never fully understand it on this side of eternity. The foregoing tentative scenario may have strengths and weaknesses, but is not another version of the Big Bang Theory. An enormous explosion could not give rise *per se* to the bewildering order observable in the Universe, and many separately rotating galaxies seem unlikely to result from the explosion force envisaged in the Big Bang Theory.

Nevertheless, this attempted scenario may satisfy those who believe that God somehow must have brought about the Universe with the use of immense explosive power. (The Creation scenario suggested by Russell Humphreys—see Chapter 9—may be a superior explanation to the tentative scenario outlined here.)

Is this scenario fundamentalist, or overly-literalistic? Neither! It simply attempts to make use of clues given to us by the trustworthy and divine principal Author of Scripture and of the reality of observable order in the Universe. It is of course speculative—but then, so are all other such Origins scenarios. Creative Origins events are unrepeatable, but we have great reassurance in knowing that the absolutely trustworthy Eye-witness (God), who can neither deceive nor be deceived, has revealed a partial account of what happened.

Even though human beings may never grasp fully the exact details of the Creation events this side of eternity, are we to be guided only by modern human experts who were not there at Creation, or rather, by what the divine Creator/principal Author of Scripture wished to reveal in *Genesis* via the human sacred writer(s)?

Some have objected that the concept of a "mature" Universe, with its apparent signs of extreme age, would amount to a deception of mankind by God. But, since fallible human beings are quite capable of deducing wrong conclusions, the *appearance* of extreme age is not proof that the Universe has existed for eons of time.

The alternative argument is that fallible human beings have mistakenly *assumed* a great age for the Universe because of the speed of light. After all, a powerful argument can be made that Adam came into existence in adult form and, if we could travel back in time to ten seconds after he was created, we could easily assume that he was aged about 25 years old. God would not have deceived us, but we would have mistakenly arrived at the wrong conclusion about Adam's age.

How can we ignore the account revealed to us by the absolutely reliable, and only, Eye-witness (God) who was there at the Creation? Any deception is in the form of Evolution Theory, with its originally supposed need for billions of years.

The Universe very likely is much more mysterious than we can even begin to imagine. Therefore it seems that, in attempting a deeper understanding of the physical form of the Universe, human beings are driven ultimately through scientific investigation to consider questions of metaphysics.[22]

Science alone is not enough to guide us about the question of "Age"; there is still so much that is not understood about the mysterious Universe, and human beings can only travel a very small distance beyond Earth. While many aspects concerning the Universe remain unresolved and subject to intense debate among scientists, much caution is warranted about placing too much trust in science alone. Rather, our trust should be in truth known from theology and Tradition.

The Reality of Metaphysics

What actually are space, time and matter, which we take for granted in everyday life? Regarding matter, within atomic particles, specialists describe the fact of short-lived appearance and disappearance of discrete subparticles which nevertheless result in constant hard matter. Wolfgang Smith (physicist) points out that the closer one looks at matter, the more it divides down to a vanishing point:

> Now atoms, supposedly, are so small as to be indivisible, and being indivisible, were held to be constant and indestructible. In a word, they were thought to be the irreducible and permanent building blocks out of which physical things are compounded. These large-scale things, moreover, possess only a more or less transient and phenomenal reality, inasmuch as their atomic constitution, as well as their internal geometry, are constantly changing. What "really exists," and what alone retains its self-identity, are the atoms. But that conception has ultimately proven to be erroneous. It turns out that neither the erstwhile atom, nor the fundamental particles into which it can be decomposed, have a true self-identity. In the words of Schrodinger, "we have been compelled to dismiss the idea that such a particle is an individual entity which in principle retains its 'sameness' forever. . . . we are now obliged to assert that the ultimate constituents of matter have no 'sameness' at all." . . . Identity *has no meaning* as a physical concept . . . it is an incurably metaphysical notion and, in fact, a name of God.[23]

---

22. Hamlyn's Encyclopedic World Dictionary defines *metaphysics* as: That branch of philosophy which treats of first principles, including the sciences of being (*ontology*) and of the origin and structure of the Universe (*cosmology*). It is always intimately connected with a theory of knowledge (*epistemology*).
23. Wolfgang Smith, *Cosmos & Transcendence*, p. 50.

This is not to argue that matter has no concrete existence; common sense informs us that steel, for example, is very real and very hard. *The Catechism of the Council of Trent* taught that "all things derive existence from the Creator's supreme power, wisdom, and goodness, so unless preserved continually by His Providence and by the same power which produced them, they would instantly return into their nothingness."[24] Thus, even though the omnipresent Creator instituted *secondary causes* by which the Universe operates, it seems that matter is somehow dependent for its continued existence upon His presence.

It is known that, while alive, our bodies are constantly undergoing a process of cellular change. Thus, in a physical way, human beings are constantly in a process of "becoming"—understood in the true sense, not to be confused with a pantheistic Evolution process of "becoming." (There is another valid sense of "becoming"—that of being drawn closer to God by seeking inner change through obedience.) It has thus been concluded that only God stays the same, while all created things have their very existence rooted in the transcendent God, who is also everywhere at once, throughout all Creation, supporting all things in their existence.

> In the words of St. Gregory, "that which is always the same, neither increasing nor diminishing, immutable to all change . . . is truly real Being." . . . The point is that relative or contingent existences cannot stand alone. They have not an independent existence, a being of their own. . . . *Indeed, the cosmos itself, in its totality, has not an existence independent of God. It is not another being, or a separate entity, standing apart from God and confronting Him, as it were. God alone Is.* [25]

Secondary Causes

Paula Haigh, in discussing the Thomistic doctrine of *secondary causes*, addresses the metaphysical reality of Being:

> *Being* is not the same for creatures as it is for God. The term *being* is used for both God and His creatures, not univocally, not equivocally or ambiguously, but analogically (i.e., with some similarities, but also with some differences). *Being*, or existence, is what God is

---

24. *Catechism of the Council of Trent*, p. 30.
25. Wolfgang Smith, *Cosmos & Transcendence*, p. 51. Emphasis added.

absolutely, but what we and all creatures only have from Him. St. Thomas Aquinas says that to exist, to be, is God's very essence or nature. He cannot *not* be. But such is not the case for every being other than God. . . .

Thus, every creature is a mixture or composite of two principles: the principle of *existence* (actuality) and the principle of *potentiality* (potency). These two principles are the basis of all change, for change is the passing over from potency to actuality and is strictly limited by the nature of the being. . . . The genetic code is the biological barrier to evolution because it is the physical expression of the precise limitations, the parameters created by God and built by Him into the very being of the kinds described in *Gen.* 1. . .

Causes are of four kinds: efficient, material, formal, and final. All of these can be secondary causes, and God created the power of their causality when He created them in the beginning. Moreover, such is the power and importance of these natural causes in the entire Universe—a power and importance given them by God—that if they were to cease to operate as secondary natural causes, the Universe would fall into chaos.

The life principle of plants and animals inheres in the material-formal constitution of the DNA code and is passed on, increasing and multiplying, by generative power alone. But the human soul is a spiritual-rational substance that the generative principles of the vegetative and lower sensitive natures are simply incapable of duplicating or infusing. The axiom "You can't give what you don't have" applies here. Only God can infuse the rational human soul into the conceptus produced by the union of man and woman.[26]

The *Catechism of the Catholic Church* declared the truth of secondary causes, that God makes use of His creatures' cooperation:

> God grants His creatures not only their existence, but also the dignity of acting on their own, of being causes and principles for each other, and thus of cooperating in the accomplishment of His plan. (306).
> *The truth that God is at work in all the actions of His creatures is inseparable from faith in God the Creator. God is the first cause who operates in and through secondary causes*: "For God is at work in you, both to will and to work for his good pleasure." Far from diminishing the creature's dignity, this truth enhances it. Drawn from nothingness by God's power, wisdom, and goodness, it can do nothing if

---

26. Paula P. Haigh, *St. Thomas Aquinas: Creationist for the 21st Century* (Nazareth, KY: unpublished manuscript, Sept., 1991).

it is cut off from its origin, for "without a Creator the creature vanishes." Still less can a creature attain its ultimate end without the help of God's grace. (308). [Emphasis added.]

Some scholars maintain that Evolution *is* secondary causes at work, but this belief is false. DNA has been designed so that only *variety within kind* is possible and Evolution *per se* cannot occur. If it could occur, there would be very few, if any, "missing links."

As will be argued in a further discussion of secondary causes in Chapter 16, any concept of Origins which relies upon *innumerable* divine interventions to "guide" and accomplish Evolution *must* be false because 1) Scripture gives no clue that God implemented the Creation this way, and 2) it would require God, paradoxically, constantly to override and frustrate His own institution of living creatures and laws of nature from fulfilling their natural properties.

The Onus of Proof Is on the Long-Age Proponents

Without presuming to know the mind of God, one can only wonder why God would have chosen an evolutionary process spanning billions of years, with a hundred million years existence and ultimate extinction of dinosaurs, before getting around to the crowning achievement of His Creation—human beings. Why billions of years of "bacteria only"?

It seems beyond reasonable doubt that the sacred writer(s) of Genesis intended to assert a literal meaning of 24-hour Creation days; therefore, they must each have been 24 hours. How could it be otherwise? And if the Creation days were only 24 hours, and Christ referred to the creation of human beings "in the beginning" (i.e., on Day 6), how can the Universe be billions of years old (in time as measured now on Earth)?

The Douay-Rheims version of the Bible contains a detailed Historical and Chronological Index (see p. 298), showing there were only approximately 4,000 years from Day 1 of the Creation to the birth of Jesus Christ. The Index states: "The Chronology followed here is according to the more general Opinions of Divines and Chronologers."[27]

In what should we believe—modern fallible science, which claims that dinosaurs lived for millions of years before suddenly, mysteriously

---

27. The Holy Bible, Douay-Rheims Version, Revised by Bishop Richard Challoner (1749-1752) (Rockford, IL: TAN, 1989).

becoming extinct some 65 million years ago; or the divinely revealed information given in Scripture? We are reliably informed in *Genesis* 2:19-20 that "the Lord God having formed out of the ground all the beasts of the earth, and all the fowls of the air, brought them to Adam to see what he would call them: for whatsoever Adam called any living creature, the same is its name. And Adam called all the beasts by their names, and all the fowls of the air, and all the cattle of the field."

This divine revelation that Adam saw and named "all the beasts of the earth," including dinosaurs, only about 6,000 years ago contradicts the modern popular idea that dinosaurs lived and died millions of years ago. After all, God could hardly expect Adam to name dinosaurs who were already long-extinct and buried as fossils in the ground!

The onus of proof does not rest with those who believe in a "young" Universe; it rests with those who insist on the long-ages concept. A meaning of 24 hours has not been disproved, and Tradition requires that we follow the instruction of Pope Leo XIII—that the literal and obvious sense of Scripture passages must be held until rigorously disproved. Therefore, *belief in the "young" Universe ought to hold pride of place, and "yom" ought to be taken in the proper sense as a natural day.*

Christian proponents of "long ages" have to prove that uniformitarianism is true out there in the gargantuan Universe when it is clearly not true on Earth or in the solar system. In addition, the idea of the Universe being billions of years old has serious conceptual problems, and these difficulties have to be explained coherently, in a comprehensive scenario, by proponents of eons of time for the age of the Universe.

The onus is also upon those who prefer the "long ages," to prove that the sacred writer(s) did not intend to convey a meaning of 24 hours for the Creation days, and to prove that all the rabbis before Christ and the greater majority of Church Fathers were all wrong in holding that *Genesis* was intended to be understood as primarily historical.

The Bible was written when myths abounded, and the sacred writer(s) of *Genesis* would hardly have been unaware of them. On the contrary, Moses himself was a very highly educated man who grew up in the court of Pharaoh and, as the writer/editor of the Pentateuch, would doubtless have been mindful of widespread pagan mythologies. And, as the principal author of Scripture, God would have ensured that *Genesis* would be free from mythology—to show that it is truly historical and deliberately to thwart future revisionist interpretations.

—Chapter 16—
PROBLEMS IN PROGRESSIVE CREATION

Credibility Problems

Progressive Creation theorists attempt to harmonize rejection of Evolution with belief in an immense age for the Universe. Following is a synopsis of their position:

> God created the Universe billions of years ago and much later specially created each kind of life form with repeated divine intervention, at great intervals of time, and set genetic variation into operation to allow later changes *within* kind.

It is argued that God continually intervened over billions of years to shape the Universe and create more complex forms of life, and that vast distances in the Universe *must* be linked to vast amounts of time. Conceptual problems arise, however, in any version of Progressive Creation.[1] (For example, one is driven simultaneously to accommodate immensely *longer* "days" of Creation and much *shorter* years of patriarchs.) Its credibility cannot be discussed in isolation from arguments relating to Origins in general, the question of age, and possible conflict with Catholic doctrine.

Hugh Ross, founder of *Reasons to Believe* (a Californian evangelical ministry) is representative of Progressive Creation beliefs. He recognizes the Mosaic authorship of the Pentateuch, total inerrancy of Scripture, and that the Bible gives no support to Evolution:

---

1. Another version of Progressive Creation is that of "vision-days," advocated by P. J. Wiseman in *Clues to Creation in Genesis*. He held that God, over six days, showed to Adam a series of visions of the actual Creation events which had taken place previously. (The "vision-days" theory seems flawed; it is subject to the objections to Progressive Creation given in this chapter. Ultimately, it only pushes the inquiry back another stage. Did the actual Creation events take place rapidly, or over billions of years?)

He specially created through fiat, miraculous means, birds, mammals, and human beings. Since the time these animal kinds were created by God, they have been subject to minor changes in accordance with the laws of nature, which God established. However, the Bible clearly denies that any of these species descended from lower forms of life.[2]

Ross argues powerfully against Naturalistic Evolution and points out that the "ratio of harmful to favorable mutations measures between ten thousand to one and a million to one" (Ross, p. 79) and "for all species, nothing like the development of new organs falls within the range of reasonable probability." (Ross, p. 80). Even if the Creation "days" were time periods of hundreds of millions of years, "these time-frames would be too brief by countless orders of magnitude for simple life to arise and become complex by natural processes." (Ross, p. 80).

However, Ross is unclear in his book about the vast gulf known to exist between Evolution Theory and the reality of genetic variation. Evolution by definition must involve the gaining of new, higher genetic information not previously possessed within each original kind of creature or plant. With respect to variety, it is well known that the observed changes over time within each basic created entity are only evidence of variety within kind, and breeders know variety has definite limits beyond which change cannot occur. Most people believe Evolution involves higher change, but even though Ross makes clear his rejection of evolutionary descent, he cites a very imprecise definition: "In its ordinary usage among scientists and non-scientists alike, especially among physical scientists, Evolution simply means 'change with respect to time.'"[3]

Surely the Origins debate is not simply over change—all sides agree that changes occur in life forms—but rather, the very nature of change. Granted that Ross acknowledges both *stasis* in the fossil record and that *natural selection* results in variety, this misleading definition gives rise to misrepresentation of "young-age" creationists. The following passages portray them as confusing variety for Evolution and contradicting themselves on their own basic arguments:

---

2. Hugh Ross, *Creation and Time: A Biblical and Scientific Perspective on the Creation-Date Controversy* (Colorado Springs, CO: NavPress Publishing Group, 1994), p. 154.
3. *Ibid.*, p. 74.

The many species of today are presumed to arise through biological Evolution from the orders and genera on Noah's Ark![4]

Such creationists brand day-age proponents, like myself, who deny any significant biological Evolution over time scales long or short, as evolutionists, while they themselves seem to concede substantial biological Evolution over very short time scales.[5]

If Ross can be so mistaken about fundamental aspects of "young-age" creationists' views, can he not also be mistaken on other aspects?

The Days of Creation

In common with theistic evolutionists, progressive creationists are intent on accommodating vast periods of time. Innumerable divine interventions are gratuitously introduced and simply asserted as factual, to make each Creation "Day" many millions of years long:

> So where does God fit in? . . . He is the one who set up all the laws and constants of physics so that hydrogen forms and, after that, burns down into galaxies, stars, planets, and life-essential elements. He also intervenes along the way, personally designing and crafting a particular galaxy, star, planets, moons, and a set of heavy elements in preparation for His Creation of life on one planet, Earth.[6] . . . Only in galaxies that are older than about 10 billion years and younger than 20 billion years will life be possible.[7] . . . *All* the relevant bodies (the universe, our galaxy, our star, our planet, and our moon) must be a few billion years old, no more, no less. . . .[8]

Ross argues that light was not created on the first "Day," but "on that day the light already created 'in the beginning' (billions of years earlier) suddenly broke through to the earth's surface." (Ross, p. 149). So the first Day was not the first day! But this would imply that the Days of Creation were only the first days recorded somehow by mankind, an idea quite beyond verification. In addition, contrary to the *Genesis* note of finality about Creation being finished on Day 6, proponents of Progressive Creation are driven to argue that Day 7 is not finished, but continues on, with God constantly intervening.

---

4. *Ibid.*, p. 73.
5. *Ibid.*, p. 83.
6. *Ibid.*, p. 132.
7. *Ibid.*, p. 137.
8. *Ibid.*, p. 138.

Such theories are really only unwarranted beliefs imposed upon Scripture; if true, they would cast the sacred writer—including God as the Principal Author of Scripture—as virtually guilty of deception, for no clue is given in *Genesis* that it was meant to be understood this way.

It would be pointless to argue alternatively that the Universe is billions of years old and that God did not create any life forms until only some six thousand years ago, for this would admit that the fossil record traces back only to a fairly recent era. It would also admit that the original dating method assumptions, which initially gave rise to the desire for many millions of years, are rendered irrelevant.

Effects of the Fall

Ross argues that proponents of young-age views are divisive. What is at stake, however, is not simply the length of the Days of Creation, but rather, the very integrity of the *Genesis* account itself, and also the traditional concept of Original Sin. A radically different concept of the Fall would have to be accepted, and it is impossible for the Church to overturn doctrine already defined as true in Tradition without also losing all credibility of its claim to being the one true Church.

The 1992 *Catechism of the Catholic Church,* referring to *Romans* 8:19-23, teaches that *the whole cosmos was affected by the Fall—not just human beings:*

> The Creation waits with eager longing . . . in hope because the Creation itself will be set free from its bondage to decay . . . We know that the whole Creation has been groaning in travail together until now; and not only the Creation, but we ourselves . . . groan inwardly as we wait for adoption as sons, the redemption of our bodies. (1046).
>
> [The visible universe will be transformed] so that the world itself, restored to its original state, facing no further obstacles, should be at the service of the just. (1047).

In contrast, Hugh Ross claims (Ross, p. 65) that the "bondage to decay" ["servitude of corruption"—Douay-Rheims Bible] referred to in *Romans* 8:19-23 relates to entropy (i.e., the tendency for things to go from order to disorder and to loss of usable energy, etc.), and not to the disobedience of Adam. He points out that some creationists hold that entropy only came into existence after Adam had disobeyed God.

Ross makes a valid point in holding the possibility that entropy and the pain of the senses have always been in existence and would have operated if Adam had been obedient. For example, heat loss caused by friction would still have occurred, plants would still have decayed, and fire would have reduced the amount of useable energy.

But the point at issue here is whether the Universe was profoundly affected by Adam's sin of disobedience, radically upsetting the tranquillity and harmony of order that would otherwise have prevailed. The Creation does indeed now groan in travail, but not because entropy existed from the beginning of Creation. If such were the case, what condition is the Earth to be restored to, especially if the Creation repeatedly deemed "good" by God was deliberately designed with the grim struggle for survival as a normal part of life on Earth?

The Flood Denied

Genesis 7:11 informs us specifically that *the Flood began "in the six hundredth year of the life of Noe, in the second month, in the seventeenth day of the month."* To deny the historical reality of the global Flood is also to deny that Noah lived the 950 years attributed to him in Scripture.

The worldwide strata configuration and phenomena such as polystrate tree trunks are not readily explained by intermittent localized catastrophes. The Flood of Noah, clearly confirmed by Jesus Christ as global in extent, thus poses profound conceptual problems for the long-ages view. The *Genesis* account places the Flood as having occurred long *after* the creation of human beings, but proponents of Progressive Creation almost invariably deny it was a true historical event and propose instead a series of catastrophes long *before* Adam was created. To admit that the Flood was global in extent would destroy the uniformitarian, long-ages interpretation of the actual strata evidence, and so they are driven to deny there ever was a global Flood.

But the evidence in the rocks in favor of a huge one-time global Flood simply cannot be ignored. The worldwide fossil evidence of the rapid death of thousands of creatures in vast death pits and the nature of worldwide sedimentary strata is consistent with a global catastrophe. And there is confirmation from Jesus Christ Himself, about a worldwide Flood destroying all except those aboard the Ark!

The Ark is amply described in *Genesis,* and the historical fact of the Flood is clearly part of God's rainbow covenant with Noah. In the *Catechism of the Catholic Church*, Noah and the Flood are referred to in covenant context (pars. 56-58); this catechism gives no indication that the Flood was anything other than a true global event in history.

In addition to his denial of the global Flood, Ross argues that multitudes of extinctions (which must have included as normal the violent deaths caused by predators ripping victims apart while they were still alive) were taking place in the immensely long periods of time *before* the Fall, between God's intervening and His creating new life forms:

> According to the fossils, more and more species of life came into existence through the millennia before modern man. The number of species going extinct nearly balanced the number of introductions, but introductions remained at least slightly more numerous. Then came human beings.[9]

The Problem of Death

As with most advocates of Progressive Creation, Ross holds that Adam and Eve were specially created "several thousand years ago" (Ross, p. 10), but this could be somewhere between 6,000 and 60,000 years ago. (Ross, pp. 140, 141). However, he finds "troubling" (Ross, p. 10) the issue of so-called hominid creatures who supposedly preceded the direct creation of Adam and Eve. He believes that right from Adam and Eve, human beings possessed "spiritual" souls—but earlier, *large-brained* primates roamed the Earth, perhaps as long ago as one million years. (Ross, p. 141).

Who were these *large-brained* creatures? Marvin Lubenow has shown conclusively in *Bones of Contention* that all the large-brained (i.e., above 700cc) so-called "hominids" were fully human beings and there never existed any human-like creatures preceding Adam and Eve. *Homo sapiens* specimens in fact date back simultaneously with the Australopithecines. In fact, the concept of so-called hominid "ancestry" is a confused and mistaken concept drawn from Evolution mythology.

Ross revises *Romans* 5:12 and *1 Cor.* 15:21 to have the texts mean that Adam's sin of disobedience brought only spiritual death; physical death was therefore always going to be mankind's inevitable fate on Earth. This belief does not address the problem of evil and stands in contrast to Jesus' Resurrection from physical death (*1 Cor.* 15:22). This cannot possibly be accepted by Catholics loyal to the Church founded by Jesus Christ. It directly contradicts Catholic teaching on human death, reiterated so strongly down the years in Tradition on this central

---

9. *Ibid.*, p. 50.

and most crucial aspect of Origins. As the 1992 *Catechism* reaffirmed unmistakably, Adam's sin of disobedience brought not only spiritual "death," but also physical death:

> Finally, the consequence explicitly foretold for this disobedience will come true: man will "return to the ground" for out of it he was taken. *Death makes its entrance into human history.* (400).
> God did not make death, and he does not delight in the death of the living. (413). Even though man's nature is mortal, God had destined him not to die. (1008).

Secondary Causes Denied

The idea of secondary causes provides another important means of gauging the credibility of both Theistic Evolution and Progressive Creation in relation to Tradition. The *Catechism of the Catholic Church* teaches that

> God grants His creatures not only their existence, but also the dignity of acting on their own, of being causes and principles for each other, and thus of cooperating in the accomplishment of His plan. (306).
> *God is the first cause who operates in and through secondary causes.* (308). [Emphasis added.]

The human family provides a good example of secondary causes at work in biological ways, an example in which free will is allowed to function and in which God does not constantly intervene by forcing individuals to behave in ways best for their salvation.

Perhaps the decision for the Body of Jesus to be conceived within Mary and then arrive humbly as a baby boy in a smelly stable not only shows the unfathomable humility of Almighty God, but also sheds light on His use of secondary causes. He could have appeared as an adult in front of multitudes and then invited all to follow Him if they chose to. Why use a human mother anyway, or why bother with being a toddler?

God's commitment to allow secondary causes to function unrestricted can be compared with His fidelity in allowing free will to prevail without interference (even though in His timeless state of being, He knows ahead of our existence that some human beings will choose sinful lifestyles and wilfully reject Him). His interventions to instruct prophets or perform wonders in Old Testament times were made known and recorded in Scripture. But where in *Genesis* or anywhere else in the

Bible are we informed that He had intervened on innumerable occasions to fine-tune His Creation, so to speak, and would continue always to do so? There is simply no indication to this effect in the Bible, and this casts very serious doubt that such interventions actually occurred.

One would expect that the only Eyewitness to the Creation events—the completely trustworthy Creator who is incompatible with deceit or error—would reveal in *Genesis* some indication of what took place; otherwise we could have no way of knowing what happened.

Does Ross or anyone else here on Earth know directly from God what He chose to do when unfolding the Universe in the mysterious Days of Creation? Is there not a contradiction in holding that God could unleash a sudden stupendous creation of the Universe—of mind-boggling proportions, almost beyond human comprehension—but then be constrained to unfold it very slowly, as though limited to the painstaking efforts required of human beings when crafting a work of art?

In any case, there would still be an element of trial and error involved on God's part if one does what Ross does—i.e., accept the alleged sequence of creatures which evolutionists do.

If Big Bang theorists are correct and the Universe is still expanding, does this mean that God is right now creating new space and time to allow it to expand? [Cf. Editor's note, p. 9]. If so, this directly contradicts the idea of the Universe's being created efficiently during the six Days of Creation as recounted in *Genesis* and then God's ceasing from that initial Creation phase and allowing secondary causes (e.g., the natural properties programmed into Nature) to function as designed.

Miracles do occur, of course, and no doubt God intervenes with grace in our lives if we seek inner change. But are we to accept as true, on the basis of subjective scientific theories which have no basis in Scripture or Tradition, that He intervenes constantly to shape new galaxies, to tailor the odd features of the solar system (such as planets and moons spinning in opposite directions), to change this creature into that time and again, or to keep on directly creating new species?

Though they see secondary causes are at work in the processes of *variety within kind*, proponents of Progressive Creation are frequently driven to abandon them, conveniently, in favor of divine intervention. (God has to be introduced by them to interrupt secondary causes intermittently, to create another batch of new creatures or expand the genome.) The concept of Progressive Creation is thus open to the charge of revisionist beliefs being imposed upon Scripture.

God's use of secondary causes is an integral part of an efficient Creation within a "young" Universe. The very interdependence of creatures, plants and atmosphere on Earth suggests that fairly rapid creative acts produced the various types of beings that exist, since their harmonious interaction requires that they be set in train rather quickly. Instead of support for the flawed notion of Progressive Creation, the superior idea of *Complementary Creation* finds an echo in the *Catechism of the Catholic Church*:

> The sun and the moon, the cedar and the little flower, the eagle and the sparrow [*not to mention myriad other life forms as well*]: the spectacle of their countless diversities and inequalities tells us that no creature is self-sufficient. (340). [Italics comment added.]

Like the mistaken concept of Theistic Evolution, it seems clear that the idea of Progressive Creation, however superficially appealing, is plagued with inherent conceptual problems which make it untenable.

—PART IV—

THE INFLUENCE OF EVOLUTION ON BELIEF SYSTEMS

—Chapter 17—

THE INFLUENCE OF EVOLUTION ON BELIEF SYSTEMS

As already noted, E. F. Schumacher was greatly concerned about the harm done by Evolutionism. He believed the evolutionists' reasoning represented a philosophical error of the most disastrous consequences. Were his fears justified?

Evolutionism has had a profound influence on various belief systems and upon man's thinking. For the Christian, there is every reason to believe that the whole notion of Evolution is one which Satan has used to bring about confusion, in an effort to turn mankind away from God. Satan seeks subtly to change one's intention from that of conformity with the laws of God, and thereby from growth in faith, to that of conforming increasingly to the spirit prevailing in the world. The idea of using Evolution Theory to achieve this purpose, by supposedly doing away with the need for God, seems brilliant, though it is erroneous.

Widespread acceptance of Evolution Theory has greatly influenced the moral standards of mankind. If atheistic Evolution is true, then "anything goes"—nudity, pornography, incest, abortion, homosexuality and so on. There can be no objective criteria for right or wrong, or for desirable or undesirable patterns of living, or for codes of law.

The Christian view is that the basic dignity of each human being derives from the fact that each one has a rational soul, made in the image of God. Each human being is destined for eternal existence and is not simply an animal which ceases to exist upon death. Both body and soul constitute the person, but because the soul persists as the body changes, it is possible to describe the soul as the "form," or the essence, of the human person. Each human being therefore has a right to knowledge of his ultimate supernatural destiny and of the fact that God, the Creator of all things, is an authority greater than man.

The idea of man as simply an evolving creature, the highest of the animals, must inevitably lead to the devaluation of the individual's fundamental rights. In many parts of the world, the rights of the individual have come to be regarded as secondary to those of the state. In "modern" societies, the very right to life itself is simply denied to pre-born human beings if abortion is desired by the mother. The widespread "blindness" throughout modern societies that regards unborn human children as less than fully human is largely attributable to pro-Evolution conditioning over many years. In most liberal democratic societies, consensus pluralism has replaced obedience to the laws of God. The Kingship of Christ is understood by relatively few, let alone considered seriously as the ideal governing philosophy of society.

As belief in Evolution Theory became ever more popular from the late 19th century onwards, the growth of evil belief systems was effectively facilitated and new possibilities emerged for theological errors to spread throughout Christianity.

Many people would be generally aware of the ruthless use to which Social Darwinism was put by unscrupulous capitalists during the 19th century era of *laissez faire* capitalism, when vast fortunes were amassed at the human expense of many workers who lived and worked in deplorable conditions. But how many appreciate the extent to which Evolution Theory has been central to various belief systems and has had a major influence within Christianity?

The appeal of powerful totalitarian belief systems in the 20th century owes much to the acceptance of Evolution as the explanation of man's existence, an acceptance that allowed evil ideas to be formulated in terms that can appeal to human reason. Take away Evolution Theory from Nazism, Communism and Secular Humanism, and these belief systems effectively lose a central premise necessary for coherence. Similarly, if Evolution Theory did not exist, the liberal theology called "Modernism" would have no scientific foundation for its revisionist propositions.*

To assert all this is not to blame Evolution Theory for the inhuman and degrading acts carried out by totalitarian forces, but rather to place the impact made by Evolution Theory in its proper context—in other words, to give it due recognition as an important causal influence allowing these "isms" to be born and grow. In the following chapters it will be argued that Evolution Theory is not incidental, but rather, is essential to the above belief systems.

---

* Modernism holds, among other tenets, that truth is relative to what the people of various generations perceive it to be. Modernism is a heresy condemned by Pope St. Pius X, 1903-1914, yet which re-emerged after Vatican Council II (1962-1965). See Appendix B for a brief look at Modernism. —*Editor, 1999.*

—Chapter 18—
NAZISM AND COMMUNISM

The 20th century has seen the rise to power in many countries of despotic systems of government which have wielded almost total control over the population. Few despotic forces in history can compare with the infamous totalitarian belief systems known as Nazism and Communism. The rhetoric used to idealize each belief system promised a better world for its peoples, Nazism placing great stress upon the leader principle and Communism upon the dictatorship of the proletariat. The power system imposed in these countries was almost invariably nothing less than brutal dictatorship, usually with an obsessive personality cult of the leader and grinding oppression of the population.

Curiously, the two systems had much in common. Nazi Germany and the USSR even made a joint peace pact at the outset of World War II. When Hitler invaded Russia, however, he unwittingly set in motion the unlamented defeat of Nazism and effectively opened the way for Stalin to occupy much of Eastern Europe and impose Communist ideology on those countries.

The pro-democratic revolution which swept Eastern Europe and the USSR in the late 1980s was bewildering, and one hopes that enlightened leadership will emerge amidst the turmoil of each country and that other countries will throw off the yoke of Communism. In the West, the many intellectuals who consistently failed to grasp the reality of Marxism-Leninism in practice and idealized the theoretical model of pure Marxism are now seen as having little credibility.

In different ways, both Nazism and Communism owe much to Evolutionism. Evolution was a powerful influence on the mindset which dominated the attitude of party leaders, and thus it helped to rationalize the brutal treatment of individuals.

Nazism ("National" Socialism)

In the decades preceding the reign of Hitler, anti-Semitism in Europe had developed to a high degree and was rationalized in religious, socio-political and biological arguments. During his youth, Hitler imbibed a fanatical hatred of the Jews and understood clearly the concept of Social Darwinism.

Based on theories of biological Evolution and natural selection, Social Darwinism held that the "survival of the fittest" applied in nature not only regarding individuals, but also in the struggle between races and nations. Those groups that were hostile to the State or were biologically inferior would have to perish if the nation were to survive.

Such ideas helped to shape the policies of the Third Reich. In pursuit of his perverted vision of eugenics and an Aryan master race, Hitler and his Nazi colleagues implemented the racist ideas of Social Darwinism, and this culminated in the "final solution"—the well organized plan to exterminate millions of Jews, Slavs, the mentally demented and other so-called inferior types.

The idea of racism as a general philosophy strongly implies that there are distinct races upon the Earth, some regarded as inherently superior to others, who have evolved from different members of a common hominid ancestry.

In a key analysis of the Nazi dictatorship, Karl Dietrich Bracher noted the rationale of Social Darwinism across pre-World War II Europe:

> The underlying idea was that, in the course of a ruthless competition and battle, a "natural" selection takes place which prevents or offsets aberrations and makes for a proper balance between population and available resources. In society, education and penal law serve as the instruments of this process of selection; according to the immutable laws of heredity, the unfit cannot be educated and therefore must be eliminated. . . . The teachings of Social Darwinism were in contradiction to the egalitarian belief in an open, mobile society and the educability of man—the basis of the democratic idea of state and society. The humanitarian idea of evolution was replaced by the concept of the planned "breeding" of an elite and the proscription of intermarriage. . . . Social Darwinism remained a sectarian philosophy; seeing man only as a biological and not a thinking, moral being, it misjudged the nature of historical and social forces, the binding values of civilization, as well as the basic differences between biological and social selection.[1]

---

1. Karl Dietrich Bracher, *The German Dictatorship* (Middlesex, England: Penguin Books Ltd., Peregrine Books, 1978), pp. 28, 29.

In translating the concept of Social Darwinism into grim reality, Hitler appears to have been influenced by the ideas advanced in the late 19th century by Frederick Nietzsche.[2] This German thinker had contended that, just as man had evolved from an ape-like ancestor, so a higher type "superman" would in time emerge—one who would be quite ruthless in exercising power over the weak. Some idea of the mentality of the Nazi leadership can be gauged from the remarks of Heinrich Himmler, made during World War II to SS leaders:

> The SS man is to be guided by one principle alone: honesty, decency, loyalty and friendship towards those of our blood, and to no one else. What happens to the Russians or the Czechs is a matter of total indifference to me. If good blood of our type is to be found among the [other] nations, we will take it, if need be by taking their children and bringing them up ourselves. Whether other peoples live in plenty or starve to death interests me only insofar as we need them as slaves for our culture; for the rest it does not interest me.[3]

Bracher also pointed out that the extermination mentality of the Nazis grew out of the biological insanity of Nazi ideology. The extermination plan was handled by a completely impersonal bureaucracy which regarded its victims as a species of inferior subhumans. Himmler regarded them as vermin, as though he were handling a biological disease.

That Social Darwinism could be put into such gruesome practice by the Nazis can in no way be blamed personally on Charles Darwin and his colleagues. However, it must be recognized that unquestioning belief in Evolution Theory was a central feature of the Nazi belief system—and not just incidental to it.

Communism ("Democratic" Socialism)

The central doctrine of Communism is the theory of Class Struggle. This theory is explained in the concept of *dialectical materialism*, which relies heavily on Evolution as an established historical fact.

One of the co-writers of the Communist Manifesto, Friedrich Engels,

---

2. It is known that Nietzsche experienced a decisive loss of faith in Christianity in 1865, after reading *Das Leben-Jesu* (1835) by David Friedrich Strauss. Strauss was among the first German thinkers who attempted to separate the "historical" Christ from the divine Christ, and later went on to write extensively from a Social Darwinist perspective.
3. Bracher, *The German Dictatorship*, p. 522. Nürnberger Dokumente cited.

wrote a booklet entitled *The Part Played by Labor in the Transition from Ape to Man*.[4] In it he claimed that exploitation began with the evolution of the first hand, millions of years ago. This explanation was accepted as factual by Marxists, and the whole Communist credo is dependent for plausibility on Evolution Theory.

Much of the indoctrination carried out in Communist "re-education" centers (concentration camps) referred to Evolution in an effort to convince people that there is no God.

According to Communist theory, everything in existence is composed entirely of matter. Matter is supposed to encompass a unity of opposite forces, with an inbuilt conflict that makes it autodynamic and self-energizing, needing no outside force. Because of this conflict—supposedly inherent in all matter—the outcome must always be orderly development of new forms of matter. Through so-called quantitative accumulations, nature is thought to be capable of "leaps" to new and higher forms of reality, and thus a Creator is not required. In *Socialism: Utopian and Scientific,* Engels wrote about Evolution:

> Nature is the proof of dialectics, and it must be said for modern science that it has furnished this proof with very rich materials increasing daily, and thus has shown that, in the last resort, Nature works dialectically and not metaphysically; that she does not move in the eternal oneness of a perpetually recurring circle; but goes through a real historical evolution. In this connection Darwin must be named before all others. He dealt the metaphysical conception of Nature the heaviest blow by his proof that all organic beings, plants, animals, and man himself, are the products of a process of evolution going on through millions of years.[5]

According to Marxist theorists, at some stage in the evolutionary process, mankind developed consciousness and then mind emerged as an intelligent, self-knowing, self-determining entity. Because man is the highest intelligence in all existence, he can, with the development of science, come to be omniscient, to know all truth about existence.

Marx held that history was on the side of the workers. As surely as all existence had evolved to the present level, it was inevitable that the exploiting class would be destroyed. The property-owning class would assist in its own demise by reducing the working class to such poverty

---

4. Marx & Engels, Selected Works (Moscow, 1975), pp. 354-364.
5. *Ibid.*, p. 407.

that a deficiency of demand would result when the workers could no longer afford to purchase the goods produced by the system.

Once the working class had grasped the laws governing their fate and had overthrown the exploiting class, it would inaugurate a new era of socialism, and the state would whither away, as society evolved to a classless community, with no need for police or armed forces.

Man was destined to find fulfillment, said Marx, *in a new world order* in which conflict would be eliminated and socio-economic harmony at last achieved. The wars, oppressions, miseries and spiritual illusions of previous ages would dissolve into the Communist peace, which was the goal of history.

Lysenkism, or Michurinism, was a genetic philosophy based on Evolutionism. It was part of the official government policy, obligatory in teaching at all levels in the USSR and its satellite countries. The idea was that acquired traits, including those acquired by training, were heritable, and thus Evolution could be guided by the state toward a better society.

The spread of Communism brought with it immense suffering. Tens of millions of human beings (about sixty million in the USSR alone) died as victims of fanatical Communists. They were clearly regarded as nothing more than creatures who were expendable in the quest for a classless society.

As with Nazism, Evolution was essential to the Communist belief system—not just incidental to it. Without the assumption of evolutionary struggle for survival against predators, the wretched notion of Class Struggle against oppressors would have lacked plausibility.

—Chapter 19—

HUMANISM

The set of ideas encompassed by Humanism has come to represent man's efforts to solve his problems and shape society apart from God. There is therefore great conflict between traditional Christianity and Humanism as to man's nature and the purpose of life. To Christians, the supernatural realm is primary and the material realm secondary, but to modern Humanists (especially atheists, as distinct from agnostics), the material realm is primary and the supernatural realm does not exist.

Humanists nevertheless recognize that man has moral obligations and that, in a purely naturalistic sense, there does exist a sort of spiritual dimension. Many such Humanists are active in bringing about social change, in accordance with their conception of society.

Humanism (defined as genuine recognition of the intrinsic value of each human being) was originally initiated when the Christian teaching of monotheism and the unselfish lifestyles of the early Christians gradually began to transform society and enable others to grasp the inherent dignity of man. Christianity brought a radical change in beliefs:

> Born in the pagan civilization of the Hellenistic-Roman world, Christianity reacted to the prevailing views, systems and assumptions, in other words, to the intellectual environment. Christian thinking was, naturally, not only a reaction to the milieu, it was primarily a new concept, a renovation of that milieu through a doctrine profoundly incompatible with pagan philosophy and science. The pagan world view, and this is true of Greek, Iranian, Chaldean, Egyptian and Indian speculation, the total pre-Christian world picture, was based on the belief that the Universe is peopled with gods and spirits, both benevolent and malevolent, that history is mechanically moving in a circular manner, always returning to the same point, and hence States and individuals are near-fatalistically tied to a pattern from which there is no escape. The Roman form of religion, moving between the *magic* which influences gods and spirits, and *pietas*, the performance

of rites honoring the ancestors, stressed man's dependence on all sorts of forces which had to be constantly appeased and propitiated.[1]

Over the centuries, however, various ideas were proposed which ironically led Humanism further toward anti-Christian views. Some scholars reasoned that all religion was much the same and that there is a "natural" religion within man. They also began to argue that God is not transcendent, but rather is an immanent reality confined within the Universe and also within man, and so man could therefore create his own moral laws. Morality increasingly came to be seen as that which is rationalized by the desires of man. The necessity of obedience to the revealed God was rejected in the process.

Over the centuries, so-called *Christian Humanism* came to be a contradiction in terms, because some Humanists distorted the true relationship between Creator and human creature. Thomas Molnar has traced the historical development of Humanist ideas and the influence of atheistic Humanism upon modern Christians. He shows how some 15th century Renaissance scholars, such as Giovanni Pico della Mirandola, began to advocate New Age type ideas which now resemble beliefs advocated by many 20th century scholars:

> The question at once arises whether what Pico intended to defend was really the Christian truth or a humanist truth? The line of argumentation of the *Oration* [*On the Dignity of Man*] is that man is privileged among creatures . . . man is a pure indeterminacy, who is allowed by God to "trace the lineaments of his own nature." This, taken to its logical conclusion, would indeed make of man a second God. . . . Further on, Pico calls upon his hearers, bishops and theologians, to lift themselves "to such ecstasy that our intellects and our very selves are united to God. . . . Borne outside ourselves, filled with the Godhead, we shall be no longer ourselves, but the very One who made us." What is the creator God if his role vis-a-vis one particular creature, man, is limited to a kind of first shaping, after which man's real creator will be man himself? It is impossible to accept the view of Fr. Henri de Lubac, for example, who sees in Pico . . . one who breaks through the timidity of certain Christians by proclaiming man as an unfinished being with unlimited self-effecting potentialities.[2]

---

1. Thomas Molnar, *Christian Humanism: A Critique of the Secular City and Its Ideology* (Chicago: Franciscan Herald Press, 1978), p. 4.
2. *Ibid.*, p. 26.

The "drift" in consciousness away from belief in a transcendent God has been traced back many centuries, long before Darwinian evolutionary ideas became popular, and it is now hard to gauge to what extent it was influenced by ancient concepts of Evolution. Nevertheless, it is now "justified" by acceptance of Evolution as a proved fact:

> The Hegelian process of the World-Spirit's gradual self-awareness is increasingly interpreted in the sense of Darwinian evolution; meanwhile the Christian [Hegelian] jubilates in his discovery of a God who is not yet, whose coming-to-be depends on him and on other progressive-minded men. The general watchword is *away from* substances which, by being permanent, appear as immobile, regressive, condemned to ossification—and *toward* whatever appears moving, becoming, a ceaseless flux. Hope, no longer a virtue, is reinterpreted as a "dimension," synonymous with progress, revolution, a restless attunement to the future. For Teilhard [de Chardin], those who are not progressive and democratic will be mercilessly crushed by evolution, which is just *now* shifting gears toward an accelerated "hominization" of our sluggish nature.[3]

The belief system known as *Secular Humanism* relies heavily on Evolution Theory for credibility. All that exists is held, *by faith*, to be self-perpetuating by a never-ending evolutionary process. Scarcely a page of any of the publications written by proponents of Secular Humanism fails to cite Evolution as an established fact central to all existence. Sir Julian Huxley praised the contribution of Darwin:

> Charles Darwin has rightly been described as the "Newton of biology": he did more than any single individual before or since to change man's attitude to the phenomena of life and to provide a coherent scientific framework of ideas for biology, in place of an approach in large part compounded of hearsay, myth and superstition. He rendered evolution inescapable as a fact, comprehensible as a process, all-embracing as a concept.[4]

---

3. *Ibid.*, p. 46.
4. Sir Julian Huxley, *Essays of a Humanist* (Pelican Books, 1969), p. 13.

Huxley also outlined the central importance of Evolution to the Secular Humanist belief system and to the rejection of religion:

> In the evolutionary pattern of thought there is no longer either need or room for the supernatural. The Earth was not created: it evolved. So did all the animals and plants that inhabit it, including our human selves, mind and soul as well as brain and body. So did religion. Religions are organs of psychosocial man concerned with human destiny and with experiences of sacredness and transcendence. In their evolution, some (by no means all) have given birth to the concept of gods as supernatural beings endowed with mental and spiritual properties and capable of intervening in the affairs of nature, including man. These theistic religions are organizations of human thought in its interaction with the puzzling, complex world with which it has to contend . . . they are destined to disappear.[5]

In the first *Humanist Manifesto* (1933), the first three affirmations acknowledge the acceptance of Evolution as a central belief:

1. Religious Humanists regard the Universe as self-existing and not created.
2. Humanism believes that man is a part of nature and that he has emerged as a result of a continuous process.
3. Holding to an organic view of life, humanists find that the traditional dualism of mind and body must be rejected.[6]

The second *Humanist Manifesto* (1973) reaffirmed belief in Evolution:

> Promises of immortal salvation or fear of eternal damnation are both illusory and harmful. They distract humans from present concerns, from self-actualization, and from rectifying injustices. Modern science discredits such historic concepts as the "ghost in the machine" and the separate soul. Rather, science affirms that the human species is an emergence from natural evolutionary forces.[7]

Atheistic Humanists thus hold that the transcendent God does not exist and that Evolution "explains" everything. In fact, they believe that

---

5. *Ibid.*, p. 82.
6. *The New Humanist* (1933), Vol. VI, No. 3.
7. *The Australian Humanist*, December, 1973.

Evolution has reached the stage where it has become conscious of itself (i.e., man, as the most intelligent being in known existence, now understands the process taking place in the Universe). Therefore it is imperative that individuals should seek to take control of the future direction of evolutionary changes. They reject absolute principles which defer to an authority higher and greater than man and instead promote subjective standards of morality and attitudes of relativism.

The permissive morality which has swept the modern world reflects acceptance of Secular Humanist standards. The concepts of marriage and the stability of the family unit have received serious challenge. This should come as no surprise, given the Evolution-consciousness of much of modern society. If man is only an evolved animal, the morals of the jungle seem more "natural" than restrictions against premarital relationships and obligations to marital fidelity. Why *not* promiscuity?

But once the Creator is banished from the ordering of society, the prospects for genuine peace and prosperity in a post-Christian world become illusory. And in their place come chaos and hardship, as society drifts further into a culture of death, a state not unlike the "kill or be killed" mentality.

There can be no doubt that the moral standards of society have been affected by Evolution-inspired Secular-Humanist attitudes. The most obvious influence of this is seen in the widespread growth of abortion on demand, but euthanasia is another issue now strongly promoted by many Secular Humanists. Their key strategy has been to change the laws of society and thereby also to influence moral standards by changing the public perception of normal moral behavior.

The practice of induced abortion is no longer regarded by many as being quite as abhorrent as it was in times past. To them, the right to life of the unborn child is secondary to the convenience of the mother. Since the legalization in many countries of what is effectively abortion-on-demand, what is now seen as legally permissible is also regarded by many as being morally acceptable.

Paradoxically, in some parts of the world, a child has the right to sue after birth for damages if injury is done to him or her in the womb, but the same human being is accorded no rights whatsoever to prevent his being completely destroyed by induced abortion while *in* the womb. The moral code of human society has been eroded so profoundly that some unborn human beings have been aborted simply because they were not of the sex desired by the parents.

With respect to the question of *in-vitro* fertilization (IVF), the very birth of test-tube babies confirms beyond doubt that human life begins at the moment of conception when the female egg is impregnated by the male sperm and the first cell of the new human person is brought into being. This fact makes nonsense of claims made that the new life in the womb may not be human for a certain period of time after conception.

As is well known, the IVF process unavoidably involves much deliberate loss of new human life at its earliest stage. Many more tiny embryos (at about the eight-celled stage) are cultivated than can be successfully implanted and carried. It is very difficult to avoid damage being done to the tiny human beings, and the fate of the "excess" embryos tends to be regarded with indifference.

The couple seeking a baby through IVF may be desperate for a child of their own, and understandably so, but in many cases their love is not disinterested love. The parents retain at all times the "right" to terminate the life of the fetus conceived through IVF if the new human life is deformed in some way.

Sadly, since atheistic Humanists believe that human beings are only animals with no supernatural destiny, induced abortion is seen as an acceptable practice in both reducing the population of the world and in pursuing one's private quest for self-fulfillment. They tend to believe that human beings only gradually develop rights, perhaps when sentience is reached (i.e., being aware of sensations, especially pain). If human beings are really nothing more than the product of Evolution, they may have little claim to priority over other creatures:

> When opponents of abortion say that the embryo is a living human being from conception onwards, all they can possibly mean is that the embryo is a living member of the species *Homo sapiens*. This is all that can be established as a scientific fact. But is this also the sense in which every "human being" has a right to life? We think not. To claim that every human being has a right to life solely because it is biologically a member of the species *Homo sapiens* is to make species membership the basis of rights. This is as indefensible as making race membership the basis of rights. It is the form of prejudice one of us has elsewhere referred to as "speciesism," a prejudice in favor of members of one's own species, simply because they are members of one's own species. The logic of this prejudice runs parallel to the logic of the racist who is prejudiced in favor of members of his race simply because they are members of his race.[8]

---

8. Peter Singer and Helga Kuhse, *"The Moral Status of the Embryo," Test-Tube Babies* (Melbourne: Oxford University Press, 1982), p. 60.

Despite the rhetoric of Secular Humanist philosophy, there can be no true liberty or human rights for the common man unless the human person is accepted as sacred from the moment of conception.

When ethical values are not grounded in the laws of God and in recognition of the uniqueness of each person, then subjective human feelings can easily be used to justify the satisfaction of one's instinctive impulses. An amoral consciousness with respect to sexual morality is thus fostered. Further, there can be no objective criteria to determine what moral standards should apply in society.

The deaths of millions of unborn human beings due to induced abortion can clearly be traced to the militant promotion of Evolutionism, rationalized free of the absolute principles that may otherwise constrain individuals from "terminating" the life of the unborn child. (One of the obnoxious features of Secular Humanism is the extensive use of euphemisms to hide reality in order to win over supporters. For example, the savage killing of unborn human beings is referred to as "interruption" or "termination" of pregnancy, apparently to assuage and desensitize troubled consciences. The evil use of euphemisms has been a tool employed by all modern totalitarian movements to hide their actual objectives and the true reality of what is being done.)

If the existence of a Creator, who has revealed laws for man to obey, is rejected, then whose moral standards are to take precedence in the ordering of society? If a dictator such as Hitler gains power and enforces his set of moral values on society, on what grounds is he to be criticized? He could argue that the views of others are only subjective anyway, and that his actions are justified by the end achieved. John Hammes noted how man-centered Humanism is incompatible with Christianity, and why it cannot answer man's deepest questions:

> First, most proponents of these views are atheistic, depending on a blind evolutionary thrust to account for man and the Universe. Second, in the rejection of God and His revelation, these viewpoints focus on man alone, man for himself, man as the highest pinnacle of life and intelligence. Consequently, one view tends to adulate man's rationality, and another his arationality. Third, these perspectives are relativistic, capable only of probability and contingency statements, rather than ones of certitude. They are also descriptive rather than explanatory. Further, a relativistic viewpoint cannot prescribe obligatory values or moral norms, and does not invite total commitment. Fourth, these interpretations of life and man are unable to explain the human condition of pain, suffering, and death. Nor can they account

for evil. Lastly, all of them are inadequate in explaining the meaning, purpose, and significance of man.[9]

The threat to civilization posed by Secular Humanism must be clearly recognized:

> Humanism should be regarded as a transition from a civilization with a strong, religion-inspired sense of reality, to an anticivilization where man's worst urges are allowed free course, in view of no other purpose than more pleasure, more hedonism, more drifting in the material and the moral universe.... From an easy-going "civilization of play," conceived as the post-war relaxation of mores, a kind of anti-puritanical reaction, a supposedly freer examination of the facts of life, etc., we have switched to the heavier, deadlier side of sin: mass-murder of unborn children, the surviving child's bodily and mental corruption in school, the live showing by the media of executions, hold-ups, atrocities, even theologians encouraging hatred, murder and removal of all restraint.[10]

One has only to reflect upon the eugenics ideology at work behind population control tactics to see just how far man-centered Humanism has moved from Christian beliefs. Acceptance of atheistic beliefs concerning the evolutionary origin of mankind has for decades helped shape the self-centered attitude of many in society who are either indifferent to the killing of unborn human beings or quite hostile to pro-life beliefs. But for many militant activists the goal has gone well beyond that of achieving abortion on demand; they believe that population growth must be controlled as though it were a disease, as though it constituted an environmental threat to life on Earth. The anti-human spectre of "helping" the process of evolution by weeding out the "weak and unfit" in society through abortion, genetic manipulation and euthanasia is likely to increase in intensity as the age of the elderly continues to increase and birth replacement rates continue to stagnate.

The antidote for man-centered Humanism lies in the rediscovery of God the Creator/Redeemer and the consequent renewed understanding of the reality of Original Sin.

---

9. John A. Hammes, *Human Destiny: Exploring Today's Value-Systems* (Huntington, IN: Our Sunday Visitor, Inc., 1978), p. 40.
10. Molnar, *Christian Humanism,* pp. 163, 164.

—Chapter 20—
CHRISTIANITY

The Impact of Evolution Beliefs

Evolution Theory has also caused great harm within Christianity. Its impact upon those who may lack a sound formation in Christian teachings has been known to destroy faith in Christ as the risen Saviour and to cause many to abandon the practice of their religion. Even when the impact is not so drastic, it can still result in a considerable degree of indifference toward belief in Christian teachings.

Although theories about Evolution can be traced back thousands of years, they were not of much significance until the 19th century. Until then, Christians had generally accepted *Genesis* as meaning what it plainly says—that God specially created the Universe from nothing and chose to implement the Creation over six days. Since uniformitarianism and Evolution beliefs had not yet been developed to any great extent, the question of whether the six days of *Genesis* were actually 24 hours each or otherwise, although troublesome, had not been anything like the contentious issue it has since become.

Along with the 19th century increase in the credibility of the Evolution Theory, seemingly valid arguments were increasingly being proposed favoring the idea of an extremely long period of time during which Evolution took place. In the then-emerging Origins controversy (long before DNA was understood and before the fossil record had been thoroughly investigated), the growing dilemma facing theologians at that time must have seemed daunting. It was thought then that Evolution needed time, and lots of it, but if Evolution were historically true, then surely the literal-as-given understanding of the Creation days *must* be mistaken!

At a superficial level, Evolution Theory must potentially have seemed shattering to convinced Christians. If Adam and Eve were not the specially created first parents, but merely symbolic representations of early mankind who had evolved over millions of years, then the doctrine of Original Sin no longer had credibility and the idea of a Redeemer no longer made sense. Scripture could not be historically true. Further, if *Genesis* were largely mythical in content—true only in a vague "religious" sense—then it would not be hard to accept the belief that Jesus Christ was only an outstanding human being, cruelly executed by the Establishment who saw Him as a threat.

Many Christian scholars of the 19th century began to seek ways of re-interpreting Scripture to harmonize it with what, at that time, seemed true in empirical science. Some proposed that the six "days" were each in fact very long periods of time, which would have enabled Evolution to take place. In the process of revision, a number of scholars eventually lost belief in Christianity, in the face of scientific arguments against *Genesis*. Thus, it comes as no surprise that, *once the credibility of the foundational* Genesis *account was placed in doubt, nothing in the Bible would be free from revision*—everything, including the New Testament, could be doubted and would have to be "re-interpreted." Even the existence of Satan and Hell came to be disbelieved by some Christians, as being incredible to the modern mind.

Evolutionist theologians held that Christ was an outstanding reformer, whose unsurpassing example had gone down in history and still challenges modern society—eulogized in death by His admirers, but definitely not divine. It made sense, they claimed, to divorce the so-called Christ of history from the Christ of faith.

The miracles worked by Christ and the Apostles were attributed by many scholars to merely natural causes. (How far could this be taken? It is hard to believe that many fishes would be naturally swimming around with a silver coin in their mouths!—see *Matt.* 17:26.) Belief in the Flood of Noah was rejected as naïve biblical literalism, and the idea of Adam and Eve being real human beings was similarly dismissed. The mistaken idea that Scripture contains errors began to be taken much more seriously, and this belief is unfortunately now widespread, even among many who consider themselves opposed to Modernism.

The doctrinal errors which said that Christ was not God were not new. In fact, almost right from Christ's Ascension into Heaven, there were those who sought to deny some part of His reality or to explain away some unpalatable teaching of His. What was new in the 19th century, however, was an apparently plausible theory (i.e., Evolution) which facilitated a drastic re-interpretation of Scripture and an apparent dethronement of God.

Over the last two centuries, much of theological and exegetical scholarship has been markedly changed because of the devastating impact of evolutionist beliefs. Many scholars now accept Evolution as historically true and are convinced that much of Scripture is myth and devoid of historical reality. This has made it possible for Humanist attitudes of relativism to find acceptance, when they should be clearly recognized as incompatible with Christian doctrine.

The impact is still being felt, and one gets the impression that many Church scholars are unaware of the full extent of modern scientific Origins findings and have yet to move beyond presuppositions handed down from mistaken 19th century science. The problem seems well-entrenched. Much material is constantly being written by Catholic intellectuals in favor of Evolution, and often the writers are patronizingly dismissive of "divisive" creationists. (Are we not repeatedly told that Galileo's detractors were at fault for not looking objectively at the evidence? How ironic, if some Catholic thinkers are unwilling to look objectively at the range of arguments within the Origins debate!)

A now-common trend is that of rejecting Darwinism but still praising Evolution, as though Evolution *per se* divorced from Darwin is viable. Fr. Stanley Jaki, O.S.B. strongly rejects both creationists and Darwinists in *The Savior of Science,* but then admits his belief in Evolution (Jaki, pp. 147ff) and gives a specific example of Evolution at work: 130 species of beetles arising on the island of St. Helena! Unfortunately, just like Darwin's finches on the Galapagos Islands, this is not Evolution; there is no new, "higher," information being added to the gene pool.

Once any version of Evolution is given credence, this *must* influence one's views on Scripture and exegesis, ultimately to the detriment of sound doctrine. Among the many examples which could be cited, consider these extracts from evolutionists—all Catholic priests:

Example 1: Writing in *A New Catholic Commentary on Holy Scripture,* Dr. Bruce Vawter, C.M. advocated acceptance of the Documentary Theory of the Pentateuch. He argued that the oldest fossil remains of man go back a million years, and therefore it is unlikely that Moses had accurate information passed down to him from early man:

> Could it be, then, that God provided miraculously for the accurate transmission of such reliable information, or that he revealed it to the author of *Genesis.* The possibility, of course, no believer in revealed religion would deny. All the evidence, however, indicates that He did nothing of the kind. . . . It does not appear possible, in other words, to effect a strong concordism between the data of *Genesis* and what

scientific archaeology and anthropology have made known to us concerning prehistory.[1]

A scientific hypothesis which envisages a gradual development of the human body need not be in conflict with the doctrine of divine creation any more than is the undoubted fact of the gradual development of the physical universe through its many eons.[2]

Example 2: Fr. Bernhard Philberth (he and his brother, Fr. Karl Philberth, are both Catholic priests/physicists) strongly favors Evolution and an age of billions of years for the Universe:

> The artificially played up contrast "evolution versus creation" is rooted in naïvety on the one hand, and demagoguery on the other. There is no genuine contrast. There is abundant evidence for a creation through development, an evolutionary creation, an ascending creation. To oppose evolution merely because it is misused as an argument against the existence of God is a scientifically suspect methodology and a morally reprehensible opportunism which is rooted in a paltry vision of God.[3]

Even though he points out (Philberth, p. 108) that, "In principle there is no fact so obvious and well-secured in this world that there is no need for proof," he cites very little in support of the "abundant evidence."

Philberth's book abounds with evolutionist *newspeak*, terms such as ontogenesis, phylogenesis, salvific homogenesis—as if such terms refer to realities and are accepted by all as credible. There is not one reference to the Magisterium or papal encyclicals, no mention of Church Fathers or Councils, and nothing from the 1992 *Catechism of the Catholic Church*. Tradition is referred to in various passages, but not by way of deferral to truths proclaimed and errors condemned by the Catholic Church concerning Origins.

Philberth outlines views which are incompatible with Tradition and can hardly be justified as sound speculation within accepted doctrine. For example, he claims that only those who choose obedience receive a rational soul from God, and then only *upon death*; the damned never receive a rational soul and eventually cease to exist (Philberth, pp. 26 and 34), after suffering the torments of Hell (which itself also self-destructs). "The negation of God on principle is the nullification of

---

1. Vawter, S.M., in *A New Catholic Commentary on Holy Scripture*, par. 146b.
2. *Ibid.*, par. 151i.
3. Bernhard Philberth, *Revelation* (BAC Australia, 1994), p. 85.

existence *per se."* (Philberth, p. 30). Finally, all memory of their existence then vanishes from the consciousness of God (Philberth, p. 36)! (Such, of course, is impossible for God, because He is omniscient, and the loss of the knowledge of something would contradict His nature.)

> Salvific homogenesis is the "ensoulment of man." An age-old problem of theology is this: "when does man receive an immortal soul?" In no onto-genetic stage such an ensoulment becomes objective, not in the union of sperm and ovum, nor in any early cell division, nor in the growth of the blastocyst, the embryo, the fetus, nor at birth, nor at the completion of the embryonic development after 21 months, *nor at any stage during a human life unto death.* . . . one cannot ascribe eternal life to the zygote; a material/chemical structure cannot be the basis of eternal life. . . . This world is neither objectified as a hellish world nor as a heavenly world. But it is objectified as such by the decision of every person rendered capable of deciding in terms of salvific homogenesis.[4]

Apparently, Adam was the first human being, several thousand years ago, to be given the possibility of choosing thus in favor of God. But who exactly were those strange earlier supposed "pre-Adamites," and how can a human being even exist without a rational soul?

> From a zoological point of view, God prepared the earthly vessel "human being" in a guided evolution to fill in with His Spirit. Pre-Adamites (before Adam) have been living for millions of years on this Earth. Obviously, Cain had descendants with a pre-Adamite and certainly *"the great heroes and famous men of long ago"* (*Gen.* 6:4) were also pre-Adamites. . . . Whether natural human beings developed from a first couple (mono-genesis), or as a manifold transition via the "threshold of hominization" (polygenesis), or as ever they were created by God, is of no importance for the salvific aspect.[5]

Philberth's terminology is very imprecise and confusing. God is constantly referred to as the "Absolute Sovereign." The Divine Trinity is referred to in such terms as "in heaven . . . God-Father and God-Spirit are face to face with the Communion of saints, which is the third person of God, as His children in Christ, His Son." (Philberth, p. 106). "The saints, who are united with God through Christ, in Christ, and as

---

4. Philberth, Revelation, p. 48. Emphasis added.
5. *Ibid.*, p. 122. Emphasis in original.

Christ, themselves receive glory, honor and power. God, the one and only God, is the 'Wholly Other' as well as the 'Communion of saints.' God unites everything in Himself, the One and the Other, He is 'All in All.' " (Philberth, p. 11).

It seems that Philberth places upon physics a higher claim to truth than that of theological reality. (Philberth, p. 37). His ideas on the mysteries of Creation suggest that he seeks too much from speculative metaphysics:

> Is the highest—the meta-cosmos of transcendence—really mirrored in the smallest—the sub-cosmos of the electron? . . . An unsettling thought springs to mind. In the infinite number of drifting-off and vanishing universes, is hell the "outer region" of God? *The hellish universes, when separating themselves from God, drift off into nothingness with an unreachable speed*; like the electric field of the electron, receding with invariance velocity c. The perpetual, renewed separation of drifting-off universes is a cruelly consistent, quasistatic event, an everlasting hell.[6]

Fr. Bernhard Philberth is hailed as an adviser to three popes and claims to have been given an out-of-this-world vision from God, who re-enacted the formation of the Universe in front of Bernhard and gave him the mathematics to go with it! Both Fathers Philberth are claimed to have been ordained as priests, *after only 6 weeks training*, because of their intellectual standing, on the instructions of the Holy Father himself.[7]

Example 3: In his book *God's Creation,* the Dutch theologian Dr. A. Hulsbosch, O.S.A. informs us that Evolution Theory must be fully embraced by the Church:

> The fact that evolution has occurred can no longer be denied. . . . We can no longer deny that, on the biological side, man originates in the animal kingdom.[8]
>
> We have transferred from a static, scientific image of the world to an evolutionary one, and the danger exists that preaching is progressively losing contact with contemporary people. The evolutionary image has inherited all the advantages of its predecessor. On this

---

6. *Ibid.*, p. 55. Emphasis added.
7. Leon Le Grand, "Philberth Under Fire," *The Message* (Vic, Australia: Centre For Peace, June, 1995).
8. Hulsbosch, *God's Creation,* p. vii.

account theology will perform a great task if it fully assimilates into its teaching evolution and all the connected facets of the cosmic reality.[9]

Hulsbosch proposed a radical change in Christian theology, one which could never be accepted by the Church, in a misguided attempt to make that theology relevant. In his view, each person is not completed while upon Earth, and the final stage of one's personal evolutionary journey is only reached upon death. Supposedly, the final evolutionary stage for those who have been good is the enjoyment of the vision of God.

Incredibly, he rejects the historical existence of the Garden of Eden and the Church's teaching of only one pair of first parents. In place of the traditional doctrine of Original Sin, he outlines a new explanation for the origin of sin in the world:

> On the former way of seeing it, sin appears as the collapse of a completed work; but in the light of evolution, sin is revealed as the refusal of man to subject himself to God's creative will.[10]
>
> Original sin is the powerlessness, arising from nature, of man in his uncompletedness as creature to reach his freedom and to realize the desire to see God, *insofar* as this impotence is put into the context of a sinful world.[11]

Before Hulsbosch's proposal could be taken seriously, difficult questions would have to be answered. Does he intend us to believe that sin is an inherited animal characteristic? What is the fate of those who have not led good lives in this world? (In his book, it appears that they completely cease to exist upon death.) Why is there no reference to Satan and Hell in the book—are we to believe they do not exist?

The theological views of Hulsbosch are clearly influenced by his belief in Evolution as a proved fact. He even dedicated his book to the infamous Fr. Pierre Teilhard de Chardin, S.J., whose views on Evolution cannot be harmonized with the teachings of the Magisterium, but have been disastrous within modern Catholicism.

---

9. *Ibid.*, p. xi.
10. *Ibid.*, p. 45.
11. *Ibid.*, p. 55.

The Errors of Teilhard de Chardin

Example 4: Fr. Pierre Teilhard de Chardin was a French priest who appeared to have devoted his whole life actively to promoting Evolutionary Theory, apparently to supplant Christianity. His intense preoccupation with it led him to propose an explanation of reality which is clearly incompatible with the teachings of the Catholic Church. But Teilhard should have been quite familiar with the Origins teaching of Vatican Council I that

> There is one, true, living God, Creator and Lord of heaven and earth, omnipotent, eternal, immense, incomprehensible, infinite in intellect and will, and in every perfection; who, although He is one, singular, altogether simple and unchangeable spiritual substance, must be proclaimed distinct in reality and essence from the world [therefore not immanent]; most blessed in Himself and of Himself, and ineffably most high above all things which are or can be conceived outside Himself. . . . This sole true God, by His own goodness and "omnipotent power," not to increase His own beatitude, and not to add to, but to manifest His perfection by the blessings which He bestows on creatures, with most free volition, "immediately from the beginning of time fashioned each creature out of nothing, spiritual and corporeal, namely angelic and mundane; and then the human creation, common as it were, composed of both spirit and body."[12]

According to Professor Wolfgang Smith, Teilhard was convinced that God could only create "evolutively,"[13] and he thus believed the Church was wedded to an outmoded scientific outlook. Teilhard believed it was vital to revise doctrine and adapt theology to harmonize with Evolution. In contrast to the objective view of reality, which holds that the Universe is comprised of real, distinct things, he proposed a very vague concept of "unification." Everything apparently is evolving through supposed *convergence* toward a future goal called Omega.

For Teilhard, God is a transcendent reality, but not in the sense as usually understood in traditional Christianity. Rather, it seems that God is secondary to the primary reality, which is the evolutionary unfolding

---

12. Vatican Council I, Denzinger 1782-3.
13. See Wolfgang Smith, *Teilhardism and the New Religion* (Rockford, IL: TAN Books and Publishers, Inc., 1988), p. 14. See also David H. Lane, *The Phenomenon of Teilhard: Prophet for a New Age* (Macon, GA: Mercer University Press, 1996). Both books discuss comprehensively the ideas of Teilhard and the influence of his pantheistic beliefs.

of all things. "God" somehow inserted Himself into this evolving force, and Christ is now the force drawing everything toward Omega. In such a concept God is clearly not the transcendent Creator of all things:

> As early as in St. Paul and St. John we read that to create, to fulfill and to purify the world is, for God, to unify it by uniting it organically with himself. How does he unify it? By partially immersing himself in things, by becoming "element," and then, from this point of vantage in the heart of matter, assuming the control and leadership of what we now call evolution. Christ, principle of universal vitality because sprung up as man among men, put himself in the position (maintained ever since) to subdue under himself, to purify, to direct and superanimate the general ascent of consciousness into which he inserted himself.[14]

He envisaged a synthesis of the Christian "God up above" with the Marxist "goal up ahead," and in so doing raised the question whether God is a still-evolving entity. Not surprisingly, his works were, and still are, rejected by the teaching Magisterium. He was prohibited by ecclesiastical fiat from publishing his theories during his lifetime and on June 30, 1962, the Holy Office issued a Monitum (or warning) against the writings of Teilhard on the grounds that they contained ambiguities and doctrinal errors. (This warning is still current, having been reaffirmed again on July 20, 1981.[15]) But Teilhard was too shrewd to leave the Church or get excommunicated, and was effective in subversion from within.

Since his ideas have been the center of so much controversy within the Catholic Church, it may be worthwhile to consider some extracts from his most definitive work, *The Phenomenon of Man*. In this book (published by others after his death in 1955), which he claims is purely a scientific treatise, he outlines all of the standard evolutionary "facts" and simply glosses over the difficult questions.

For Teilhard, the origin of the Earth was purely accidental:

> Some thousands of millions of years ago, not, it would appear, by a regular process of astral evolution, but as the result of some unbelievable accident (a brush with another star? an internal upheaval?) a fragment of matter composed of particularly stable atoms was

---

14. Pierre Teilhard de Chardin, *The Phenomenon of Man* (London: William Collins Sons & Co., Ltd. and Fount Paperbacks, 1980), p. 322.
15. See *l'Osservatore Romano*, July 20, 1981.

CHRISTIANITY

detached from the surface of the sun. Without breaking the bonds attaching it to the rest, and just at the right distance from the mother-star to receive a moderate radiation, this fragment began to condense, to roll itself up, to take shape. Containing within its globe and orbit the future of man, another heavenly body—a planet this time—had been born.[16]

The Earth was probably born by accident; but, in accordance with one of the most general laws of evolution, scarcely had this accident happened than it was immediately made use of and recast into something naturally directed.[17]

The origin and reproduction of cells provides no problem to Teilhard:

If their spontaneous generation took place only once in the whole of time, it was apparently because the original formation of the protoplasm was bound up with a state which the general chemistry of the earth passed through only once.[18]

At first sight reproduction appears as a simple process thought up by nature ... what was at first a happy accident or means of survival, is promptly transformed and used as an instrument of progress and conquest.[19]

For the evolution of man's consciousness, he coined the novel but false concept of *noogenesis*, and equated Evolution with truth:

... the engendering and subsequent development of the mind, in one word *noogenesis*. When for the first time in a living creature instinct perceived itself in its own mirror, the whole world took a pace forward.[20] Man discovers *he is nothing else than evolution become conscious of itself*, to borrow Julian Huxley's striking expression.[21]

Is evolution a theory, a system or a hypothesis? It is much more: it is a general condition to which all theories, all hypotheses, all systems must bow and which they must satisfy henceforward if they are to be thinkable and true. Evolution is a light illuminating all facts, a curve that all lines must follow.[22]

---

16. de Chardin, *The Phenomenon of Man,* p. 73.
17. *Ibid.,* p. 80.
18. *Ibid.,* p. 162.
19. *Ibid.,* p. 115.
20. *Ibid.,* p. 73.
21. *Ibid.,* p. 243. Emphasis in original.
22. *Ibid.,* p. 241.

Since Teilhard was a Jesuit priest, it is incredible that nowhere in the book does he discuss the reality of sin. The appendix on evil has no mention of Satan and Hell, and the word Creation does not even rate inclusion in the index. The basic flaw in Teilhard's ideas has been pointed out by Fr. G. H. Duggan, S.M. in *Teilhardism and the Faith*:

> Teilhard's fundamental error was to seek for something more elementary than *being* as the basis of his metaphysics. He thought he had found it in the concept of unification, but he was mistaken. It is true that the reality of created things can, in a certain sense, be accounted for by a process of unification, since they are composite and so may be said to result from the conjunction of their components. Similarly, since they are oriented towards an end that is distinct from themselves, and their attainment of their full perfection consists in their union with this end, their development could be described also as a process of unification. But created being is composite and oriented towards an end distinct from itself, not in so far as it is being, but in so far as it is created. And what is true of created being *qua* created is not true of being as such. "Being is being" or, more concretely, "Beings exist." This is the primordial truth attained by metaphysical intuition, and any attempt to find a more fundamental truth is foredoomed to failure. Reality must exist before any kind of union or unification is possible—*esse* ["to be"] is metaphysically prior to *uniri* ["to be unified"]; and cannot be either identical with it or posterior to it.[23]

Regarding the integrity of Teilhard, it is noteworthy that, according to the prominent evolutionary scientist, Stephen Jay Gould, he was almost certainly a co-conspirator in the Piltdown Man hoax. This hoax was found to be a portion of a human cranium combined with a piece of the lower jaw of an orangutan. The bones were stained with chemicals to give the appearance of great age, and the orangutan teeth were filed to resemble human wear and match the human teeth in the upper jaw. The fraud was not uncovered for 40 years, during which time Piltdown Man was widely cited as proof of human Evolution.

On the other hand, John Evangelist Walsh, in *Unraveling Piltdown: The Science Fraud of the Century and Its Solution*, contends that Charles Dawson was the sole perpetrator of the hoax and depicts Teilhard as a naïve victim of Dawson. Teilhard made a major discovery at Piltdown in 1913—a canine tooth of the lower jaw, apish in appear-

---

23. G. H. Duggan, *Teilhardism and the Faith* (Cork: The Mercier Press, 1968), p. 33.

ance but worn in a human fashion—and this could have been "planted" by Dawson.

Against this, Gould, in a lengthy 1980 analysis,[24] raises a disturbing argument—that Teilhard must have had inside knowledge of the affair. He argues that Piltdown Man should have been cited by Teilhard as outstanding proof—virtually the only proof—for his theme of multiple, parallel lineages within human Evolution. And yet, quite uncharacteristically for such a passionate evolutionist, in all of his writings Teilhard remains silent about Piltdown Man, except for a short article in 1920. Apart from this article, Gould was only able to find half a dozen small references to Piltdown Man, none even as long as one sentence, in the whole twenty-three volumes of Teilhard's complete works. Gould also records his outrage that the Paris edition of Teilhard's works had expurgated all Piltdown references without so stating.

Teilhard's astonishing silence about Piltdown Man tends to confirm that he had intimate knowledge of the hoax. Gould believes that Teilhard knew all about it and tried to indicate in a veiled way in the 1920 article that Piltdown Man was phony. Teilhard's strange reaction to the exposure of the hoax (noted also by Walsh), especially in a series of letters written to Kenneth Oakley (who with others had exposed the hoax in 1953), only served to draw suspicion upon his own role in the affair. Gould lets Teilhard off quite charitably by suggesting that it may have been an elaborate joke which got out of hand. He notes, however, that distinguished careers were ruined by the hoax, and it is not good enough simply to dismiss the matter as of little consequence.

That such a prominent evolutionary scientist as Prof. Gould should come to this conclusion is extremely embarrassing for supporters of Teilhard. His criticism is reminiscent of that of another well-known scientist, Nobel Prize-winner Sir Peter Medawar, who declared that Teilhard's "visionary" pseudo-scientific ideas were "pious bunk."

What was it that made Teilhard so captivating to so many Catholics, both clergy and laity? Was there not something amiss in the scholars' understanding of Origins over the past 150 years, that his ideas could be so widely accepted as very learned? No doubt, his appeal was precisely because his ideas *appeared* to be very learned, in an era when all the facts against Evolution had not yet been fully

---

24. Stephen Jay Gould, "The Piltdown Conspiracy," *Natural History*, Volume 89, No. 8, pp. 8-28, American Museum of Natural History, August, 1980.

marshalled. Writing about the devastating criticism made by Medawar of Teilhard's use of ambiguity, Wolfgang Smith shows how the *illusion of Evolution* was employed by Teilhard:

> And there is something else that very much needs to be pointed out: in what appears to be his central criticism (i.e., that "Teilhard habitually and systematically cheats with words"), Medawar has no doubt hit the nail on the head. The word "cheat" is of course very strong, and should not be interpreted in its prime sense, which is "to deceive by trickery, to defraud, to swindle"; it should be understood, rather, in its milder sense, which is "to fool" or "to beguile," without any implication of ill intent. What concerns us, in any case, is the fact that a certain systematic misuse of language is to be found throughout the writings of Teilhard de Chardin, and that this does indeed tend to fool and beguile the reader. Medawar has put it very well when he observes that "It is the style that creates the illusion of content." Precisely; and the single most effective and most frequent applied "device" is the abuse of metaphor.[25]

Dietrich Von Hildebrand, the outstanding German scholar (as well as courageous and outspoken opponent of Hitler) and author of the spiritual treasure *Transformation in Christ*, pinpointed an important reason for the fascination with Teilhard:

> Many people are impressed by a thinker who constructs a new world out of his own mind, in which everything is interconnected and "explained." They consider such conceptions the most eminent feat of the human mind. Accordingly, they praise Teilhard as a great synthetic thinker. In truth, however, the measure of a thinker's greatness is the extent to which he has grasped reality in its plenitude and depth and in its hierarchical structure. If *this* measure is applied to Teilhard, he obviously cannot be considered a great thinker.[26]
>
> Teilhard's thought is hopelessly at odds with Christianity. Christian revelation presupposes certain basic natural facts, such as the existence of objective truth, the spiritual reality of an individual person, the radical difference between spirit and matter, the difference between body and soul, the unalterable objectivity of moral good and evil, freedom of the will, the immortality of the soul, and, of course, the existence of a personal God. *Teilhard's approach to all of these*

---

25. Wolfgang Smith, *Teilhardism and the New Religion*, p. 57.
26. Dietrich Von Hildebrand, *Trojan Horse in the City of God: The Catholic Crisis Explained* (Manchester, NH: Sophia Institute Press, 1993), p. 302.

> *questions reveals an unbridgeable chasm between his theology fiction and Christian revelation.*[27]
>
> Teilhard's Christ is no longer Jesus, the God-man, the epiphany of God, the Redeemer. . . . In his basic conception of the world, which does not provide for original sin in the sense the Church gives to this term, there is no place for the Jesus Christ of the Gospels; for if there is no original sin, the redemption of man through Christ loses its inner meaning.[28]

The fact that Teilhard's pseudo-scientific theological fantasy was so captivating for so long (he is still held in high regard by some Catholics who otherwise would be regarded as "conservative" about doctrinal beliefs), and not readily grasped by most as obvious heresy, suggests that a significant degree of confusion concerning Origins has prevailed within Catholic institutions during the last two centuries.

Whether such things as rationalist Higher Criticism and the quest for the historical Christ came before the impact made by belief in uniformitarianism and Evolution beliefs or vice versa, is immaterial. The reality is that the biblical distortions imposed upon the *Genesis* account since the mid-19th century helped greatly to facilitate the Modernist revolution which wreaked untold damage within the Church.

The attempt by modern dissenting theologians to revise even the notion of the Magisterium itself, in quest of making the Church relevant in the modern world, should come as no surprise.

If doubts about the historical credibility of Scripture passages are seen as acceptable in modern biblical scholarship, then why not also allow dissent from teachings proclaimed by the Magisterium? Why not even revise the concept of the Magisterium itself, away from the commission given it by Jesus Christ?

Modernism

Another strong influence of Evolution Theory can be seen in the heresy of Modernism or, as better known today, *neo*Modernism. (See Appendix B for a brief look at Modernism.) This belief system, with its theories of immanence, subjectivism and historical criticism of Scripture, embraces belief in Evolution as a fact central to all existence.

---

27. *Ibid.*, p. 284.
28. *Ibid.*, p. 286.

Modernism holds, along with Marxism and Secular Humanism, that religion itself is only a product of man's evolutionary consciousness. Thus, Modernism makes little sense without Evolution as an integral part of its package. As one would expect, this heresy is quite incompatible with Catholic doctrine. Early in the 20th century, Pope St. Pius X (1903-1914) pointed out the insidious nature of Modernism when he described its central proposition:

> [Modernism declares that] the religious sentiment, which through the agency of vital immanence emerges from the lurking-places of the subconscious, is the germ of all religion, and the explanation of everything that has been or ever will be in any religion. This sentiment, which was at first only rudimentary and almost formless, gradually matured, under the influence of that mysterious principle from which it originated, with the progress of human life, of which, as has been said, it is a form. This then, is the origin of all religion, even supernatural religion, it is only a development of this religious sentiment. Nor is the Catholic religion an exception, it is quite on the level with the rest; for it was engendered, by the process of vital immanence, in the consciousness of Christ, who was a man of the choicest nature, whose like has never been, nor will be . . .
>
> There are many Catholics, yes, and priests too, who say these things openly; and they boast that they are going to reform the Church by these ravings. . . . We have reached the point when it is affirmed that our most holy religion, in the man Christ as in us, emanated from nature spontaneously and entirely. Than this there is surely nothing more destructive of the whole supernatural order.[29]

Despite the efforts of Pope St. Pius X, the cancer of Modernism grew quietly but steadily in the first half of the 20th century and then burst forth with astonishing impact at the time of Vatican Council II.

As well-documented in many books, a fairly sudden and open revolution began to convulse the modern Catholic Church, beginning in the 1960s. Around the world in many countries, packed Sunday Masses began to diminish; hours spent waiting to enter the confessional dwindled to only a few minutes; and open dissent by clergy, teachers and laity became the norm in many places.

Parishes were split down the middle wherever dissenting clergy (themselves mostly trusting devotees of well-known dissenting theolo-

---

29. Pope St. Pius X, *Pascendi Domenici Gregis* (Encyclical "On Modernism"), par. 10.

gians) chose to deny, or even mock, previously cherished Catholic beliefs such as prohibition of contraception. The unsuspecting laity, who for the most part were poorly informed about the basis of their beliefs, were easy targets and were caught unprepared for such profound and unexpected challenges to their faith.

To those in the pews, the apparently doctrinally united Catholic Church had somehow lost cohesion, and confusion had set in; and this was baffling to most. Even the scholarly dissenters themselves must have been astonished at the scale and speed of their re-conversions.

In North America alone, some 10,000 priests and 50,000 nuns left their vocations in the ten years between 1966 and 1976, marking an unprecedented collapse of faith within the Catholic Church. The impact made by Modernism is not hard to discern, and neither is the related Origins confusion. Sadly, many nuns were won over to radical feminist theology (some even became strongly pro-abortion) and experienced a revolutionary conversion away from Catholicism:

> Usually summed up today as "the spirit of Vatican II," this body of neo-modernist opinion has penetrated deeply into the Church. It represents the substance of the new theological "liberal consensus," which is shared even by some prelates. Liberation theology and process theology are among its expressions. But nowhere has its impact been greater than among Catholic women religious, where its poisonous spiritual fruits include rabid feminism, goddess-Wicca and New Age thought. When women religious understood themselves as brides of Christ, their consecration expressed not what they *did* but what they *were*. When they lost their faith, they were convulsed by a crisis of identity so profound that it threatens to destroy the very possibility of consecrated life. *Their rapid disintegration has no precedent in the entire history of the Church.* . . . Many swallowed the neo-modernist reinterpretation of Scripture and catechetics, the new morality and new psychology, already prevailing among avant-garde professors.[30]

That Modernism could only result in great harm to the Church was pointed out by Cardinal Siri, Archbishop of Genoa. Again, the impact of evolutionary beliefs is evident :

---

30. Donna Steichen, *Ungodly Rage: The Hidden Face of Catholic Feminism* (San Francisco: Ignatius Press, 1992), p. 258. Emphasis added.

> Modernism, now as at the beginning of the century, with new designations and shades of meaning, first implicitly and then explicitly, damages the principle of Revelation which is replaced by the elaborations of a "religious sense" in the subconscious. Today almost more than in its beginning, it pushes towards a quasi "transcendental" agnosticism and a "dogmatic evolutionism," in order to destroy every notion of objectivity in Revelation and in acquired knowledge.[31]

The Modernist cancer still continues to flourish, with strong emphasis on subjective feelings and experiential catechetics, and it is hard to see how the current dissent can be effectively countered until the credibility of the *Genesis* account is restored. Meanwhile, the number of practicing Catholics continues to dwindle, and in many Catholic schools around the world large numbers of the young are being denied their right to knowledge of truth about doctrine—in Catholic schools of all places, which should ensure that Catholic doctrine is taught!

Despite the Modernist revolution, Pope St. Pius X's condemnation of Modernist theology is still relevant:

> Methods and doctrines brimming over with errors, made not for edification but for destruction, nor for the formation of Catholics but for plunging of Catholics into heresy; methods and doctrines that would be fatal to any religion.[32]

The body of St. Pius X has not suffered disintegration (the body was not embalmed) and lies today in a glass coffin in St. Peter's Basilica for all to see. The miraculous preservation of his body shows that he must have been very pleasing to God to be so honored. Perhaps it is a sign from God that all forms of this heresy are to be fully comprehended and utterly rejected, and traditional teachings proclaimed instead. A renewed interest in Special Creation—to re-establish the idea of objective truth and counter the modern fascination with subjective consciousness—is of crucial importance in trying to overturn the grip of Modernism.

---

31. Cardinal Joseph Siri, *Gethsemane: Reflection on the Contemporary Theological Movement* (Chicago: Franciscan Herald Press, 1981), p. 49.
32. Pope Pius X, *Pascendi Domenici Gregis*, par. 37.

Subjective Consciousness and Dissenting Theology

Dissenting theology, which is so opposed to traditional Christian beliefs about Creation, tends to place great emphasis on subjective views. The quest to demythologize Scripture (i.e., to remove all traces of supernatural concepts) is accompanied by belief that genuine reality is an ongoing, restless process of change, where man is constantly challenged to find his true "self."

Of particular concern is the effect that this mentality can have on "ordinary" people. It engenders an attitude of mind which, over a period of time, can lead the individual to a position where basic Christian teachings are no longer seen as being very important. As this "drift" in consciousness gradually takes place, the individual becomes convinced that love is the only reality to pursue and tends to pay less attention to other important aspects of the Christian religion. Thus, he can become confused about relativist attitudes and lose sight of the central truths of Christianity.

Consider, for example, the effect that exposure to the errors of "situation ethics" theology can have on a person who has been misled by others who themselves no longer believe in absolute principles:

> Situational ethicists have made a positive contribution in alerting us once more to the dangers of legalism and Pharisaism. But they have also departed from the path of traditional Christian morality in several ways, which may be summarized under four heads. First, situational ethics denies the binding obligatory power of negative moral absolutes in all moral acts. This denial of intrinsic evil leaves only extrinsic evil, circumstantially determined. Second, situational ethics assumes unimpaired human judgment and claims that love alone can result in correct moral decisions, without any need of negative moral absolutes, or moral law. Third, situational ethics, in making intentions, end, and purpose the sole criteria of moral legitimacy, rejects the traditional Christian principle that no end, no matter how worthy, can ever justify evil means. Fourth, situational ethics, by defining love in terms of subjective intent, falls prey to relativism and thus ends in moral chaos.[33]

If meaning and values were only feelings which emerged from within the individual, rather than qualities emanating from an authority

---

33. Hammes, *Human Destiny*, p. 139.

higher and greater than man, then subjective desires could come to dominate the individual.

We know that human beings are imperfect creatures, quite capable of making mistakes. If a person is intent on creating his own truth through personal experiences, then subjective judgments could easily be shaped by self-centered desires. On the other hand, recognition of a moral authority to which the individual is responsible provides an important challenge for the individual. The ever-present tendency to pride within all human beings is also an important reason why respect for moral authority is necessary: to help learn humility.

There are many modern Christians who feel disposed to attack traditional Christian teachings, and perhaps there is a certain excitement associated with doing so. But it appears that the underlying motivation is that of changing Christianity to conform with what they perceive to be the mentality of the modern world.

It is not possible conveniently to group all the various viewpoints of those who attack traditional Christian teachings into one category. Nevertheless, this author believes a general comment has validity: the Modernist, Humanist ideas prevalent in dissenting theology are really modern-day heresies. Perhaps they should be considered as varieties of the heresy of idolatry, where man has been elevated to God-like status.

The great upheaval which has occurred within Catholicism in recent times has clearly had much to do with the acceptance by Christian scholars of both uniformitarianism and Evolution Theory.

On the one hand, the magnitude of man's technical achievements has made many atheistic and agnostic scientists believe that man really is the master of this planet, and there is no doubt that science is continuing to discover new horizons of man's knowledge. On the other hand, the contribution of many religious "experts," with their dissection of Scripture and their explaining away of things they find disquieting or unacceptable to the modern mind, has only brought more confusion among Christians.

Is it not a paradox that so many of those who seem disposed to query almost everything in Scripture seem loath to query anything of Evolution? In this confused world, if Evolution is taken for granted as proved, is it any wonder that theories can be seriously proposed and find acceptance which proclaim that "God is dead" or that He is only a still-evolving entity, that Satan and Hell do not exist, and that the Bible contains much that is only myth? Fr. Paul Crane, S.J. contends that acceptance of Evolution as fact has helped bring about the rejection of the doctrine of Original Sin:

> Those who see the Church's primary task as that of assisting the self-completion of man are forced to disregard or dismiss altogether the doctrine of Original Sin. They have to do this because what the doctrine says so rightly is that man cannot achieve perfection on his own terms, through his own efforts alone, unassisted by the Grace of God. The doctrine as it stands is, therefore, a repudiation of those Christian secularists who see the Church as concerned, necessarily and primarily, with promoting the self-completion of man. . . .
>
> He has proceeded, in consequence, to dismiss it; substituting in its place something he refers to as "the difficulties of the human condition." Or to hold up as true the cruder versions of the theory of evolution, which leave no room for Grace and, therefore, no room for man's fall from it; which means no room for Original Sin. Anything but admit it. This is the method of the autonomous Christian and the admittance is increasingly widespread, no matter how outworn and suspect Darwinism has now become, both outside the Church circles as well as within them.[34]

Once a person loses sight of the doctrine of Original Sin, he loses sight of the fact that he needs God's grace in order to overcome his inherited fallen human nature. This problem is then compounded by loss of grace. Clearly, the "social" gospel of today's dissenting clergy places little emphasis on the supernatural or the doctrine of Original Sin, and Christian endeavor can easily translate into little more than emulation of Christ the Man as the model for social reform. The emphasis on consensus, community Christianity, is not hard to comprehend when so many Christians doubt the existence of Heaven and Hell and are loath to acknowledge the miracles worked by Christ.

While compassion and charity toward the underprivileged are noble values, unless God is seen as an authority greater than man, whose revealed laws are to be taken seriously, then the subjective conscience can become the rationalizing influence for one's motives in seeking social justice. Objective truth can sometimes be elusive, and it is possible for the conscience to be ill-informed and mistaken—and one's actions can thereby unintentionally bring about harm, rather than good.

If, through the influence of Evolution Theory, one loses sight of the fact that the basic dignity of each human being is derived from the possession of an immortal, rational soul, it is quite possible to support

---

34. Paul Crane, S.J., *Christian Order* (London, March, 1982), p. 215.

changes in society (e.g., abortion law "reform") which do not respect the sanctity of life and rights of human beings, both born and unborn.

Being central to a number of belief systems, Evolution Theory has been "winning" on a number of fronts in the attack upon Christianity.

The great rejection of Communism in the USSR and Eastern Europe brought welcome relief from totalitarian power, but only after many decades of terror. The cross borne throughout the nightmarish decades of Communism has been very real in terms of imprisonment, lost jobs, lack of housing, loss of educational opportunities and having one's children taken away. To be a Christian in Communist countries has been indeed to carry a very heavy cross, and neo-Modernist theories are useless to people living in such conditions.

While all this has been happening, the dissenting Humanist clergy in the West have offered, in effect, a form of Christianity with a very vague supernatural dimension. While they do indeed stress the need for personal sacrifice, their concept seems primarily motivated by worldly considerations. Social justice will always be a pressing concern for Christians, but the motivation should primarily be that of seeking to love and honor God. Ideally, the desire for social justice ought to be a natural consequence resulting from a heightened view of reality and compassionate concern for the plight of others.

On the one hand, therefore, belief in Evolution Theory made possible the Communist belief system which has tormented Christians of modern times no less than the martyrs of old, and yet it also made possible the drift in consciousness of Christians in the West away from a clear grasp of unchangeable Christian teachings.

One unfortunate effect of the evolutionary tone of things has been a tendency for many Christians, even "conservative" believers, to be wary of believing that instances of divine intervention are recorded in Scripture. There is a strange tendency to want to ascribe these events to purely natural forces coinciding at the time. (Paradoxically, innumerable instances of divine intervention, implicit in Theistic Evolution concepts of Origins, tend to be accepted without question.)

It is possible that God may have used natural physical laws in some instances (e.g., the destruction of Sodom and Gomorrha and the moving star seen by the three Wise Men at the birth of Christ), but other events seem clearly beyond such natural forces. What explanation, other than supernatural phenomena, can be given for such important incidents as the rolling away of the rock by angels from Christ's tomb, Peter's being led out of prison by an angel, water changed into wine, or the miracles of the loaves and fishes?

A further general observation which is of importance in considering the effect of Evolution beliefs upon Christian thinking is the fascination displayed by some individuals for "modern" morality. To them, it seems that whatever is perceived as acceptable to the modern enlightened secular mentality tends also to be regarded as automatically better than standards of morality held dear in the past, as if truth can be determined solely by majority opinion. The dissenting theology which seeks to modify the revelation of God to suit the mentality of modern secular society only results in further obscuring of genuine truth. It cannot succeed in making this revised form of Christianity relevant to society.

In the light of all the turmoil that has arisen over the Origins debate and the alleged irrelevance of creationist ideas, a certain irony can now be seen. As Evolution theories encounter ever more difficulties and the case against Evolution becomes more confirmed and better understood, it can be argued that those who support Special Creation are actually in the vanguard of modern thought.

Evolution Theory is not being conveniently blamed here for all the ills of Christianity, but the price paid for flirtation with the myth of Evolution has nevertheless been too high for the Church founded by Jesus Christ. It is a matter of giving due recognition to the enormous harm which has been done by its influence.

—PART V—

THE SEARCH FOR MEANING IN LIFE

— Chapter 21 —

THE SEARCH FOR MEANING IN LIFE

Mankind lives in an imperfect world where such factors as ignorance, confusion, poverty and the abuse of power can predominate and trap the individual in an existence where feelings of unhappiness and loneliness are hard to escape. While most people can cope with the problems of life, some are not so fortunate. Sometimes the frustration and anxiety of the person are let loose in the form of violence, and escape from reality is often sought by taking drugs. There are, of course, many factors involved in all this, and it seems that many of the problems in society are locked into an apparently endless chain of cause and effect. And so human beings will continue to seek a coherent explanation of reality and to ask questions such as, "Who am I?" "Where am I going?" "What meaning is there in my life?"

The question of Origins is relevant to the search for meaning in life, for it is central to the most fundamental questions concerning reality. Are human beings only the product of forces of nature operating randomly over an immense period of time, with no real meaning and no future, or are they the creation of an omnipotent, loving God?

In seeking for a personal philosophy of life, modern man is confronted with a wide spectrum of ideas, some of which are religious in orientation and others purely naturalistic. As societies are becoming more pluralistic, it is becoming more difficult to find common standards by which truth may be safeguarded. Indeed, the very notion of truth has itself been called into question: Is truth an objective reality or a subjective reality? Does it depend on majority vote, or is it something beyond the whim of man?

The depth of the thirst for answers about one's existence can be observed in the increasing attraction of cults. But some of these cults offer explanations which are bizarre in the extreme and lacking in meaningful answers. It is worthwhile, therefore, to probe some of the problems which can be encountered in the search for meaning in life.

It will be argued that Evolution Theory tends to increase the level of confusion in society, whereas belief in Special Creation can have an enlightening effect.

—Chapter 22—
RELIGION AND MEANING

The Importance of Meaning

Mankind has made spectacular technological progress, but the same cannot be said for his ability to live in harmony with his fellow man. Many diseases have been conquered, machines have taken over many monotonous or extremely tiring tasks, the moon and planets are no longer out of reach, and people tend to live longer. In spite of all this, however, the disposition of individuals is still clearly influenced by the effects of Original Sin.

Problems such as greed, selfishness and laziness are a fact of life and tend to place great strain on political, social and economic systems. The situation is aggravated by the population drift from country areas to the cities, a phenomenon seen all around the world. All too often, city life has become a nightmare of squalor for many in society, who long for real justice. As cities increase in population *while moral values decline*, social disharmony tends to increase well out of proportion.

Together with the sheer complexity of industrialized societies and the problems encountered with "bigness" (e.g., big government, big business, big unions, big bureaucracies) is the effect on the individual of feeling helpless, of being borne down, of "not belonging"—in other words, the problem of personal alienation.

As mentioned earlier, one of the features of modernity is the tendency toward pluralization. Within democratic societies, all sorts of religious beliefs, philosophies and lifestyles compete in very uneasy co-existence. In consequence, the revolution in moral standards, with its disintegrating effect upon family bonds, has brought added tension to the ordering of society.

A clearly observable phenomenon has developed in modern societies. The individual tends to lead two lives—the "public" life and the "private" life. And people tend to find meaning and contentment only in the private world (e.g., in having close friends, a stable family milieu, the religious fraternity or club to which they belong, and other private interests that help individuals to cope with the often overwhelming forces encountered in "public" life). Needless to say, if one's private world begins to suffer great stress, then the individual will experience even more alienation.

Arguing that pluralization has a secularizing effect upon the role of religious beliefs, the authors of *The Homeless Mind* have noted the historical importance of religion in providing the individual with a coherent world-and-life view which enabled people in the past to grasp readily a sense of meaning in life:

> Through most of empirically available history, religion has played a vital role in providing the overarching canopy of symbols for the meaningful integration of society. The various meanings, values and beliefs operative in a society were ultimately "held together" in a comprehensive interpretation of reality that related human life to the cosmos as a whole. Indeed, from a sociological and social-psychological point of view, religion can be defined as a cognitive and normative structure that makes it possible for man to feel "at home" in the Universe. This age-old function of religion is seriously threatened by pluralization . . . the plausibility of religious definitions of reality is threatened from within, that is, within the subjective consciousness of the individual. . . . Increasingly, as pluralization develops, the individual is forced to take cognizance of others who do not believe what he believes and whose life is dominated by different, sometimes by contradictory, meanings, values, and beliefs. As a result, quite apart from other factors tending in the same direction, *pluralization has a secularizing effect*. That is, pluralization weakens the hold of religion on society and on the individual.[1]

But mankind cannot dispense with religion if genuine meaning is to be found. Helping troubled people to find meaning in life in a confused, alienating world is a great challenge today for therapists, and the question of Origins is of vital importance:

> The interpretations of the Adam and Eve story has both theoretical and practical implications for psychology. Theoretically speaking, with regard to a psychological model of man, is he to be considered

---

1. Peter L. Berger, Brigitte Berger and Hansfried Kellner, *The Homeless Mind: Modernization and Consciousness* (Middlesex, England: Penguin Books Ltd., 1974) p. 75. Emphasis in original.

a child of evolution or a child of God? Is he created in the image of God or is he animal alone? Is he ordained to final termination in death, or to a glorified individual immortality? Practically speaking, the interpretations of the Adam and Eve story bear on the deepest existential problems encountered in counseling and psychotherapy. Carl Jung once declared that the basic problems of almost all of his patients over thirty-five years of age was finding a religious outlook on life.... Our age is characterized by anxiety, meaninglessness, and a search for enduring values ... problems relating basically to the human condition. *Whether or not one can give satisfactory and authentic answers to today's questions on the meaning of guilt, suffering, and death rests ultimately on one's interpretation of the Fall of man.*[2]

Not surprisingly, the influence of evolutionist beliefs can be observed in various value systems offered to modern man. B. F. Skinner, the behavioral psychologist, integrated the theory of biological evolution with the concept of environmental determinism.

He believed in Evolution and proposed that human beings take control of its future course. His concept necessarily involves belief that man's conscious awareness—one's very thought processes—originated in the random, selective environment of nature. In effect, he declared that natural selection is the efficient cause of all progress:

> Before the 19th century the environment was thought of simply as a passive setting in which many different kinds of organisms were born, reproduced themselves, and died. No one saw that the environment was responsible for the fact that there *were* many different kinds (and that fact, significantly enough, was attributed to a creative Mind). The trouble was that the environment acts in an inconspicuous way: it does not push or pull, it *selects*. For thousands of years in the history of human thought the process of natural selection went unseen in spite of its extraordinary importance. When it was eventually discovered, it became, of course, the key to evolutionary theory.[3]

In light of the evolutionary milieu of the last two centuries, Western societies are becoming like the ancient Greeks who had no idea of a Creator. With Christianity suffering greatly from inner turmoil and

---

2. Hammes, *Human Destiny: Exploring Today's Value-Systems,* p. 258. Emphasis added.
3. B. F. Skinner, *Beyond Freedom and Dignity* (Middlesex, England: Penguin Books Ltd., 1979), p. 22.

division, many individuals have turned to Eastern religions such as Buddhism and Hinduism in search of truth and the real "self." In reaching out to these people we need, like St. Paul, to inform them about Creation.

The New Age Movement

There are scores of books which feature the various wares of the *New Age* movement. A vast range of ideas are embraced under this movement, and it provides a socio-religious worldview which is more or less shared by many individuals on various intellectual and cultural levels. Its themes include Earth environmentalism, theosophy, Eastern mysticism, occult spiritualism, immanence and cosmic consciousness. Extreme versions include *neo*pagan beliefs and witchcraft.

The widespread allure of New Age ideas suggests that human beings will always be fascinated with questions about the meaning of life. Clearly, man does have an inner drive to comprehend his location in time and in the cosmos and to find answers for such things as the existence of good and evil. The truth about Origins ultimately is very important to a person's "public" and "private" life.

The appeal of New Age concepts (especially in modernized Western societies, which now give minimal assent to Christian teachings) is that they seem to offer a spiritual dimension giving meaning to human life, against the alternative backdrop of selfish marketplace materialism and disidentifying lifestyles. But New Age concepts are really not new. In reality, they are only newer versions of ancient mystical Eastern concepts of religion which are opposed to Judaism, Christianity and Islam.

The counterfeit "Christ" of the New Age Movement is not the transcendent Creator/Redeemer of mankind. Their "Christ" is an exemplary "Way-shower" to the God within each person's evolutionary consciousness. This idea of the "immanent God within" must be countered with the true story of the omnipotent Creator-God. The very idea of "Christ crucified" does not fully make sense without knowing the story of what brought on the need for Christ to pay the ransom for mankind.

In commenting on the insidious danger of the New Age movement to Christian beliefs, Donna Steichen has described its features. Pantheistic, evolutionist beliefs lie close below the surface:

> As a phenomenon in American society, the movement is new, but its ideas . . . are rooted in an ancient underground search for mystical

illumination: in the Hermetic tradition, gnosticism, alchemy and the Kabbalah. It re-emerged down the centuries . . . in Masonry, Rosicrucianism, the transcendentalist concept of an "Oversoul," the "collective unconscious" of Jungian psychology, Whitehead's process theology and Teilhard de Chardin's vision of spiritual evolution to an Omega Point, summed up by Aldous Huxley as "the perennial philosophy." Today, adherents look expectantly for a "quantum leap" in spiritual evolution when a "critical mass" of the human race comes to recognize "the God within" and realize that "All is One." To hasten the process, they, like their gnostic predecessors, are trying to invent a new religious mythology. They see themselves as prototypes of the higher species that will build a "planetary society" in unspecified ways.[4]

The Jung Cult

One of the most influential thinkers within New Age consciousness is Carl Gustav Jung (1875-1961), whose impact still lingers long after his death. The movement known as Jungism still attracts many followers because of a perceived spirituality. But it is an anti-Christian spirituality which relies upon Naturalistic Evolution and *neo*paganism.

Consider the background of post-Enlightenment, late 19th century Europe to which Jung was exposed: *Fin-de-siècle* ("end-of-century") movements in Germany and Switzerland were tinged with such things as occult revival, mystical eroticism and return to paganism. Many Protestant ministers had been attracted to the writings of David Friedrich Strauss, and Tübingen scholars denied the divinity of Jesus Christ and sought to uncover the "historical" Christ. Many evolutionists had by then abandoned Christian doctrine and attacked organized religion; the concept of uniformitarianism appeared to lend credibility to the idea that human beings and other life forms came about by slow evolutionary transformation over millions of years, all of which directly challenged the *Genesis* account. In addition, opposing schools of *vitalistic* and *mechanistic* notions of Evolution (both incompatible with Christian doctrine) were being proposed.

The son of a Protestant minister, Jung eventually rejected Christianity in 1912 and embraced the polygamous beliefs practiced by Dr. Otto Gross. Jung went on to promote his own cult religion of *rebirth* and *individuation*, and told the story of his own deification in 1925.

---

4. Donna Steichen, *Ungodly Rage*, p. 212.

Most importantly, Jung was profoundly influenced by the writings of Ernst Haeckel, a German professor of zoology and zealous 19th century evolutionist, and especially by Haeckel's so-called *Biogenetic Law*.[5] Haeckel's infamous and erroneous concept of "*ontogeny recapitulates phylogeny*" was shown early in the 1900's to be completely false, and it now has few supporters. Nevertheless, the idea of *recapitulation* has served as an important vehicle in evolutionary mythology and it made a great impact upon Jung personally.

In a frank treatment of Jung's life and ideas, Richard Noll shows how, although Jung's views changed markedly during his life, Evolution beliefs remained central to his various theories about existence and meaning. Spiritualism and evolutionary biology formed the earliest basis of his psychoanalytical theories, which were intended to replace Christianity:

> [In] late 1909 Jung first began to hypothesize that the unconscious mind had a deeper "phylogenetic" or racial layer beyond the memory store of personal experiences, and that it was from this essentially vitalistic biological residue that pre-Christian, pagan, mythological material emerged in dreams, fantasies, and especially psychotic states of mind. Jung's final theories, those of a trans-personal collective unconscious (1916) and its archetypes (1919), marked a transition away from an already tenuous congruence with the biological sciences of the 20th century and instead returned to ideas popular during his grandfather's lifetime—the age of Goethe.
>
> By this time a metaphysical idea that was somewhat implicit in Jung's earlier thought became explicit: namely, that all matter—animate and inanimate—has a kind of "memory." Such ancient ideas, ironically, are what Jung is best known for introducing as modern innovations. Indeed these essentially transcendental concepts are so widely spread in our culture through their connections to psychotherapeutic practice, New Age spirituality, and neo-paganism that they continue to be the subject of innumerable workshops, television shows, bestselling books, and video cassettes, and they form the basis of a brand of psychotherapy with its own trade name: Jungian analysis.[6]

---

5. Known otherwise as *recapitulation theory*, this "law" holds that an organism's embryological development (i.e., its ontogeny) repeats (or recapitulates) the stages of the adult form of its ancestors. Haeckel is known to have falsified some of the pictorial evidence for his theory, and as a result was severely rebuked by his peers.
6. Richard Noll, *The Jung Cult: Origins of a Charismatic Movement* (Princeton, NJ: Princeton University Press, 1994), p. 6.

Wolfgang Smith has pointed out that Jung certainly perceived the psyche in evolutionist terms. According to Jung, the earlier stages of psychic prehistory can be compared to the anatomical prehistory of life forms. But whereas the fossil record is a collection of dead fossils, the earlier stages of psychic prehistory of mankind are still with us in the form of the collective unconscious:

> Moreover, if our individual consciousness has evolved—both in a phylogenetic and an ontogenetic sense—out of the collective unconscious, then this wondrous being is quite literally the parent of us all, and the giver of life. In a word, it began to dawn upon Jung that what he had come upon was nothing less that the numinous source from which all the religious conceptions of mankind have sprung, and to which they ultimately refer.[7]

As for the adverse impact of Evolution Theory upon modern belief systems, it is sobering to reflect upon how Jung finally rationalized away any lingering Christian prohibitions against mistresses:

> What did Jung discover about himself in [Otto] Gross? . . . Perhaps the natural state of humans who were civilized only in the last few thousand years after a million or so of evolution was indeed the primal polygamy of our ancestors . . . only suited for tribal life in a small *Gemeinschaft* of hunters and gatherers . . . this notion of biologically based polygamous impulses from an ancestral past as a major determinant of human social behavior [is] gaining scientific ascendancy in the work of sociobiologists and "evolutionary personality psychologists" in the 1990's.[8]

Jung correctly discerned that the modern crisis of lack of purpose or meaning within individuals is religious in character, but incorrectly sought to supplant the True Religion revealed by the transcendent Creator. Despite their stark contradiction to Catholic doctrine (e.g., to Original Sin), Jungian theory and mystical New Age concepts are still widely influential in Catholic educational institutes, with exaggerated emphasis being made on personality types and the meaning of feelings.

Jung was not the only thinker profoundly influenced by Evolution. Within the psychoanalysis movement which arose later in the 19th cen-

---

7. Wolfgang Smith, *Cosmos & Transcendence*, p. 116.
8. Noll, *The Jung Cult*, p. 158.

tury, evolutionist beliefs were central to theories developed by Sigmund Freud (1856-1939) and his colleagues, disciples and foes.

Freud and Evolution Beliefs

As Wolfgang Smith has also pointed out, in the late 19th century it was only a short step from Darwin to Freud:

> Given that the human species derives from sub-human ancestors, it follows that its mentality, too, has evolved out of a sub-human rudiment: the rational from the non-rational, the self-conscious from the instinctual. Now if that be the case, it is but natural to suppose that the bestial psyche still exists in us, concealed behind or beneath the conscious mentality, as a living vestige of the animal stage. And so we arrive essentially at the Freudian id, the psychic substratum which Freud takes to be "the core of our being."[9]

Among Freud's colleagues, disciples and foes, *belief in recapitulation theory and Lamarckian-acquired characteristics were of crucial importance* in stimulating theories about supposed ancestral behavior now affecting modern human beings.

The ideas of staunch evolutionists (such as Jean-Baptiste Lamarck, Alfred Russel Wallace, Ernst Haeckel, Charles Darwin, George John Romanes and Herbert Spencer) made an enormous impact upon late 19th century society, permeating to all levels and affecting daily beliefs. Darwin was also a major early figure in the growing science of the mind, later developed more fully by Freud and others. His belief that man is descended from lower animals led Darwin to delve into questions such as the evolution of intelligence and the origin of instincts and emotions—hence his 1872 book *The Expression of the Emotions in Man and Animals*, published soon after *The Descent of Man*. Frank Sulloway has shown how much Freud owed to Darwin:

> Darwin's [*M* and *N*] notebooks touch repeatedly upon unconscious mental processes and conflicts; upon psychopathology (including double consciousness, mania, delirium, senility, intoxication, and a variety of other psychosomatic phenomena); upon the psychopathology of everyday life (for example, forgetting and involuntary recall);

---

9. Wolfgang Smith, *Cosmos & Transcendence*, p. 92.

upon dreaming (Darwin records three of his own dreams and subjects them to partial psychological analysis); upon the psychology of love and the phenomena of sexual excitation . . . and upon the evolution of the aesthetic sense, of morality, and of religious belief.[10]

Thus, the influence of belief in Evolution upon the early developing psychoanalysis movement cannot be over-emphasized. In addition to Freud and Jung, many other theorists (such as Wilhelm Fliess, Thomas Clouston, Wilhelm Bolsche, Shobal Clevenger, Cesare Lombroso, Sandor Ferenczi, James J. Atkinson and G. Stanley Hall) developed themes which derived from recapitulation/Lamarckian perspectives. Darwin's influence was all-important:

> It is certainly fitting that the influence of Charles Darwin, the man whose evolutionary writings did so much to encourage young Freud in the study of biology and medicine, should have been so instrumental in turning psychoanalysis into a dynamic, and especially a genetic, psychobiology of mind. Indeed, perhaps nowhere was the impact of Darwin, direct and indirect, more exemplary or fruitful outside of biology proper than within Freudian psychoanalysis. Yet it was not until Freud had freed himself from the quest for a neurophysiological theory of mind that he finally began to reap the full benefits of this Darwinian legacy within psycho-analytic theory. By then—the late 1890s—Darwin's influence upon Freud's scientific generation had become so extensive that Freud himself probably never knew just how much he really owed to this one intellectual source. Darwinian assumptions (1) pervaded the whole nascent discipline of child psychology from which Freud drew, and to which he in turn contributed so much; (2) reinforced the immense importance of sexuality in the contemporary understanding of psychopathology; (3) alerted Freud and others to the manifold potentials of historical reductionism (the use of the past as a key to the present); (4) underlay Freud's fundamental conceptions of infantile erotogenic zones, of human psychosexual stages, and of the archaic nature of the unconscious; and (5) contributed a number of major psychical concepts— like those of fixation and regression—to Freud's overall theory of psycho-pathology. Finally, the simultaneous influence of Lamarckian notions served to convince Freud that psychoanalysis, with its insight into conscious and unconscious psychical adaptation, was itself a cul-

---

10. Frank J. Sulloway, *Freud, Biologist of the Mind: Beyond the Psychoanalytic Legend* (Cambridge, MA: Harvard University Press, 1992), p. 241.

minatory achievement in evolutionary theory. . . . Freud, toward the end of his life, recommended the study of evolution be included in every prospective psychoanalyst's program of training.[11]

Much emphasis was placed in the early psychoanalysis movement upon human behavior being explicable primarily in sexual/Evolution concepts.[12] (Elizabeth Thornton[13] argued that Freud's fascination around 1893 with sexuality and his personality change are also attributable to an unwitting addiction to cocaine, undertaken initially to treat migraine attack.) Indeed, without a naturalistic evolutionary underpinning, it is hard to see how Freudian psychoanalysis could even exist:

> As a committed psychobiologist in his overall approach to the human mind, Freud knew that proximate (that is, psychological and physiological) as well as ultimate (evolutionary) explanations were necessary for a complete theoretical understanding of the subject. *Despite the speculative nature of his excursions into phylogenetic theory, Freud fully believed that some such prehistoric drama had to have occurred if his various psycho-analytic claims about repression, sexuality, and neurosis were to possess universal truth.*[14]

Unfortunately for Freud and others in the psychoanalysis movement, both recapitulation theory and Lamarckian acquired characteristics came to be widely seen as erroneous scientific beliefs and were eventually abandoned by almost everyone, although such notions still linger in Evolution mythology. Freud, however, refused the urging of colleagues to accept this bad news and instead clung thereafter to belief in recapitulation theory and acquired characteristics. This sad situation in old age must have been intensely disappointing to him.

---

11. *Ibid.,* p. 275.
12. Such beliefs were cited to give plausibility to the grotesque ideas of bisexual practices and the so-called *Oedipus complex* (the supposed unfulfilled desire for sexual gratification with one's parent of the other sex, and harking back to primal ancestry killing of father by sons for possession of females in the evolving animal harem).
13. See E. M. Thornton, "O Ye of Little Faith," *Daylight Origins Society Magazine,* Spring, 1996.
14. Sulloway, *Freud, Biologist of the Mind,* p. 391. Emphasis added.

— Chapter 23 —

THE PROBLEM OF EVIL

Any discussion of the meaning of life would be incomplete without consideration of the question of evil. One's view of evil and suffering must influence his attitude on life. The harsher aspects of life can bring sorrow to the individual and cause feelings of dismay. Unless some meaning in suffering can be discerned, a person may be drawn to cynicism or despair. Consider problems which confront us:

Natural disasters or terrible accidents take place, involving much loss of life and human suffering. The individual, or someone very dear to him or her, may suffer from a serious illness or disease, and great pain may be associated with it. Honest mistakes can be made by well-intentioned but imperfect individuals, and yet other people may suffer great harm as an unintended result. Misunderstandings occur among family members or among friends and tend to be exacerbated by ignorance and confusion. People often get manipulated and "used" by others, and the underprivileged in society may suffer loneliness and deprivation through no fault of their own. A loved one may suddenly die, and we are left bewildered and find that words are very inadequate tools with which to express the feelings of the heart.

Problems such as these can bring distress, and one may feel inclined to rage against it and to wonder why. In addition, decent human beings are repelled by the barbarity to which some others descend. Examples of barbarity abound throughout history, and only one example need be cited:

Commenting on the potential dimensions of evil-doing in the context of the nightmare of Marxism-Leninism in Russia, the Russian writer Alexander Solzhenitsyn raised the possibility that evil-doing is much more dreadful than commonly supposed. Speculating on the possibility that some condemned prisoners were fed alive to wild animals in the Petrograd and Odessa city zoos by the CHEKA (Communist Secret Police) between 1918 and 1920, he notes that

Evidently evil-doing also has a threshold magnitude. Yes, a human being hesitates and bobs back and forth between good and evil all his life. He slips, falls back, clambers up, repents, things begin to darken again. But just so long as the threshold of evil-doing is not crossed, the possibility of returning remains, and he himself is still within reach of our hope. But when, through the density of evil actions, the result either of their own extreme degree or of the absoluteness of his power, he suddenly crosses that threshold, he has left humanity behind, and without, perhaps, the possibility of return.[1]

The Reality of Evil

The nature of evil is a mystery which has long puzzled mankind, and various belief systems have proposed different ideas about it:

> Among the world's living religions, Hinduism accepts evil and suffering as part of the human condition, from which one attempts to escape in some future pantheistic unity with Brahma. For Buddhism, suffering is consequent to selfish human desires, which must be suppressed to achieve happiness. Buddha, an atheist, has no religious explanation for the existence of evil in the world. Taoism does not explain the existence of evil and suffering, but rather accepts it and attempts to escape it, as well as world involvement in general. Confucianism, being more a practical philosophy that avoids speculative theology, offers no adequate explanation for evil and suffering. Shinto, immersed in nature deities and primarily a religion of the natural world, has become intermingled with Buddhism and Confucianism. Its solution to evil and suffering lies in self-purification.
>
> Judaism and Christianity believe the devil, a fallen angel, to be evil personified. Original man succumbed to the devil's temptation and was responsible for the present human condition of suffering, guilt, and death. However, all human beings, through personal sin, contribute to evil and suffering. Zoroastrianism and Islam have similar concepts of the devil and of personal responsibility for moral evil.[2]

Atheistic explanations cannot account for the existence of evil, except perhaps to assert that it originated in the evolutionary struggle for existence, or is perhaps a product of evolving consciousness.

---

1. Alexander Solzhenitsyn, *The Gulag Archipelago* (Collins/Harvill Press and Fontana, 1974), Vol. 1, p. 175.
2. Hammes, *Human Destiny*, p. 74.

Christianity, however, offers a clearer understanding of evil, sin, suffering and death. But what actually is "evil," and what is "sin"?

Evil is always a lack of something, the absence of due good; it has no positive existence. The total reality of evil is, of course, very much a mystery to our limited human comprehension, but we know from Revelation that it is a very awesome reality. We know also that God never permits evil unless good can somehow come from it.

Catholic doctrine holds that Satan is not, strictly speaking, "evil personified," but rather a fallen being, one who rebelled through pride and rejected the choice of the vision of God. While now suffering torment in Hell, Satan and the other fallen angels are consumed with hatred of human beings and seek to draw all mankind away from God.

But God will not allow the forces of evil to tempt a human being beyond his or her capacity and, through the prayers and penance of the individual and other members of Christ's Mystical Body, the grace received from God can assist human beings to cope with life's trials.

In *Fundamentals in the Philosophy of God*, Fr. A. J. Benedetto, S.J. notes the various distinctions and divisions which are warranted in analyzing the reality of evil. Some extracts from Chapter 12, *"The Problem of Evil,"* are presented here:

> From the metaphysical point of view, evil is non-being, but it is not non-being in the sense of mere absence or limit or finitude, but in the sense of the negation, in a being, of something that this being, according to its nature, should have. Evil, properly considered, is a privation of good; it is the absence of some good or degree of being that should be present.
>
> The first and most important division to make with respect to evil is the division into moral and physical evil. Moral evil is sin, that is, the deliberate violation of God's moral law. It is defined as the absence or lack of the agreement and conformity that should be present between human conduct and the rule or norm of what that conduct ought to be. Physical evil is a defect, or privation of a perfection, in a being, marring it in its natural integrity or in the exercise of its normal activities or in both. Physical evil thus consists in the destruction or impairing of material substances and in the pain (suffering), bodily or mental, of sensitive and intellectual substances.
>
> It is important to make this division. For sin, and not suffering, is the prime analog of evil. Indeed, sin or moral evil is the only real total evil in the Universe. All other so-called evils can, under certain circumstances, be the positive object of a good will. But sin can never be willed positively, not even as a means to a good end. Moral evil

cannot be rationalized into a system that would meet with God's unqualified approbation. It can exist if God permits it—that is, if He does not prevent its occurrence, even though He has the power to do so and has certainly forbidden it—but it is caused only through the insolence of man defying God's prohibition. Physical evil, on the other hand, can be positively willed by God as a morally good, or at least as a morally indifferent, means to a (physically or morally) good end. Physical evil, therefore, unlike moral evil, is not an altogether unmitigated evil.

Sin is the choice of something forbidden by God as incompatible with the attainment of man's last end. It is the voluntary grasping after a good which, in the circumstances, is only a seeming good—"seeming," since nothing that is willed in conflict with right order can be a true good. Sin is thus the rejection of the supreme good; it is the alienation of a moral being from his intended destiny. God cannot positively will such an alienation; how sin can occur is the most mysterious and the most problematical element in the whole "problem of evil." However, almost inextricably interwoven with the mystery of sin is the problem of the eternal sanction against unrepented mortal sin, the sanction which is eternal loss, Hell, damnation.[3]

A Time of Testing

Christian teachings declare that each person's life on Earth is a time of probation, a time of testing. Man is presented with opportunities to develop such qualities as patience, humility, compassion, unselfishness, fortitude—in short, all those qualities which help the person to be more like God. The physical evils referred to by Fr. Benedetto may be considered as being allowed by God for the moral good of man.

Sometimes God, in His wisdom, may allow a person to meet with misfortune or lead him into a frustrating experience, and the reason may be to humble him and curb a tendency to pride. Or perhaps God may provide the individual with the challenge to be a great source of inspiration to other human beings.

Human suffering is, therefore, very painful, but explicable in the light of Revelation. One's character can be purified and strengthened by suffering, and acceptance of suffering can enable a person to reach a

---

3. Fr. A. J. Benedetto, S.J., "The Problem of Evil," *Fundamentals in the Philosophy of God* (New York: MacMillan Co., 1963), pp. 284-317.

high level of holiness. Suffering offered up brings blessings from God onto other members of the human race, and so strengthens the Mystical Body of Christ. Man must also contend with the effects of Original Sin, as noted in the Vatican II document *Gaudium et Spes*:

> It cannot be denied that men are often diverted from doing good and spurred towards evil by the social circumstances in which they live and are immersed from their birth. To be sure, the disturbances which so frequently occur in the social order result in part from the natural tensions of economic, political and social forms. But at a deeper level they flow from man's pride and selfishness, which contaminate even the social sphere. When the structure of affairs is flawed by the consequences of sin, man, already born with a bent toward evil, finds there new inducements to sin, which cannot be overcome without strenuous efforts and the assistance of grace.[4]

Even as great a person as St. Paul recorded his exasperation with this residual inner problem of sin: "For to will, is present with me; but to accomplish that which is good, I find not. For the good which I will, I do not; but the evil which I will not, that I do. Now if I do that which I will not, it is no more I that do it, but sin that dwelleth in me." (*Romans* 7:18-20).

Modern man is most certainly still affected by the Fall. We know, however, that the grace of God is available to each individual, to enable him or her ultimately to transcend the "spirit of the world." It seems clear, in fact, that God desires for each person to seek humility and to be open for God to use him or her in transforming the world.

Another sobering aspect to consider is the cost involved in disobeying the laws of God. They cannot be ignored with impunity. In view of the widespread modern problem of AIDS, the instruction given in Scripture to observe the natural law is especially pertinent: "Fly fornication. Every sin that a man doth, is without the body; but he that committeth fornication, sinneth against his own body. Or know you not, that your members are the temple of the Holy Ghost, who is in you, whom you have from God; and you are not your own? For you are bought with a great price. Glorify and bear God in your body." (*1 Cor.* 6:18-20).

Mankind may have freedom to reject God's wisdom, but there is a price to pay. Quite apart from the dreaded viral disease of AIDS, there

---

4. Vatican Council II, Pastoral Constitution on the Church in the Modern World, *Gaudium et Spes*, par. 25.

is the general problem of venereal disease. The alarming increase of herpes, a "new" type of venereal disease, can clearly be traced in the USA to the growth of permissive lifestyles. Since the nervous system of the body is affected, any cure is regarded as being a long way off.

Notwithstanding the fact that Satan has been temporarily allowed a great deal of say in the prevailing "spirit" of the world, God is still in control of things. We know from the Church's teaching that Satan and the other evil spirits, as well as those human beings who die in a state of willful rejection of God, will forever remain in Hell, which we understand to be a timeless state of lonely isolation from God, where one suffers not only the loss of the vision of God but also suffers terrible pain of the senses and mutual hatred of others in Hell.

Jesus Christ leaves us no illusions about the reality of evil, and the Gospels abound with His thoughts in this respect. The temptations to which Christ was subjected by Satan provide ample evidence of the existence of evil spirits. Christ's stern warnings to those who lead others into sin demonstrates the grim reality of Hell: "But he that shall scandalize one of these little ones that believe in me, it were better for him that a millstone should be hanged about his neck, and that he should be drowned in the depth of the sea." (*Matt.* 18:6). The references to the "weeping and gnashing of teeth" of those cast into Hell (*Luke* 13:28, *Matt.* 13:50) are very explicit.

Pope John Paul II reiterated the sober, traditional Catholic teaching on Hell in his book *Crossing the Threshold of Hope*:

> ... *eternal damnation* as the consequence of man's rejection of God.
> ... resurrection of the body is to be preceded by a *judgment* passed upon the works of charity, fulfilled or neglected. As a result of this *judgment*, the just are destined to eternal life. There is a destination to eternal damnation as well, which consists in the ultimate rejection of God, the ultimate break of the communion with the Father and the Son and the Holy Spirit. *Here, it is not so much God who rejects man, but man who rejects God.*[5]

The Providence of God

God has been described as omnipotent, omniscient, and omnipresent (i.e., He is all-powerful, all-knowing, and everywhere at one time). God created time itself and knows everything that has happened so far and

---

5. Pope John Paul II, *Crossing the Threshold of Hope*, pp. 70, 72.

everything that is still to happen.

All things have been designed to reflect God's greatness, and since He is infinitely perfect, both error and sin are incompatible with His perfection. The very fact of God's greatness provides a clue to the existence of evil, sin and suffering. Since man's will is of its nature free, justice requires that the consequences of a free choice cannot be overruled; otherwise, free will would not truly have been granted to mankind. Also, as Fr. Benedetto observes,

> Were God to abstain from giving existence to a soul because He foresaw that that soul would choose the path of evil, the perversity of the creature would have prevailed against the goodness of the Creator, and human wickedness [would] have compelled God to modify His purposes. God, despite the infinite resources of His wisdom and power, would refuse to create because of foreseen evil—this would be like surrendering in advance to evil.[6]

Thus, if God were so to act, then the omnipotent God would no longer be truly omnipotent—His infinite qualities would have become limited to some extent by His created subjects.

Against this fact, it may be argued still further, how can God remain omnipotent when He later acts to save us after we have sinned? Has not sinful man in some sense "compelled God to modify His purposes"? The answer is that His omnipotence is not affected at all when He allows each person the opportunity for reconciliation with Him. There is no compulsion forced upon God by His created subjects to make Him act in this way. The existence of evil has still not worked to negate the perfection of the Creator. Ultimately, the choice to be reconciled with God after a person commits sin still rests with the individual.

Another aspect that requires comment in any discussion of evil is that of the element of chance in the affairs of life. All events which take place in the world, even those which appear to be purely fortuitous, are incorporated into God's plan for the Universe. Everything serves a purpose according either to God's permissive Will or to His compulsive Will. Everything is mysteriously foreseen by God. That He is in charge is clear from Scripture: "Are not two sparrows sold for a farthing? and not one of them shall fall on the ground without your Father. But the

---

6. Benedetto, "The Problem of Evil," p. 310.

very hairs of your head are all numbered." (*Matt.* 10:29-30).

In the context of the consequence of Original Sin in darkening man's intellect, the Christian can still grasp the fact that, although one does not perfectly understand all this now, there will come a time when he will understand the harsher aspects of life. Our view in this life is that of the rear of the tapestry; the view from the front is yet to be seen by us. St. Paul touches upon this in his first letter to the Corinthians: "We see now through a glass in a dark manner; but then face to face. Now I know in part; but then I shall know even as I am known." (*1 Cor.* 13:12).

If a person meets with an accident, then there is a reason for it which will be made clear at the Last Judgment, when "God shall wipe away all tears from their eyes: and death shall be no more, nor mourning, nor crying, nor sorrow shall be any more, for the former things are passed away." (*Apoc.* 21:4).

In the final analysis, Christianity is the only religion which claims not only that God revealed something of Himself to mankind, including a partial account of the creation of the Universe, but that He also entered into the world as a human being (in the Person of Jesus Christ) and then allowed Himself to be humiliated, to suffer physical torment and be put to death, all out of love for His estranged subjects and in order to satisfy the infinite justice of God, which was offended by Original Sin. (Only God could have made reparation that has an infinite value, but only as man could He have suffered, since in His divine nature he is implacable [incapable of undergoing change, e.g., suffering]. Hence the need for God the Son to become man to make the needed reparation to God the Father and thus heal the rift between God and man caused by Original Sin.) This has to be the greatest story ever told!

— Chapter 24 —

EXISTENTIALISM

The Subjective Viewpoint of Reality

A discussion of Existentialism at first glance may not seem very relevant to the Origins debate. However, this philosophy has a significant bearing on the debate between scholars over the credibility of Scripture, and so an analysis is warranted. This chapter will also consider the impact that Evolution-inspired Existentialism can have on religious education.

Existentialism has been a strong influence on the attitude of many modern students and scholars toward the very definition of truth. Does truth consist of unchangeable values fully harmonious with God, or is truth a relative phenomenon, something which can be determined by each person according to his or her disposition?

Being a humanistic philosophy, Existentialism attempts to explain the existence of man from a personal, subjective point of view. Because of this approach, contemporary Existentialism cannot readily be defined in neat categories. There are, it seems, as many versions of Existentialism as there are existentialists.

Viewed in the light of the Origins debate, a crucial point about the credibility of Existentialism can be made: *Since the central proposition of Existentialism is that man has no specific nature, the concept makes sense only if man is the product of Evolution by chance processes. If such Evolution did not occur, this concept of man's nature is utterly flawed.*

As already shown, Evolution Theory is nothing more than science fiction, and Theistic Evolution is plagued with such weaknesses that it should be rejected as untenable. But the concept of Special Creation reinforces the traditional Christian view that objective truth does exist, and that each human being has been given a specific nature, i.e., one made in the image of God by the Creator.

Just as the laws of nature clearly have objective existence, objective truth can be shown to exist (e.g., two plus two can never equal other than four). Therefore the idea that man must create his own truth is fallacious. Rather, truth exists outside of man and has to be discovered.

Existentialist attitudes facilitated the growth of the cult of the "self" and became an important rationale with which to rationalize sin and to "demythologize" Scripture. The overall effect has been to diminish modern man's awareness of the omnipotent God and to assist the drift in consciousness away from belief in absolutes.

If one believes that human beings are only the result of chance processes, then there is no such thing as destiny, meaning or significance, and existentialist philosophy might appear to be valid. On the other hand, if man is the product of Special Creation, then Existentialism is invalid. Thus, the truth of the Origins debate is most pertinent to the credibility of Existentialism.

The Beliefs of Existentialists

In pondering the meaning of life, existentialist philosophers tend to draw a distinction between *Being* and *Existence*. *Being* refers to reality (i.e., the very fact of being alive) whereas *Existence* refers to possibility (i.e., the potential inherent in being alive).

The existentialist interpretation of human existence locates the individual in the midst of a world of objects. Man finds himself thrown into existence, hampered by many limitations, which restrict the opportunity for him to do as he may want, and he faces the inevitable prospect of death. However, his life is constituted of possibilities, from among which he may choose and thus project himself. In so doing, the individual "creates" his authentic self. This central theme has become known as *"existence precedes essence."* ("Essence" in this context means the "nature" of a being.)

Atheistic versions of Existentialism reject the traditional distinction between body and soul. The body is seen as a lived-through experience, lacking definite essence or nature that would determine a mode of being and acting. Man lives alienated from others and "homeless" in a purposeless Universe, and he must "create" himself and his own truth by exercising his free will. In so doing, *truth becomes an individual, subjective attitude, rather than objective reality*.

In the existentialist view, man is held to achieve his essence only in the exercise of free choice and through the experiences gained in life. Most importantly, he gains knowledge primarily by personal experiences. Truth becomes established simply by the individual's affirming that he has experienced it. Compliance with absolute principles is superseded by the subjective choices made according to the disposition of the individual.

Theistic versions of Existentialism acknowledge the existence of God, a transcendent Being separate from human beings, but they reject the idea that His existence can be deduced by the use of human reason. One can only gratuitously accept by an act of faith that God does exist. In this view, acceptance of His existence comes only from revelation and not from any observation by man of the wonders of the Universe. Such belief does not accord with the teachings of the Catholic Church, which hold instead that man can know the existence of God by the use of reason, which is of vital importance, and that faith and reason are inseparable. Vatican Council I declared that the Catholic Church

> Holds and teaches that God, the Beginning and End of all things, can be known with certitude by the natural light of human reason from created things.[1]

The 1992 *Catechism of the Catholic Church*, quoting Vatican Councils I and II in almost identical words, reaffirmed this teaching:

> The Church holds and teaches that God, the first principle and last end of all things, can be known with certainty from the created world by the natural light of human reason. (36)

A human being can use reason to deduce an awareness of knowledge (e.g., two plus two equals four) and to deduce that one's conscience can discern the existence of good and evil. Therefore it must be possible for man to deduce (or reason to) the existence of a Creator. (The existence of laws in nature points overwhelmingly toward the existence of a Creator.)

Atheistic existentialists recognize that man is something more than simply a creature. With his reflective capacity, man is capable of much more than the lower animals. Man has been defined by existentialists as a "constant possibility"; Heidegger's *Dasein* (or There-Being) and Jean-Paul Sartre's *Being-For-Itself* denote man as always capable of becoming more than he is at his present stage. His possible "self" is always in a state of "becoming," until death ends the process.

---

1. Vatican Council I (April 24, 1870), *Dogmatic Constitution concerning the Catholic Faith, The Sources of Catholic Dogma*, Henry Denziger's *Enchiridion Symbolorum* [1785] (30th Edition), translated by Roy J. Deferarri (1957).

Tragically, atheistic Existentialism has no room for Hope. Death is the unavoidable reality which is ever present and which will end the process of making choices. The best that one can do is to face this inevitable demise with resolute determination. Courageously confronting one's ultimate extinction at least provides some self-respect.

Regarding self-respect, atheistic existentialists believe that man is only living responsibly when he chooses to live an authentic existence. Only when man rises above self-imposed restrictions, or behavior which conforms to phony popular opinion (and thus when he is true to himself), does he really make a choice for an authentic existence. When he surrenders personal responsibility to the anonymity of the crowd or to collectivism, he is guilty of *inauthentic* existence or "bad faith," according to Sartre.

On the other hand, theistic Existentialism, with its acknowledgment of God and of the existence of a life after death, holds that the authenticity of one's existence is dependent upon one's openness toward other persons. Responsibility for, and availability to, other human beings is considered essential for the enrichment of one's existence.

The Cult of the Self

Since the primary preoccupation of existentialist philosophy is centered on the "inner" man, it is important to look closely at this area. The emphasis placed by existentialists on consideration of the "self" highlights for the individual an acute awareness of himself and his own life on Earth. This awareness is of a particular and individual nature (e.g., my existence, your existence, rather than our existence). Other human beings can tend to be seen simply as impersonal objects. It is considered important that the person "finds" or "actualizes" himself in the encounters or relationships that he has with other people.

Based on the premise that man is inherently good and that evil influences can be attributed to society in general, "selfism" is in reality a secular substitute-religion. Each person is encouraged to find his "actualization" in whatever way that appeals. Compliance with objective principles is rejected in favor of subjective values derived from one's experiences. This selfist viewpoint is obvious in the current fascination with humanistic psychology on the part of certain secular educationalists and many Christian clergy. In contradiction to the central Christian teaching that man's nature is inherently flawed as a result of Original Sin, humanistic psychology declares that man is good and that evil comes from outside of man.

Greed, selfishness and aggression are attributed essentially to the social environment. The impact of this type of thinking upon catechetics has been disastrous. Ironically, instead of emphasis on the need for repentance and obedience to God's Will, the message from many modern Christians is in effect that one should put his trust in the subjective conscience. The American psychologist, Paul C. Vitz, has noted some important features of this preoccupation with the self:

> An important existentialist concept is "becoming," the process of self-development or fulfilling one's potential. This process unfolds by way of the self's choosing its own course of self-fulfillment. Acts of choice bring the self from its initial existence into an actualized self, with a nature or essence created by its choice. Thus, the self first exists (i.e., "I am") but without any *a priori* nature or essence. Instead, through acts of choice the self's essence is created. These choices are courageously made in the face of the self's awareness of non-being and its experience of *Angst* [dread, anxiety, anguish]. Guilt arises through failure to develop the self's potential, through blocking or ignoring one's chance to become one's potential. Transcendence is the name of the important capacity of the existential self to surpass or climb beyond the prior level of self-development. Thus, as self-potential is developed, each new stage is a transcending of the earlier stages, and this process often is called "becoming."[2]

Implicit in the concept of *becoming* is the idea that human beings are evolving creatures whose nature is progressing "upwards." This view necessarily opposes the idea of the Fall of man.

The traditional Christian view, on the other hand, holds that the faithful individual may look forward to salvation in the next life and thus can enjoy a measure of contentment in this life amidst the turmoil of a fallen world. But for selfists, the notion of salvation tends to be perceived rather as "integration," when the individual finds self-discovery, self-acceptance and self-actualization through creative activity and pursuit of happiness.

The self-centered lifestyle of many individuals, seeking optimum gratification with as little delay as possible, has resulted in serious family dislocation and personal distress in Western countries. This type of

---

2. Paul C. Vitz, *Psychology as Religion: The Cult of Self Worship* (Hertfordshire, England: Lion Publishing, copyright William B. Erdmans Publishing Co., Grand Rapids, MI, 1979), p. 24.

lifestyle can only be attempted effectively in times of material prosperity, but even then, the distress of those who attempt to live this way is very real.

Since it is an illusory concept of human existence, selfism is resulting in a deepening crisis of identity for individuals. If absolute principles are held to be non-existent, then one can only speculate and offer propositions based on personal experiences, and consequently any private choices will tend to reflect attitudes of relativism. Once persons or actions are given no intrinsic value as such, and the only value involved is that which the individual decides to assign to it, then such moral attitudes tend to be regarded as "value-free."

To claim that a moral attitude can be "value-free" is erroneous; rather, it signifies an attitude of indifference. If a mother decides to impute worthlessness to her unborn fetus, then the killing of this new human being cannot be regarded as "neutral" or "value free." It is clearly a judgment of indifference as to the life or death of that unborn person, and this judgment in itself constitutes a value.

Professor Vitz also noted the reality of selfism for modern man:

> Hendin's brilliant descriptions portray exactly where so much of selfism ends: the self as subject frantically trying to gain control over others—the objects—in order to build its own self as subject. As more and more people have their "consciousness raised"—that is, as they are "liberated" and take on the role of subjects—the competition becomes fierce. Life has become a game where there are only two states: winning and losing, sadist and masochist. . . .
>
> If you show weakness, such as a need for love, you get slaughtered; if you withdraw to a machine-like, emotion-free competence and develop complete identification with career, you are isolated and starved for intimacy and love. Perhaps there is some relief in temporarily losing the self in sexual or other sensations and afterwards counting each new experience as a score for the self, but a lonely deathlike living is inescapable.[3]

In contrast to this background of selfishness and emptiness, Viktor Frankl pointed out that unselfish attitudes are good for the individual. When self-satisfaction is made a primary end in itself, it is most unlikely to be attained. The person who cherishes values above and beyond the self is more likely to find self-satisfaction as an unintended side effect. Frankl noted that when the *"I Ought"* dimension is intro-

---

3. *Ibid.*, p. 120.

duced, the effect is to complement the subjective aspect of human existence, *Being,* with its objective counterpart, which is *Meaning.*[4]

Inherent Weaknesses

Existentialism contains serious inherent weaknesses. If the existence of absolute principles is rejected, there can be no objective criteria or standards with which to establish a viable basis for proposing a particular point of view as being more worthy of acceptance than another. If no universal, objective criteria exist regarding truth, and instead only what each person perceives to be the truth, then on what basis can any personal choice be morally justified in the mind of the atheistic existentialist? Any consideration of metaphysics as it is normally studied would have to be rejected, because the matters under consideration would be beyond one's personal experiences. Were morality to consist only of subjectively imputed worth given to a particular action, the criterion could amount to little more than expediency. Thus, atheistic Existentialism may in certain instances be opposed to such noble ideals as compassion and altruism.

Another weakness of Existentialism relates to the basis upon which conclusions may be drawn. For example, how is it possible for an existentialist to make a claim that, say, life is meaningless? What standards apply? A consistent existentialist attitude should reject all criteria, if no objective principles are thought to exist. In theory, of course, one could never begin existential life because the process of choosing could not get started. There would be no standard applicable on which to base the first choice.

In practice, the individual existentialist has no option but to accept the existence of certain values or standards. If an existentialist acknowledges that some degree of meaning and purpose can be observed in the altruistic lifestyles of inspiring individuals, or if he is repelled by the savage atrocities committed by totalitarian regimes, then he must also acknowledge that a set of objective principles does exist. He must then agree with Albert Camus that, while life may not be completely comprehensible to him, it is not without meaning. In so doing, the existentialist is moving away from the subjectivity that is central to existentialism.

---

4. Viktor E. Frankl, *Psychotherapy and Existentialism* (New York: A Touchstone Book, published by Simon & Schuster, 1976), p. 54.

Another problem arises from the fact that existentialists cannot avoid the use of reason in arriving at their beliefs. Man's very thought processes, his logic or means of reasoning, proceed from essences which are not the result of either personal experiences or of choice. This fact shows that *existence and essence are in fact complementary*; one does not proceed from the other or exist without the other.

This point is verified when one considers that the body of the person can only be accepted by the existentialist as a "given," a reality over which he has no choice. For example, the skin pigment, the color of hair and eyes, the height of the body, the shape of the ears and so on, are factors beyond the choice of the individual. The existence of these characteristics in no way depends upon the efforts of the person.

Existence and essence may be distinct notions, but they are inseparable in all beings. The essentially complementary nature of essence and existence has been pointed out by Fr. Francis J. Lescoe:

> When we consider the case of any given concrete individual, the question of the priority of essence over existence or vice versa becomes totally meaningless. This is so because, while it is true that the act of existing does enjoy a kind of primacy within being, yet both essence and existence are concomitant principles. With regard to created, finite being, neither its essence nor its existence is prior; both actively constitute a composite being. The essence determines or limits the existence which a particular entity exercises. Since essence and existence are absolutely co-related, acting in the manner of co-principles, there is no way in which one of the principles can rightly be said to exercise a priority over the other.[5]

Since it is only over one's thought processes that the existentialist can claim to have complete choice, he then finds himself drawn into the field of the spirit and questions of metaphysics. But how does one explain the existence of one's knowledge of right and wrong without also granting that a spiritual entity exists in every human being?

The Impact on Christianity

Existentialism was utilized with great effect in the attack on traditional Christian beliefs, because it provided an effective vehicle with which to attempt to "demythologize" Scripture. The famous reinter-

---

5. Fr. Francis J. Lescoe, *Existentialism: With or Without God* (New York: Society of St. Paul, 1974), p. 283.

preter of Scripture, Rudolf Bultmann, was greatly influenced by Martin Heidegger's atheistic existential analysis of Being. He felt the only way to make Scripture intelligible and acceptable to modern man was to demythologize it and revise it.

Bultmann proposed a drastic change in Christian consciousness, but with little semblance to traditional Christianity. The doctrine of Original Sin was revised, with a completely changed meaning; the miracles of Christ were rejected; and a distinction was introduced between the Christ of history and the Christ of faith. With *Genesis* regarded as unscientific, the Creation account would be re-interpreted with entirely subjective concepts about existence. The truth of Scripture would be thus only what is deemed to be meaningful to the experiences of modern man. In his analysis of Bultmann's beliefs, John Macquarrie outlines and endorses the logic of this existential re-interpretation of *Genesis:*

> Adam's dependence and creatureliness is both the writer's and mine when I read his story and understand it as something which touches my existence. Adam's disobedience is the writer's disobedience and my disobedience; Adam's fall is the writer's fall and my fall. . . . So also with the account of the creation of the world of nature. It does not teach me how to understand the world considered as an object for theoretical study, as science teaches me, but it does teach me to understand the world as a constant factor in my experience as being-in-the-world, that the world is good and for my use, but that at the same time I lose myself in the world in preferring the creature above the Creator, as Adam did. Thus understood, there can be no possible conflict between the teaching of these accounts and the teaching of science and cosmology, and thus understood it is clear also that these accounts of the creation touch a level of truth more fundamental and important for my existence than any theory of cosmic origins could be.[6]

Thus, *Genesis* is held to be no more than a "story," the moral of which can only be discerned after Scripture is revised by modern scholars.

If one believes that errors really do exist in Scripture, then such subjective rationalizations as proposed by Bultmann and other liberal scholars are likely to be uncritically accepted by naïve Christians. In the view of the demythologizers, a great deal of Scripture consists of

---

6. John Macquarrie, *An Existentialist Theology: A Comparison of Heidegger and Bultmann* (Middlesex, England: Penguin Books Ltd., 1980), p. 61.

myths and legends—symbolic stories which must be understood existentially in the light of personal experiences.

A major weakness in their theology, however, is that as the case against Evolution Theory becomes stronger, the *Genesis* account is increasingly more difficult to dismiss as mythology. In fact, the assertions of the demythologizers are really gratuitously based assumptions, lacking in any real evidence.

The erroneous ideas within Existentialism have clearly contributed to the modern turmoil within Catholicism. In an important 1970's study of the inroads made by this form of philosophy into Christian beliefs, an Australian Catholic scholar noted its impact upon catechetics:

> It is quite clear that this corrosive philosophy of existentialism, by denying the possibility of a philosophy of God, and by infiltrating an extreme relativism and subjectivism into the minds of the new Catholic intellectual elite, has wreaked havoc in the sphere of catechesis. At the very time when Catholic school children, in keeping with the mood of the wider secular culture, are questioning the existence of God and of an objective moral order, the catechists are refusing to speak—indeed, they are philosophically incapable of speaking—about these things. When our civilization is "leaving the light of its ancestral religion and sliding back into the dark" and when it has despaired even of natural religion, the existence of God and of an objective moral order, the existentialist catechists can speak only of "Christian living." They cannot see that their catechesis fails because it begs the very questions about which contemporary culture is most in doubt.[7]

The erroneous notion of theistic Existentialism—that man only attains his authentic self by choosing in favor of commitment to God by a "leap of faith"—is perhaps best exemplified in saying that "faith cannot be taught, it can only be caught." This saying is only partly true. Faith is definitely a gift from God and is a mystery. Nevertheless, one can grow in the knowledge of God and gradually move from an incomplete awareness to a more complete understanding of Him, and thus be "touched" by Him. Unless the individual were to undergo something like the experience of St. Paul on the road to Damascus, how else but by prayer, study and the example of others can one grow in faith?

The liberal theologians who promote existential reinterpretations of

---

7. Gary Scarrabelotti, *Existentialism and Catechetics* (Melbourne: John XXIII Fellowship Co-operative Limited, 1977), p. 14.

Scripture are in effect placing man above the Church. If a modern Christian sees himself as autonomous and able to redefine standard teachings which have prevailed for nearly two thousand years, then he obviously believes that there is no absolute truth for the Church to guard. Revisionist liberal theology in effect subjects to human judgment the revealed doctrines which are of their very nature beyond and independent of human thought. It promotes the idea that man can determine what is truth, rather than acknowledge and accept the eternal authority of God in revealing it.

In *Iota Unum*, a comprehensive examination of the collapse of faith in 20th century Catholicism, Professor Romano Amerio analyses the present confusion in catechetics. (Muddled but very persuasive evolutionary concepts, such as the "process of becoming," have been widely disseminated, and this now affects the contents of catechetics/homilies.)

> The Christian religion recognizes that apart from the natural light there is also a supernatural light that illuminates the human spirit from above, and makes it capable not indeed of *seeing* truths surpassing the natural sphere, but of assenting to them without seeing them, and of appropriating them to itself. For modern pedagogy on the other hand, reality is self-creation, truth self-knowledgeable, and teaching self-teaching, because it believes that the true and the good and every other spiritual value are inherent in the spirit itself, in short that the divine is immanent in man.
>
> ... In fact the proper and formal goal of teaching, including catechetics, is not to produce experience but knowledge. The teacher draws the pupil to proceed from one state of knowing to another by means of a dialectical process of the presentation of ideas. *Thus, the immediate end of catechesis is not an existential and experiential meeting with the person of Christ (to think it is, is to confuse catechesis with mysticism), but is rather the knowledge of revealed truths and of the preambles to them.* The modernist origin of pedagogy is obvious to anyone who knows that the fundamental philosophical principle of modernism was that sentiment, or feeling, contains all values within itself and is superior to all theoretical knowledge, the latter being merely an abstraction from the concrete reality which is experience.[8]

---

8. Prof. Romano Amerio, translation (1996) from the Second Italian Edition by Rev. Fr. John P. Parsons (Canberra), *Iota Unum: A Study of Changes in the Catholic Church in the XXth Century* (Kansas City, MO: Sarto House, 1996), pp. 291, 292.

Pope John Paul II and Phenomenology

Any discussion on the controversy surrounding the question of *existence* and *essence* would be incomplete without some consideration of the thought of Pope John Paul II.

The concept of *phenomenology* (i.e., the study of the objects of consciousness) was proposed by Edmund Husserl as a means of describing in an objective manner the facts of one's personal consciousness. The intention was that the essences of these objects could then be analyzed. However, various existentialist philosophers developed markedly differing concepts of phenomenology. Martin Heidegger, a disciple of Husserl, rejected any notion of stable essences and made *existence* the center of his philosophy.

The concept of phenomenology can have a useful place in pondering the reality of life, *provided that objective principles are retained in the analysis*. William A. Marra demonstrated this in his book *Happiness and Christian Hope*, in which he highlighted some of these "insights into the given."[9]

In analyzing the question of happiness, he drew a distinction between the experiences which produce feelings of happiness by direct contact (e.g., as when a child has her forehead stroked by her mother) and those that are motivated by the awareness of a quality outside of oneself (e.g., when one is in the presence of an especially liked friend).

The first type can be seen as caused, immanent experiences which are capable of subdivision into instincts, moods and sensations. The second type are spiritual in nature and can include love for the other person and joy when that love is requited. Though both types produce feelings of happiness, they each belong to quite different levels of reality. (At the most sublime level of *reality* lies the prospect of never-ending happiness with God.)

Prof. Karol Wojtyla (prior to his election as Pope John Paul II), together with other scholars at Catholic University of Lublin, Poland, developed a comprehensive set of ideas, defined as "existential personalism," regarding man's self-realization. He sees value, *only up to a point*, in the methods of phenomenology. In the experiences of life through which an individual passes, one can observe that there are moral values which exist and which can be discovered and recognized. A set of objective principles is therefore essential, to enable

---

9. William A. Marra, *Happiness and Christian Hope: A Phenomenological Analysis* (Chicago: Franciscan Herald Press, 1979).

human beings to discern *truth. Phenomenology can only be of limited use.*

The future Pope defined love as being a disposition toward goodness; thus, *truth and love are inseparable realities.* Man can only fully "realize" himself by seeking to conform to absolute principles. In this way the person becomes truly free to love his fellow man.

> Only truth about oneself can bring about a real engagement of one's freedom in relation with another person. It is a giving of oneself, and giving of oneself means exactly to limit one's own freedom for the sake of another person. The limitation of one's freedom could be something negative and painful, were it not for love, which transforms it into something positive, happy and creative. . . . Will strives towards goodness, and freedom is a prerequisite of will. Freedom, therefore, is for love, since it is through love that man participates in goodness. This is the basis for its principal position in the moral order, in the hierarchy of values, and the hierarchy of proper longing and desires of man. Man needs love more than he needs freedom, since freedom is only a medium, whereas love is a purpose. *Man desires, however, true love, because only when it is based on truth can an authentic engagement of freedom be made possible.*[10]

But how can one discern *authentic truth* amidst the present disharmony within Catholicism? In keeping with the Pope's above declared support for widespread knowledge of truth, a genuine commitment to objective truth necessarily requires also that the full truth about Creation/Evolution/Origins be disseminated to all. What better way than by issuing an updated, comprehensive encyclical on all matters relating to Origins?

---

10. See Karol Wojtyla, *Love and Responsibility,* quoted by Fr. Andrew N. Woznicki in *A Christian Humanism: Karol Wojtyla's Existential Personalism* (New Britain, CT: Mariel Publications, 1980), p. 28. Emphasis added.

―Chapter 25―

CREATION REDISCOVERED

One does not have to be a prophet to see that the world of today is one of stress and tension, even though multitudes thirst for peace. Living in the shadows of nuclear destruction, totalitarian enslavement and personal alienation, modern man also has had to cope with rapid technological changes and increasing levels of unemployment. In addition, deteriorating family structures are giving rise to increasing incidence of loneliness, heartache and financial hardship.

Despite the work of Redemption achieved by Christ, mankind continues to remain in bondage to sin as a consequence of Original Sin. Unfortunately, at a time in history which has been likened to a new pre-Christian era, the reality of the inherited inner sinfulness of human beings seems poorly understood, even by many Christians. On the other hand, the ideal of resisting one's self-centered desires will always be relevant and has been recognized as vital for the well-being of individuals, despite their religious beliefs:

> What would be the result if man had the opportunity to satisfy completely each of his needs and drives? Assuredly, the results of such an experiment would in no way consist of an experience of deepest fulfillment, but on the contrary, of a frustrating inner void, of a desperate feeling of emptiness. . . . Only as man withdraws from himself in the sense of releasing self-centered interest and attention will he gain an authentic mode of existence.[1]

Thus, thoughtful human beings can discern that a certain enrichment of character can be gained from unselfish behavior. Since Christianity teaches that the Creator is the very epitome of unselfishness, it makes sense, therefore, to have a clear understanding of one's true relationship to the Creator. Only then will the sacrifice that is part

---

1. Frankl, *Psychotherapy and Existentialism,* pp. 42, 46.

of living a Christian lifestyle be fully comprehended. Only through God can man truly "find" himself:

> It is only through an awareness of God, the Absolute, that man can be aware of himself; only in that way can he appreciate the fact that he has been made in God's image. When one has that awareness of God, all other things fall into place and are seen in true perspective. Human origin, purpose and destiny are then comprehensible; the ultimate meaning of one's personal existence is grasped. It was said earlier that relativism stemmed from a prideful rejection of God. The recognition of God, on the other hand, leads to the opposite attitude, that of humility. More important, it leads to faith. It has always been recognized that humility and faith are both inversely related to pride; the greater those virtues are, the less will be human pride.[2]

Paul Vitz confirms the truth of this statement, in recalling his own difficulties in rediscovering Christianity:

> In spite of my rejection of self-theory, large parts of me remain which are still thoroughly indoctrinated with it. Particularly difficult are religious ideas like penitence, humility, accepting my dependence on God, praying for help. In my heart and in my mind I know these are good, true, and necessary for spiritual life. I know they are needed to curb pride and purge arrogance. But my yet luxuriant, overblown ego baulks at and rejects being labeled "a miserable offender," wonders about the need for penitence, and occasionally bristles at metaphors referring to me as a sheep, child, or obedient servant. Equally foreign is the concept of judgement.
>
> I know that many others who are not as ready as I to listen are turned away by these words at once. The justification for these concepts has been lost, and re-education is desperately needed. We need updated orthodox theology. We need sermons on radical obedience, on the mysticism of submissive surrender of the will, on the beauty of dependency, on how to find humility.[3]

What modern man most definitely does not need is a watered-down version of Christianity, heavily laden with erroneous theological concepts. Unfortunately, many Christian scholars—in a futile effort to make Christianity "relevant"—have proclaimed a Christian message which is so emasculated it only succeeds in making itself seem irrelevant. Aware of this factor and of the intimidation by Modernist forces

---
2. Hammes, *Human Destiny*, p. 205.
3. Vitz, *Psychology as Religion*, p. 124.

in society, Vitz points out the need for a revitalized Christian response:

> Conservative Christians too often intuitively recognize the nature of these conflicts without being able to articulate their position with much sophistication. Meanwhile, the liberal churches have often enthusiastically embraced selfism and humanistic psychology, without regard to its hostility to Christian teaching. It seems high time to transcend both reactions with a post-modern, intellectually sound counter-response, regaining for the Church the large, legitimate religious issues it has surrendered to secular ideologies like selfist psychology.[4]

There is no doubt that substantial numbers of young people have left the Catholic Church precisely because they know very little about Catholic doctrine; the so-called "experiential" method of catechetics has failed. Though they may be poorly informed about the real story of Christianity, many of them nevertheless can sense they have been given a waffle which is lacking in substance.

What is to be done? God the Creator must be rediscovered! If individuals are to be told the true story which will enable them to see past the meaninglessness of secular society and to grasp the Christian vision of the dignity of man, then clarification of Origins beliefs is essential. The 1992 *Catechism* notwithstanding, a revitalized comprehensive teaching of all matters relating to Origins doctrine is overdue within the Catholic Church.

The pressing urgency of this need has been recognized with clarity by Fr. David Becker, editor of *Watchmaker*, a Catholic Origins Society magazine. In highlighting the struggle of *supernaturalism versus naturalism* in modern societies, and showing that the Origins controversy is a *faith* issue, Fr. Becker argues that the continuing consignment of *Genesis* to mythology by learned scholars is central to the continuing crisis within the Catholic Church:

> "You evolutionistic scientists can practice your brand of science and we won't bother you; and we will practice our biblical scholarship in such a way that evolutionistic theories will never contradict the Word of God, for we will treat the Word of God as myth, legend and story, which when properly demythologized and interpreted by us, will yield a truth that is purely religious. By this neat compartmentalization we can live together in concord!"
>
> But the deal is a Faustian bargain. For one who accepts it, the pay-

---

4. *Ibid.*, p. 109.

off is meager, immunity from being called a "fundamentalist," and the price is high, capitulation on the truth of the Word of God. The Book of *Genesis* is foundational to the Christian faith, not as a scientific textbook, but as an *historical* book. The historicity of *Genesis* is crucial to the credibility of the whole structure of Christian dogma. The Christian faith and the Church stand or fall as the historical foundation, *Genesis*, stands or crumbles.

Our experience in the Church over the past thirty years confirms that as soon as you open the door to evolution, revolution comes in, and the theological dominoes begin to fall, as fundamental dogmas are denied: Adam and Eve as real people, marriage as [a] heterosexual and monogamous union, the prohibition of divorce, Original Sin, the need for redemption, the Immaculate Conception, the Incarnation, the Redeemer, the divinity of Christ, the saving blood sacrifice of Christ, the Mass as Sacrifice. The rejection of these basic dogmas of the Faith is very real and widespread in the Church today, and this rejection is inspired by the evolutionary mindset that dominates our schools and seminaries.[5]

Thus, it is vital to recognize the existence of the Creator and to see oneself as subject to God. But it is not enough simply to concentrate on the idea that there is a Creator and that there was a Creation. This limited idea has prevailed too long already and only addresses part of the story; it is essential to understand and proclaim the rest of the story.

The assaults on traditional Christian beliefs which exploded in the second half of the 20th century may be traced to a range of causes. In the opinion of this writer, the present diminution in Christian conviction has come about primarily because the true story of Creation, especially the section dealing with Original Sin, has been largely "lost," and the notion of God the Creator, in effect, has been downgraded. In its place has come, almost implicitly, the weird notion of God the Evolutor.[6]

Perhaps the most obvious external factor which facilitated the 20th

---

5. Fr. David Becker, private correspondence, September, 1994.
6. As an aside, is seems appropriate to point out here that unless the doctrine of Original Sin comes to be more clearly understood, it is hard to see how genuine peace can be achieved on Earth. No attempt has been made in this book to address the question of peace, for this necessarily touches upon a wide range of issues such as social justice, human rights, defense and foreign affairs, resources and environmental policies, and man's stewardship of the planet. Suffice it to say that the Christian notion of peace—"tranquility of order"— is hard to reconcile with evolutionary concepts such as "kill or be killed" and "survival of the fittest," and it can be fully comprehended only when the reality of Original Sin is understood.

century crisis in Catholic beliefs was the progress made in medical technology, which in turn made possible the humanist revolution of contraception and abortion on demand. Other factors were those of increasing materialism and the dramatic impact made by the now almost instantaneous nature of the worldwide communications media.

The situation was not helped by the fact that many Catholics do not study seriously the basis of their religious beliefs. Not surprisingly, large numbers were unprepared for the fairly sudden and yet profound moral challenges which exploded in the 1960s against traditional Christian teaching. Instead of seeing the humanist challenge for what it really is, many Catholics accommodated themselves to its values. Instead of seeing the fallacies of the extreme subjective mentality promoted by liberal theologians, many were swayed by its novelty.

The harm now done to Christianity is immeasurable by human comprehension. Tragically, at a crucial time in history when learned Christian scholars were especially needed by the laity to promote incisively and enthusiastically the central teaching of the Church, many of them turned out to be agents of change, steeped in evolutionist influenced theology, who expended their talents in trying to overturn unchangeable doctrines. The many years spent by these scholars imbibing such ideas as the "demythologizing" of Scripture had finally produced its bitter fruits.

Many eminent scholars, who could have been developing modern insights into the central teachings revealed by God, seemed to have lost sight of these very teachings. Instead of encouraging the laity regarding, say, the need for trust in the Creator and trust in His teachings handed down from the Apostles, these same scholars instead became dissenters, causing confusion. The effect of their efforts has been to weaken the very credibility of Christianity and to cause others to lose interest in it.

Ironically, Evolution Theory now stands exposed as false—the biggest mistake made in science and the most enduring *myth* of modern times, but the effect of naturalistic beliefs upon modern society has been quite profound; a point well noted by Michael Denton:

> The entire scientific ethos and philosophy of modern western man is based to a large extent upon the central claim of Darwinian theory that humanity was not born by the creative intentions of a deity but by a completely mindless trial and error selection of random molecular patterns. The cultural importance of evolution theory is therefore immeasurable, forming as it does the centerpiece, the crowning

achievement, of the naturalistic view of the world, the final triumph of the secular thesis, which since the end of the Middle Ages has displaced the old naïve cosmology of *Genesis* from the Western mind.

The 20th century would be incomprehensible without the Darwinian revolution. The social and political currents which have swept the world in the past eighty years would have been impossible without its intellectual sanction. It is ironic to recall that it was the increasingly secular outlook in the 19th century which initially eased the way for the acceptance of evolution, while today it is perhaps the Darwinian view of nature more than any other that is responsible for the agnostic and skeptical outlook of the 20th century. . . .

Considering its historic significance and the social and moral transformation it caused in Western thought, one might have hoped that Darwinian theory was capable of a complete, comprehensive and entirely plausible explanation for all biological phenomena from the origin of life on through all its diverse manifestations up to, and including, the intellect of man. That it is neither fully plausible, nor comprehensive, is deeply troubling. One might have expected that a theory of such cardinal importance, a theory that literally changed the world, would have been something more than metaphysics, something more than a myth. *Ultimately the Darwinian theory of evolution is no more nor less than the great cosmogenic myth of the 20th century.*[7]

In view of all the turmoil which has taken place concerning the debate over objective and subjective viewpoints of reality, it is easy to overlook the true status of man—that of a subject of God. There can be no mistaking the fact that promotion of naturalistic philosophy has greatly contributed to many people's losing sight of, or being frustrated from recognizing, the truth of God as Creator. And Theistic Evolution has failed because it seeks accommodation with erroneous Evolution Theory and cannot effectively counter Evolutionism.

Nevertheless, we know with certainty that God is still in control of the Universe and the true doctrine on Creation is still there—awaiting rediscovery in a fuller sense and awaiting further proclamation. The Catholic Church has long claimed to be the Guardian of Truth, and the objective truth about Creation/Evolution/Origins ought to be proclaimed by the teaching Magisterium, despite popular opinion.

The Catholic Church was commissioned by Jesus Christ to "teach

---

7. Denton, *Evolution: A Theory in Crisis,* p. 357. Emphasis added.

all nations," and thus to influence the spiritual/moral standards of each generation. As Pope John Paul II pointed out, Christians now face the extremely difficult task of transforming society anew:

> About 150 years after Descartes, all that was *fundamentally Christian* in the tradition of European thought *had already been pushed aside*. This was the time of the Enlightenment . . . when *pure rationalism held sway*. The French Revolution, during the Reign of Terror, knocked down the altars dedicated to Christ, tossed crucifixes into the streets, introduced the cult of the goddess Reason. On the basis of this, there was a proclamation of *Liberty, Equality, and Fraternity. The spiritual patrimony and, in particular, the moral patrimony of Christianity were thus torn from their evangelical foundation. In order to restore Christianity to its full vitality, it is essential that these return to that foundation.*[8]

How *can* Christian teachings be fully instilled anew in largely amoral modern societies which esteem subjective consciousness, if the vital message of Creation remains distorted? At this point in time, even for many Catholics who are appalled at the modern collapse of faith in the Church, the idea of Special Creation has still not been considered seriously within the Catholic Church. (It has not been tried and found wanting; rather, it has been thought irrelevant and not tried at all.)

With the present drift away from respect for moral teachings and authority, the Church founded by Jesus Christ can only stand to gain by comprehensively re-examining the question of Origins. This is especially so, since Paul VI stated (June 30, 1972) that "the smoke of Satan has entered the temple of God," and Cardinal Ratzinger laments the malaise into which Catholicism has drifted and is greatly concerned about the virtual disappearance of Creation theology.

The concept of Special Creation seeks to bring God the Creator back to center stage in society, and it offers the prospect of relief from confusion. It brings welcome clarity about basic teachings, and thus can provide a very incisive thrust in spreading the good news of Christ. Further, since it is perennially important to each new generation, a clearly defined concept of Origins also sheds light on the type of catechetics most likely to transmit accurately the central teachings

---

8. Pope John Paul II, *Crossing the Threshold of Hope*, p. 52. Emphasis added.

of Christianity from one generation to the next. As Cardinal Ratzinger himself has pointed out, "catechetics must get back to being, not one opinion among others, but a certainty drawing on the Church's faith, the substance of which far surpasses accepted opinion."[9]

A crucial aspect of the importance of Origins is the impact upon the all-important element of certainty and the feeling of assurance or security that goes with it, which any belief system must offer the believer in order to attract and maintain his or her allegiance. It has been shown in the material outlined earlier that belief in Evolution is a common denominator of Nazism, Communism, Secular Humanism, Existentialism and Modernism. Thus it enables those belief systems to appear "closed," and to purport that they are a total explanation of reality—in effect, to claim a mortgage on truth and to offer certainty to the adherent. But, since Evolution Theory can be shown to be untenable, these belief systems are not closed at all, and certainly they can no longer be offered for credible belief.

On the other hand, would the rediscovery of Creation strengthen belief in Christianity? Would it contribute to greater awareness of and reverence for the essential dignity of all human beings, and thus have beneficial impact on society generally? To answer questions such as these in the affirmative, the concept of Special Creation offers a comprehensive explanation of reality, and it powerfully reaffirms the indispensable doctrine of Original Sin. Special Creation is fully harmonious with both Scripture and modern science and can be as easily grasped by the layman as by the professor.

Against the back-drop of a troubled and often harsh society, when a person can see that he or she is a unique creature, loved personally by an omniscient and compassionate Creator and Redeemer, the effect is totally uplifting and consoling, and this consolation and assurance helps the individual to cope better in this transitory and often difficult life.

The rediscovery of the true story of Creation therefore remains an essential prerequisite for both the reconversion of society and the restoration of Catholicism as the True Faith and perennial philosophy. This is not to suggest and should not be misrepresented as being a return to some sort of Camelot, to a mythical halcyon era of the past. Rather, it is to recognize that the Church must speak to the troubled

---

9. Joseph Cardinal Ratzinger, interviewed by Vittorio Messori, *The Ratzinger Report* (San Francisco: Ignatius Press, and John XXIII Fellowship Co-op Ltd., Australia, 1985), p. 144.

world of each century in the most enlightened terms possible. And the story of Creation—just as it is given in *Genesis*—will always be relevant to her "Good News" message to a fallen world.

·

—Appendix A—
Pontifical Biblical Commission on *Genesis* (1909)[1]

In 1909 the *Pontifical Biblical Commission* declared its ruling on the historical character of the first three chapters of *Genesis*. The actual statements of the Commission (translated from Denzinger) are as follows:

1. False Exegesis:
   Whether the various exegetical systems which have been proposed to exclude the literal historical sense of the first three chapters of the book of *Genesis*, and have been defended by the pretense of science, are sustained by a solid foundation?
   Reply: In the negative.

2. Historical Character of the First Three Chapters:
   Whether, when the nature and historical form of the book of *Genesis* does not oppose, because of the peculiar connections of the first three chapters with each other and with the following chapters, because of the manifold testimony of the Old and of the New Testaments; because of the almost unanimous opinion of the Holy Fathers, and of the traditional sense which, transmitted from the Israelite people, the Church always held, it can be taught that the three aforesaid chapters of *Genesis* do not contain the stories of events which really happened, that is, which correspond with objective reality and historical truth; but are either accounts celebrated in fable drawn from the mythologies and cosmogonies of ancient peoples and adapted by a holy writer to monotheistic

---

[1]. Pontifical Biblical Commission on *Genesis* 1-3 (June 30, 1909) [nos. 2121-8], *The Sources of Catholic Dogma, Henry Denzinger's Enchiridion Symbolorum* (30th edition 1954, revised by Karl Rahner, S.J.), translated by Roy J. Deferrari (St. Louis, MO: B. Herder Book Co., 1957; reprint Powers Lake, ND: Marian House, n.d.).

doctrine, after expurgating any error of polytheism; or allegories and symbols, devoid of a basis of objective reality, set forth under the guise of history to inculcate religious and philosophical truths; or, finally, legends, historical in part and fictitious in part, composed freely for the instruction and edification of souls?
Reply: In the negative to both parts.

3. Historical Character of Certain Parts:
Whether in particular the literal and historical sense can be called into question, where it is a matter of facts related in the same chapters, which pertain to the foundations of the Christian religion; for example, among others, the Creation of all things wrought by God in the beginning of time; the Special Creation of man; the formation of the first woman from the first man, the oneness of the human race; the original happiness of our first parents in the state of justice, integrity, and immortality; the command given to man by God to prove his obedience; the transgression of the divine command through the devil's persuasion under the guise of a serpent, the casting of our first parents out of that first state of innocence; and the promise of a future restorer?
Reply: In the negative.

4. Interpretation:
Whether in interpreting those passages of these chapters, which the Fathers and Doctors have understood differently, but concerning which they have not taught anything certain and definite, it is permitted, while preserving the judgment of the Church and keeping the analogy of faith, to follow and defend that opinion which everyone has wisely approved?
Reply: In the affirmative.

5. Literal Sense:
Whether all and everything, namely, words and phrases which occur in the aforementioned chapters, are always and necessarily to be accepted in a special sense, so that there may be no deviation from this, even when the expressions themselves manifestly appear to have been taken improperly, or metaphorically or anthropomorphically, and either reason prohibits holding the proper sense, or necessity forces its abandonment?
Reply: In the negative.

6. Allegory and Prophecy:
   Whether, presupposing the literal and historical sense, the allegorical and prophetical interpretation of some passages of the same chapters, with the example of the Holy Fathers and the Church herself showing the way, can be wisely and profitably applied?
   Reply: In the affirmative.

7. Scientific Expression:
   Whether, since in writing the first chapter of *Genesis* it was not the mind of the sacred author to teach in a scientific manner the detailed constitution of visible things and the complete order of Creation, but rather to give to his people a popular notion, according as the common speech of the times went, accommodated to the understanding and capacity of men, the propriety of scientific language is to be investigated exactly and always in the interpretation of these?
   Reply: In the negative.

8. The word yom (day):
   Whether in that designation and distinction of six days, with which the account of the first chapter of *Genesis* deals, the word (dies) can be assumed either in its proper sense as a natural day, or in the improper sense of a certain space of time; and whether with regard to such a question there can be free disagreement among exegetes?
   Reply: In the affirmative.

Msgr. John E. Steinmueller, in *A Companion to Scripture Studies*, noted that Pope St. Pius X clarified the authority of the then Pontifical Biblical Commission. Presumably, the clarification by Pius X in 1907 is applicable to the declarations given on June 30, 1909. Msgr. Steinmueller wrote:

> In this century the *Pontifical Biblical Commission* has played an important role in the history of the Catholic Bible. It was instituted on Oct. 30, 1902, by Leo XIII to promote and direct Biblical studies. About five years later (Nov. 18, 1907) St. Pius X in his *Motu Proprio* determined the authority of its decisions. From these it follows: (1) that its decrees are neither infallible nor unchangeable; (2) that they enjoy the same authority as the decrees of the other Sacred Congregations; (3) that external as well as internal consent is required; (4) that this assent need not be absolute and irreformable; (5) that the formal object of these decrees is the security or nonsecu-

rity of any doctrine, that is, it does not stress so much the truth or falseness of a Biblical interpretation as it safeguards a revealed doctrine by declaring such and such an interpretation is unproven, untimely, and tends to weaken the teaching of the Church.[2]

---

2. Msgr. John E. Steinmueller, *A Companion to Scripture Studies* (Houston, TX: Lumen Christi Press, 1969), Vol. 1, p. 300.

—Appendix B—
Modernism—A Snapshot[1]

Principal Errors:
1. God cannot be known and proved to exist by natural reason.
2. External signs of Revelation, such as miracles and prophecies, do not prove the divine origin of the Christian religion and are not suited to the intellect of modern man.
3. Christ did not found a Church.
4. The essential structure of the Church can change.
5. The Church's dogmas continually evolve over time so that they can change from meaning one thing to meaning another.
6. Faith is a blind religious feeling that wells up from the subconscious under the impulse of a heart and a will trained to morality, not a real assent of the intellect to divine truth learned by hearing it from an external source.

Background:
   The heresy of Modernism was inspired by tendencies prevalent in liberal Protestantism and secular philosophy. It was influenced by Kant's subjectivist ideas, by the evolutionary metaphysics of Hegel, by liberal Protestant theologians and biblical critics (such as Schleiermacher and von Harnack), by the evolutionary theories of Darwin, and by certain liberal political movements in Europe. Its centers were in France, England, Italy and Germany. Two of its leading figures were Fr. Alfred Loisy, a French theologian and Scripture scholar, and Fr. George Tyrell, an Irish-born Protestant who became a Catholic, though he was dismissed from the Jesuits in 1906.

Condemnation:
   By Pope St. Pius X in the *Syllabus Condemning the Errors of the Modernists—Lamentabili Sane* (July 3, 1907) and the *Encyclical on the*

---
1. Reprinted from *Christian Order,* London (February, 1996).

Doctrines of the Modernists—Pascendi Dominici Gregis (September 8, 1907).

Modernist Errors: (Extracts taken from *Lamentabili Sane*)
- The Magisterium of the Church, even by dogmatic definitions, cannot determine the genuine sense of the Sacred Scriptures.
- Since in the Deposit of Faith only revealed truths are contained, in no respect does it pertain to the Church to pass judgment on the assertions of human sciences.
- When the Church proscribes errors she cannot exact any internal assent of the faithful by which the judgments published by her are embraced.
- Divine inspiration does not extend to all Sacred Scripture so that it fortifies each and every part of it against an error.
- In many narratives the Gospel writers related not so much what is true, as what they thought to be more profitable for the reader, although false.
- The divinity of Jesus Christ is not proved from the Gospels, but is a dogma which the Christian conscience has deduced from the notion of the Messias.
- When Jesus was exercising His ministry He did not speak with the purpose of teaching that He was the Messias, nor did His miracles have as their purpose to demonstrate this.
- It may be conceded that the Christ whom history presents is far inferior to the Christ who is the object of faith.
- Christ did not always have the consciousness of His Messianic dignity.
- The Resurrection of the Saviour is not properly a fact of the historical order, but a fact of the purely supernatural order, neither demonstrated nor demonstrable, and which the Christian conscience gradually derived from other sources.
- It was foreign to the mind of Christ to establish a Church as a society upon Earth to endure for a long course of centuries; rather, in the mind of Christ the kingdom of Heaven together with the end of the world was to come presently.
- The organic constitution of the Church is not immutable, but Christian society, just as human society, is subject to perpetual evolution.
- Simon Peter never even suspected that the primacy in the Church was entrusted to him by Christ.
- The progress of the sciences demands that the concepts of

Christian doctrine about God, Creation, Revelation, the person of the Incarnate Word and Redemption be readjusted.
- Present-day Catholicism cannot be reconciled with true science unless it be transformed into a kind of non-dogmatic Christianity, that is, into a broad and liberal Protestantism.

BIBLIOGRAPHY

Aardsma, Gerald A. "Geocentricity and Creation," in *Impact,* No. 253. El Cajon, California: Institute for Creation Research, July, 1994.
Allis, Oswald T. *The Five Books of Moses.* Phillipsburg, NJ: Presbyterian and Reformed Publishing Co.
Amerio, Prof. Romano. *Iota Unum: A Study of Changes in the Catholic Church in the XXth Century.* Kansas City, MO: Sarto House, 1996.
Andrews, E. H. *God, Science & Evolution.*
Aquinas, St. Thomas. *Summa Theologica.* Vol. 1. Westminister, MD: Christian Classics.
Austin, Steve. *Mount St. Helens: Explosive Evidence for Catastrophe!* (video). N. Santee, CA: Institute for Creation Research.
Baker, Sylvia. *Bone of Contention.* Hertfordshire, England: Evangelical Press, 1981.
Barbarito, Archbishop Luigi (then Apostolic Pro-Nuncio, Australia), correspondence, 1983.
Barnes, Thomas G. "Origin and Destiny of the Earth's Magnetic Field," *ICR Technical Monograph.* No. 4. San Diego: Creation-Life Publishers,* 1973.
Barr, Prof. James, in a personal letter dated April 23, 1984 to David C. Watson and quoted by Dr. Clifford Wilson in *Visual Highlights of the Bible,* Vol. 1. Australia: Pacific Christian Ministries, 1993.
Barrosse, Fr. Thomas, CSC. *God Speaks to Men: Understanding the Bible.* Notre Dame, IN: Fides Publishers, Inc., 1964.
Becker, Fr. David. "The Divine Pedagogy," in *Watchmaker* newsletter of Morning Star Catholic Origins Society, P.O. Box 189, Shade Gap, PA 17255, April, 1995. (Now c/o St. Stephen's, 303 Lincoln Way E., McConnellsburg, PA 17233.) Also, private correspondence, September, 1994.
Behe, Michael J. *Darwin's Black Box: The Biochemical Challenge*

---

* Note: It is our understanding that Creation-Life Publishers is no longer in operation and that inquiries for their books should be directed to Institute for Creation Research (a Protestant ministry), P.O. Box 2667, El Cajon, CA 92021 or to Master Books, P.O. Box 727, Green Forest, Arkansas 72638-0727. —*Publisher,* 1999.

to Evolution. New York: The Free Press, 1996.
——"Molecular Machines: Experimental Support for the Design Interface," in *Watchmaker*. Shade Gap, PA: Catholic Origins Society, Jan.-Feb. 1996.

Benedetto, Fr. A. J., S.J. "The Problem of Evil," in *Fundamentals in the Philosophy of God*. New York: MacMillan Co., 1963.

Berger, Peter L., Berger, Brigitte and Kellner, Hansfried. *The Homeless Mind: Modernization and Consciousness*. Middlesex, England: Penguin Books, Ltd., 1979.

Bergman, Jerry. "Advances in Integrating Cosmology: The Case of Cometesimals," in *Creation Ex Nihilo Technical Journal*. Creation Science Foundation (now called Answers in Genesis), Acacia Ridge D.C., Qld., Australia/Answers in Genesis (a Protestant ministry), Florence, KY. Vol. 10, Part 2, 1996.

Borel, Emil. *Elements of the Theory of Probability*. New Jersey: Prentice Hall, 1965.

Boulet, Fr. André. *Création et Rédemption*. Chambray-les-tours Cedex, France: C. L. D. (Publisher), 1995.

Boudreaux, Edward A. "Hydrogen-Helium Ionic Model of Solar Evolution." Department of Chemistry, University of New Orleans, September 14, 1990.

Bracher, Karl Dietrich. *The German Dictatorship*. Middlesex, England: Penguin Books, Ltd., 1978.

Brenner, Joe. *The Pilot's Guide to the Universe*. Scottsdale, Arizona: unpublished manuscript, 1993.

Brown, Fr. Raymond E. *The Critical Meaning of the Bible*. London: Cassell Ltd., Geoffrey Chapman Book, 1982.

Brown, Walt. "The Foundations of the Great Deep." *In the Beginning*. Phoenix, AZ: Center for Scientific Creation, 1995.

Carroll, Warren H. *A History of Christendom*, Vol. 1. Front Royal, VA: Christendom College Press, 1985.

The Catechism of the Council of Trent. Rockford, IL: TAN Books and Publishers, Inc., 1982.

Chesterton, G. K. *The Everlasting Man*. Garden City, New York: Image Books, 1955.

Chui, Chris. *Did God Use Evolution to "Create"?* Canoga Park, CA: Logos Publishers, 1993.

Clark, Marlyn E., M.S. *Our Amazing Circulatory System . . . By Chance Or Creation?* San Diego, CA: Creation-Life Publishers, 1976.

Cohen, I. L. *Darwin Was Wrong—A Study in Probabilities*. Greenvale, NY: New Research Publications, 1984.

Crane, Paul, S.J. *Christian Order*. London, March, 1982.
Cummins, Harold and Charles Midlo. *Finger Prints, Palms and Soles: An Introduction to Dermatoglyphics*. Philadelphia: The Blakiston Company, 1943.
D'Abrera, Bernard. *Butterflies of the Neotropical Region, Part VI Riodinidae*. Black Rock, Victoria, Australia: Hill House, 1994.
Darwin, Charles. *The Origin of Species*. Middlesex, England: Penguin Books Limited, 1979.
Davies, Paul. *The Mind of God*. London: Penguin Books Ltd., 1992.
Davis, Percival, *et al*. *Of Pandas and People: The Central Question of Biological Origins*. Dallas, 1989.
Dawkins, Richard. *The Blind Watchmaker*. Essex, England: Longman House, 1986.
de Chardin, Pierre Teilhard. *The Phenomenon of Man*. London: William Collins Sons & Co. Ltd. and Fount Paperbacks, London, 1980.
Denton, Michael. *Evolution: A Theory in Crisis*. London: Burnett Books Limited, 1985. (U.S. distributor: Woodbine House, Bethesda, MD.)
Dillow, Joseph C. *The Waters Above: Earth's Pre-Flood Vapor Canopy*. Chicago: The Moody Bible Institute of Chicago, 1981.
Dobzhansky, Theodosius. *Evolution*.
Douay-Rheims Bible (1899 edition). Rockford, IL: TAN Books and Publishers, Inc., 1989.
Duggan, G. H. *Teilhardism and the Faith*. Cork: The Mercier Press, 1968.
Eldridge, Niles. *Discovery*. Melbourne: ABC Television, July 29, 1982.
Encyclopaedia Brittanica, Macropaedia, Vol. 7, 1979, 23.
Ewing, Fr. J. Franklin. "Human Evolution—1956," in *Anthropological Quarterly*, Washington, DC: Catholic University of America Press, October, 1956.
Fehlner, Fr. Peter Damian. "In the Beginning," Vol. XXXIII, in *Christ to the World*. Rome: Via di Propaganda, 1988.
Frankl, Viktor E. *Psychotherapy and Existentialism*. New York: A Touchstone Book, Simon and Schuster, 1976.
Fuller, Rev. Reginald C., *et al. A New Catholic Commentary on Holy Scripture*. London: Thomas Nelson and Sons Ltd., 1975.
Funk & Wagnalls New Encyclopaedia, Vol. 15. New York: Funk & Wagnalls, Inc., 1978.
Gaudium et Spes. Vatican Council II, Pastoral Constitution on the Church in the Modern World.

Gentry, Robert V. *Creation's Tiny Mystery*. Knoxville, TN: Earth Sciences Associates, 1988.

Gish, Duane T., *Evolution, The Fossils Say No!* San Diego, CA: Creation-Life Publishers, 1980.

Gould, Stephen Jay. *Everyman*. Melbourne: ABC Television, September 2, 1982.

——"The Piltdown Conspiracy," in *Natural History*, Vol. 89, No. 8. American Museum of Natural History, August, 1980.

Grebe, John J. "DNA Studies in Relation to Creation Concepts," in *Why Not Creation?* Grand Rapids, MI: Baker Book House, 1970.

Green, William Henry. *The Higher Criticism of the Pentateuch*. Grand Rapids, MI: Baker House Books.

Grigg, Russell. "Ernst Haeckel: Evangelist for Evolution and Apostle of Deceit," in *Creation Ex Nihilo*. Creation Science Foundation Ltd. (now called Answers in Genesis), Acacia Ridge D.C. 4110, Qld., Australia, 1996. (U.S. distributor: Answers in Genesis, P.O. Box 6330, Florence, KY 41022.)

Hahn, Scott. *St. Paul's Letter to the Hebrews*. West Covina, CA: St. Joseph Communications, audio tapes.

Haigh, Paula P. *St. Thomas Aquinas: Creationist for the 21st Century* (unpublished manuscript). Nazareth, KY: 1991.

Ham, Ken and Snelling, Andrew and Wieland, Carl. "What About Continental Drift?" in *The Answers Book*. Acacia Ridge D.C., Qld., Australia: Creation Science Foundation (Answers in Genesis), 1990. (U.S. publisher: Master Books, P.O. Box 727, Green Forest, AR 72638-0727.)

Hammes, John A. *Human Destiny: Exploring Today's Value-Systems*. Huntington, IN: Our Sunday Visitor, Inc., 1978.

Hardon, Fr. John A., S.J. *The Catholic Catechism*. London: A Geoffrey Chapman Book published by Cassell Ltd., 1977.

Harrison, Fr. Brian W., O.S. "Bomb-Shelter Theology," in *Living Tradition*. Rome: Sedes Sapientiae Study Center, May, 1994.

——"Did The Human Body Evolve Naturally? A Forgotten Papal Encyclical" in *Living Tradition*, Nos. 73-74, Jan.-March 1998.

Hitching, Francis. *The Neck of the Giraffe: Or Where Darwin Went Wrong*. London: Pan Books Ltd, 1982.

Holt, Roy D. "Evidence for a Late Cainozoic Flood/post-Flood Boundary," in *Creation Ex Nihilo Technical Journal*, Vol. 10, Part 1. Acacia Ridge D.C., Qld., Australia, 1996.

Hulsbosch, Fr. A., O.S.A. *God's Creation: Creation, Sin and Redemption in an Evolving World*. London: Sheed and Ward.

Humphreys, Russell. *Starlight and Time: Solving the Puzzle of Distant Starlight in a Young Universe*. Colorado Springs, CO: Creation-Life Publishers, Master Books (now in Green Forest, AR), 1994.

Huxley, Sir Julian. *Essays of a Humanist*. Pelican Books, 1969.

Jastrow, Robert. "Have Astronomers Found God?" in *New York Times Magazine*, June 25, 1978.

Johnson, J. W. G. *The Crumbling Theory of Evolution*. Los Angeles: Perpetual Eucharistic Adoration, Inc., 1986.

Johnson, Paul. *Modern Times*. London: Orion Books Limited, 1992.

Johnson, Phillip E. *Darwin on Trial*. Washington, DC: Regnery Gateway, 1991. (Reprinted by InterVarsity Press, Westmont, IL.)

———*Reason in the Balance: The Case Against Naturalism in Science, Law & Education*. Downers Grove, IL: InterVarsity Press, 1995.

Jones, Fr. Alexander, S.T.L, L.S.S. *Unless Some Man Show Me*. London: Sheed and Ward, 1961.

Julien, Pierre Y. and Berthault, Guy. *Fundamental Experiments on Stratification*. (Video.) Monaco.

Kazmann, Raphael G. *Geotimes*, Jan. 9, 1979, quoted in Gentry: *Creation's Tiny Mystery*.

Kitchen, K. A. *The Bible in Its World: The Bible and Archaeology Today*. Exeter, England: The Paternoster Press, 1977.

Koestler, Arthur. *The Sleepwalkers*. London: Penguin Books, 1986.

Kramer, Fr. William C.PP.S. *Evolution & Creation: A Catholic Understanding*. Huntington, IN: Our Sunday Visitor, 1986.

Lane, David H. *The Phenomenon of Teilhard: Prophet for a New Age*. Macon, Georgia: Mercer University Press, 1996.

Leakey, Richard. *The Making of Mankind*. Melbourne: ABV2 Television, April 24, 1983.

LeGrand, Leon. "Philberth Under Fire," in *the Message*. Vic, Australia: Centre for Peace, June, 1995.

Lejeune, Jerome. "On the Mechanisms of Speciation." Paris: Proceedings of the Sessions of the Biological Society, 1975.

Lescoe, Fr. Francis J. *Existentialism: With or Without God*. Staten Island, NY: Society of St. Paul, 1974.

Lewin, Roger. "Evolutionary Theory Under Fire," in *Science*, 210 (Nov. 21, 1980):883-887.

Lewis, C. S. *Christian Reflections*. Glasgow: William Collins Sons & Co. Ltd., 1985.

Long, Fr. Valentine. *Upon This Rock*. Chicago: Franciscan Herald Press, 1982.

Lubenow, Marvin L. *Bones of Contention: A Creationist Assessment of*

Human Fossils. Grand Rapids, MI: Baker Book House Company, 1992.

Macquarrie, John. *An Existentialist Theology: A Comparison of Heidegger and Bultmann.* Middlesex, England: Penguin Books Ltd., 1980.

Marra, William A. *Happiness and Christian Hope: A Phenomenological Analysis.* Chicago: Franciscan Herald Press, 1979.

McCarthy, Msgr. John F., O.S. *The Science of Historical Theology: Elements of a Definition.* Rockford, IL: TAN Books and Publishers, Inc., 1991.

——"Not the Real Genesis 1," in *Living Tradition.* Rome: Sedes Sapientiae Study Center, March, 1994.

McKee, Fr. John. *The Enemy Within the Gate.* Houston: Lumen Christi Press, 1974.

Molnar, Thomas. *Christian Humanism: A Critique of the Secular City and Its Ideology.* Chicago: Franciscan Herald Press, 1978.

Moore, Ruth and the editors of *Life.* Nederland: Time-Life International, 1964.

Morris, Henry M. and Parker, Gary E. *What Is Creation Science?* San Diego, CA: Creation-Life Publishers, 1983.

——*The Troubled Waters of Evolution.* San Diego: Creation-Life Publishers, 1980.

Most, Fr. William. *Free From All Error.* Libertyville, IL: Prow Books/Franciscan Marytown Press, 1990.

Muggeridge, Anne Roche. *The Desolate City: Revolution in the Catholic Church.* New York: Harper & Row, Publishers, Inc., 1990.

Mukerjee, Madhusree. "Explaining Everything." In *Scientific American* (January 1996).

Nelson, Ethel R., Broadberry, Richard E. and Ginger Tong Chock, *God's Promise to the Chinese.* Dunlap, TN: Read Books, 1997.

Noll, Richard. *The Jung Cult: Origins of a Charismatic Movement.* Princeton, NJ: Princeton University Press, 1994.

Norman, Trevor and Setterfield, Barry. *The Atomic Constants, Light and Time.* Technical report. Flinders University of South Australia, August, 1987.

O'Connell, Canon J. P., ed. *The Roman Martyrology,* Typical Edition approved by Pope Benedict XV. MD: The Newman Press, 1962.

O'Reilly, Sean. *Bioethics and the Limits of Science.* Front Royal, VA: Christendom Publications, 1980.

Ott, Dr. Ludwig. *Fundamentals of Catholic Dogma.* Cork: The Mercier

Press, Ltd., 1955; Rockford, IL: TAN, 1974.

Paley, William. *Natural Theology: Or Evidences of the Existence and Attributes of the Deity Collected from the Appearances of Nature.* 1802.

Philberth, Fr. Bernhard. *Revelation.* Australia, 1994.

Pope John Paul II. *Fides et Ratio* encyclical, September 18, 1998.

——*Crossing the Threshold of Hope.* London: Random House, 1994.

Pope Leo XIII. *Providentissimus Deus* encyclical, November 18, 1893. *Arcanum Divinae Sapientiae* encyclical.

Pope Paul VI. *Credo of the People of God*, Section 16.

Pope St. Pius X. *Pascendi Dominici Gregis* encyclical, September 8, 1907. *Lamentabili Sane*, July 3, 1907.

Pope Pius XII. *Humani Generis.* Denzinger, Wanderer Press trans.

——Address to the First International Congress of Medical Genetics, September 7, 1953.

Radday, Yehuda T. and Shore, Haim. *Genesis: An Authorship Study in Computer-Assisted Statistical Linguistics.* Report by Rome Biblical Institute Press, 1985.

Ratzinger, Joseph Cardinal, interviewed by Messori, Vittorio. *The Ratzinger Report.* Ormond, Australia: John XXIII Fellowship Co-Op Ltd., 1985, and San Francisco: Ignatius Press.

Raup, David M. "Conflicts Between Darwin and Paleontology," in Field Museums of Natural History, Vol. 50, No. 1, January, 1979.

ReMine, Walter James. *The Biotic Message.* St. Paul, MN: St. Paul Science, 1993.

Ross, Hugh. *Creation and Time: A Biblical and Scientific Perspective on the Creation-Date Controversy.* Colorado Springs, CO: NavPress Publishing Group, 1994.

Ruffini, Cardinal Ernesto. *The Theory of Evolution Judged by Reason and Faith.* New York: Joseph F. Wagner, Inc., 1959. English translation by Fr. Francis O'Hanlon, Melbourne.

Sarfati, Jonathan. "Life from Mars?" in *Creation Ex Nihilo Technical Journal.* Creation Science Foundation (now called Answers in Genesis), Acacia Ridge D.C., Qld., Australia/Answers in Genesis, Florence, KY. Vol. 19, No. 1, Dec., 1996 and Vol. 10, Part 3, 1996.

——"Exploding Stars Point to a Young Universe," in *Creation.* Creation Science Foundation (now called Answers in Genesis), Acacia Ridge D.C., Qld., Australia/Answers in Genesis, Florence, KY. Vol. 19, No. 3, June-Aug., 1997.

Scarrabelotti, Gary. *Existentialism and Catechetics.* Melbourne, Australia: John XXIII Fellowship Cooperative Limited, 1977.

Schumacher, E. G. *A Guide for the Perplexed*. London: Sphere Books Limited, 1978.

Sennott, Br. Thomas Mary. *The Six Days of Creation*. Cambridge: The Ravengate Press, 1984.

Sheehan, Fr. Michael. *Apologetics and Catholic Doctrine, Part II*. Dublin: Gill, 1962.

Simpson, George Gaylord. *Tempo and Mode in Evolution*. New York: Columbia University Press, 1944.

——"The Biological Nature of Man," in *Science*, Vol. 152, April 22, 1966.

Singer, Peter and Kuhse, Helga. "The Moral Status of the Embryo," in *Test-Tube Babies*. Melbourne: Oxford University Press, 1982.

Siri, Joseph Cardinal. *Gethsemane: Reflections on the Contemporary Theological Movement*. Chicago: Franciscan Herald Press, 1982.

Skinner, B. F. *Beyond Freedom and Dignity*. Middlesex, England: Penguin Books, Ltd., 1979.

Slusher, Harold S. "Age of the Cosmos," in *ICR Technical Monograph*, No. 9. San Diego, CA: Institute for Creation Research, 1980.

Smith, Fr. Richard F., S.J. "Inspiration and Inerrancy" in *The Jerome Biblical Commentary*, Section 66:75. Prentice-Hall, Inc., 1968.

Smith, Wolfgang. *Cosmos and Transcendence: Breaking through the Barrier of Scientific Belief*. Peru, IL: Sherwood Sugden and Co., 1990.

——*Teilhardism and the New Religion*. Rockford, IL: TAN Books and Publishers, Inc., 1988.

Snelling, Andrew A. "Creating Opals," in *Creation Ex Nihilo*, Vol. 17, No. 1, Dec., 1994.

Snelling, Andrew and Malcolm, David, "Earth's Unique Topography," in *Creation Ex Nihilo*, Vol. 10, No. 1, 1987.

Solzhenitsyn, Alexander. *The Gulag Archipelago*, Vol. 1. Collins/Harvill Press and Fontana, 1974.

Spieker, Edmund M. "Mountain-Building and Nature of Geologic Time-Scale," in *Bulletin of the American Association of Petroleum Geologists,* Vol. 40 (August 1956): 1803.

Steichen, Donna. *Ungodly Rage: The Hidden Face of Catholic Feminism*. San Francisco: Ignatius Press, 1992.

Steinmuller, Msgr. John E., S.T.D. *A Companion to Scripture Studies*, Vol. 1. Houston, Texas: Lumen Christi Press, 1969.

Strickling, James, "A Statistical Analysis of Flood Legends," in *Creation Research Society Quarterly*, Vol. 9, No. 3. Terre Haute, IN, 1972.

Sulloway, Frank J. *Freud, Biologist of the Mind: Beyond the Psychoanalytic Legend*. Massachusetts/London: Harvard University Press, 1992.

Taylor, Charles V. *The Oldest Science Book in the World*. Australia: Assembly Press Pty Ltd., 1984.

Taylor, Paul S. *The Illustrated Origins Book*. Mesa, Arizona: Films for Christ Association Inc., 1989.

Thornton, E. M. "O Ye of Little Faith," in *Daylight Origins Society Magazine*, Herts, AL3, 6BL UK, Spring, 1996.

Vitz, Paul C. *Psychology as Religion: The Cult of Self Worship*. Hertfordshire, England: Lion Publishing, copyright William B. Eerdmans Publishing Co., Grand Rapids, MI, 1979.

Von Hildebrand, Dietrich. *Trojan Horse in the City of God: The Catholic Crisis Explained*. 1967; republished: Manchester, NH: Sophia Institute Press, 1993.

Watters, Thomas R. *Planets—A Smithsonian Guide: The Story of Our Solar System—From Earth to the Farthest Planet and Beyond*. New York: Macmillan USA, 1995.

Whitcomb, Jr., John C. *The World that Perished*. London: Evangelical Press, 1974.

Wilder-Smith, Arthur E. *A Basis for a New Biology*. Telos International NR, 1976.

Wilders, Peter. "Evolution: End of the Story?," in *Christian Order*, London (April 1990): 253.

Williams, Emmett L., Ed. *Thermodynamics and the Development of Order*. Norcross, Georgia: Creation Research Society Books, 1981.

Wilson, Clifford. *Visual Highlights of the Bible*, Vol. 1. Boronia, Victoria, Australia: Pacific Christian Ministries, 1993.

Wiseman, P. J. "New Discoveries in Babylonia about Genesis." 1936, republished by Wiseman, D. J. in *Clues to Creation in Genesis*. London: Marshall, Morgan and Scott, 1977.

Wojtyla, Karol. "Love and Responsibility," quoted by Woznicki, Fr. Andrew N. in *A Christian Humanism: Karol Wojtyla's Existential Personalism*. New Britain, CT: Mariel Publications, 1980.

Woodmorappe, John. *Noah's Ark: A Feasibility Study*. El Cajon, CA: Institute for Creation Research, 1996.

Wysong, R. L. *The Creation-Evolution Controversy*. Midland, MI: Inquiry Press, 1980.

Young Age of the Earth, The. (Video.) Alpha Productions, 1994.

INDEX

A

Aardsma, Gerald, 235
abortion, xxiv, 41, 43, 187, 205, 235, 282, 292-295, 311, 316, 358
Abraham and Melchisedech, 86
Abraham—called to sacrifice Isaac, 250
absolute principles, 43, 92, 184, 238, 239, 240, 291, 294, 295, 313, 342, 345, 347, 353
Acheulean hand ax, 108
acquired characteristics, 27, 287, 330, 332
Adam
— did not arrive as a baby boy, 183
— could not have had animal "parents," 183, 184, 188-195
— created as adult from non-living matter, x, xvi, xvii, 46, 58-60, 183, 192, 193, 264
— hiding in Garden of Eden, 219
— lived less than 10,000 years ago, 248
— named "all the beasts of the earth" including the dinosaurs, 61
— one person––not a group, 46, 197, 198
Adam and Eve
— could not have come from animal parents, 183, 190-199
— created as adults in state of high perfection, 58-60, 85, 192, 193
— created in state of holiness and justice, 177, 178, 188, 189
— gave rise to great human diversity, 24, 158-10
— had brilliant intelligence, 181
— individuals (our first parents), xviii
— made male and female from the beginning of Creation, 256
— worshipping at east gate of Garden of Eden, 84
aestivation, 68

Age of Universe
— appearance of old age, 155, 238, 264
— conceptual problems, 245, 269
— no credible third position, 242, 243
— onus of proof, xxviii, 268, 269
— question of age, 13, 18, 21, 47-49, 64, 79, 107, 108, 110, 132, 136-165, 241-269
— relevance of age question, xxviii, 64, 136, 137, 141, 241, 242, 268, 269
— three distinct beliefs, 136
aging effects, 85, 180
alienation, 63, 85, 323, 324, 336, 354
allegorical sense, 194, 209
altruism, 167
Allis, Oswald T., 228
Amerio, Romano, 351
amino acids, 6, 7, 30, 31, 35, 120, 125, 134
amphibians, 111-116
anagogical sense, 209
analogous structures, 56
analogy of faith, 209
ancestry, 18, 22, 30, 37, 85, 105, 112, 116, 129, 147, 275, 284
Andrews, E. H., 17, 28, 43
angels, xvi, xxix, 44, 46, 47, 63, 91, 92, 222, 257, 262, 316, 334, 335
angels, power of good, 91, 92, 222, 222-7, 316
animals, birds—and the Ark, 67, 68
animals, etc.—traveling around Earth, 67-69
Anthropic Principle, 14, 47, 166
anthropomorphism, 64, 137
apologetics, xiii, xviii
Aquinas, St. Thomas, xvi, 59, 190, 209, 210, 257-262, 266, 267
Aquinas—on the works of Creation, 261, 262
Arcanum Divinae Sapientiae encyclical, 192, 193

archaeology, 9, 144-147
Archaeopteryx, 113, 115
Ark of Noah
— feeding creatures on, 62, 68
— general notes, 62-68, 83, 86, 248-252, 261, 272, 274
— how many creatures were aboard, 67, 68, 249
— order aboard the Ark, 68
— size of modern ship, 249
artificial selection, x, 126
asexual reproduction, 61
assert, intention of sacred writer to, 211-218, 236, 242, 252, 253, 257-264, 268
asteroid belt of rocks, 51, 52, 83
atmosphere, Earth's early, 69, 122
atmospheric pressure, pre-Flood, 69, 74, 75
Augustine, St., 76, 179, 184, 219, 257-260
Austin, Steve, 67, 143
australopithecines, 13, 107-109, 144, 275
avian lung data, 114, 115, 158-8
Ayers Rock (Uluru) data, 72

B

Babel, Tower of, 84
Baker, Sylvia, 128, 129, 140
Baptism, 189
baramin, 54
Barbarito, Luigi, 202
Barnes, Thomas G., 153
Barr, James, 260, 261
Barrosse, Thomas, 244, 259, 260
Barrow, John, 169
Basking Shark, 143
bats, 70, 111, 115, 138
Beatific Vision of God, 46
Becker, David, 255, 256, 356, 357
"becoming," state of, xviii, 47, 266, 289, 343, 345, 351
behavior of predatory creatures—affected by the Fall?, 62, 63, 67, 89-92, 178-181
Behe, Michael J., 15-17
being, xviii, xix, 4, 47, 162, 266-268, 289, 306, 342-353
beliefs, crucial to Origins debate, xxi, xxvii, 3, 5, 8-11, 39-44, 202, 295

Bellarmine, St. Robert, 231, 233
Benedetto, A. J., 335-339
Beresovka mammoth, 72-74
Berger, Peter *et al.*, 324
Bergman, Jerry, 159, 160
Berthault, Guy, 20, 102, 142
bias, human, 50, 163, 211, 233
Big Bang concept—no center and no boundary, 161
Big Bang theory, 8, 12, 50-53, 133, 155-159, 161, 165, 166, 169, 170, 203, 254, 263, 264, 277
big institutions, etc., 323
biochemistry, 14-16, 43, 120, 198, 204
biochemistry—Lilliputian challenge to Darwin, 15-17
"Biogenetic Law," 116, 328
birds' behavior, 57
birds' evolution?, 27, 57, 113-115, 158-8
birth rates, stagnation of, 295
black holes/white holes discussion, 161-164
bloodshed, xx, 90, 91, 178, 179
Blythe, Edward, 23
Bonaventure, St., xv, xvi, 260
border ceremony, Chinese, 84
Borel, Emil, 33
bottleneck effect, 23
boundary around Universe?, 78, 161-164, 203
Boudreaux, Edward, 156
Boulet, Andre, 180
Bracher, Karl D., 284, 285
brain data, hominid/human, 107, 167
breeding, x, 126
Brenner, Joe, 75-77, 81, 82
Brincard, Henri, 181
Broom, Robert, 70
Brown, Driver, Briggs, 37
Brown, Raymond E., 221, 222
Brown, Walt, 71-75, 78, 107
Buckland, William, 19
Buddhism, 326, 334
Bultmann, Rudolf, 187, 242, 349
butterflies, 57, 58, 158-6, 158-7, 167

C

Calaveras skull (human), 107
Cambrian Period, 18, 100, 141
Camus, Albert, 347
canopy, water vapor, 75-83

canyon formation, rapid, 66, 67, 143
Carbon-14 dating, 79, 138, 140, 153
Carlsbad Caverns, 138
carnivores/herbivores, 178-180
"carriers" of Modernism, xxii, xxx
Carroll, Warren H., 92
Cartwheel Galaxy, 222-6
Castanedolo skeletons (human), 107
cataclysmic global Flood, 65-71
catastrophism, xxiii, 19, 64-75, 82, 103, 109, 110, 138, 141-144, 179, 274
Catholic Church
— authority from Christ to declare on both Scripture and doctrine, 185, 207, 208
— catechetics, 311, 344, 350-352, 356-361
— *Catechism of the Catholic Church* (1992), xxvi, xxxi, 177, 178, 180, 184, 189, 190, 196, 198, 199, 206, 236, 248, 249, 267, 268, 273-278, 299, 343, 356
— *Catechism of the Council of Trent*, 46, 88, 218, 222, 249, 266
— collapse of belief, xxi, xxvii, 189, 202, 238, 239, 311, 314, 351, 360
— doctrine declared in Tradition cannot be overturned, 188
— human evolution—impermissible/irreconcilable with doctrine, 191, 197
— restoration of Catholicism, xxi, 361
— sense of the sacred, loss of, xxiii
Cat's Eye Nebula, 222-6
Cause, First, 45, 58, 132, 133, 169, 262, 267, 276
cave-paintings, man's "signature," 146
Cercle Scientifique et Historique (CESHE), 20
chance, xxiii, 5, 6, 7, 10, 18, 22, 25-36, 43, 57, 58, 60, 120-123, 127, 165, 166, 181, 187, 342
change, xv, xviii, 23, 25, 30, 44, 126, 242, 265, 271, 277, 313
Chesterton, G. K., 146
Christ, Jesus
— accepted Old Testament as real history, 217, 218, 248, 274
— atonement, voluntary, 63, 64
— claimed to be God, 218
— Creator, 45, 181, 217, 228, 248, 256, 261
— doubt cast on His words, 250, 256
— genealogies, 242, 246-248, 261, 268
— Incarnate Saviour, xxv, xxvi
— involved as divine Principal Author of Scripture, 217, 218, 256
— Kingship of Christ, xxiii, 45, 282
— Mystical Body of, 335, 337
— New Age "Christ," 326
— referred to Genesis/Flood/Moses, 228, 248-252, 274
— Redeemer, 45, 63, 87, 217, 237, 248, 256, 309
— "second" Adam, xvii, 45, 217, 222, 222-7, 236
— separating Christ of "history" from the divine Christ of faith, 285, 309, 327, 349
— trustworthy, reliable Creator who cannot deceive, xxx, 9, 186, 217, 240, 254-256, 264, 276, 358
— Word made flesh, 45
— wept over Lazarus, 181
chromosomes, 56, 61, 119, 123-125
Chui, Chris, 155, 156
cilium, 16, 17
circular (evolution) reasoning, 57, 143
civilization, 41, 284, 295, 350
cladistics, 17, 116
Clark, Marlyn E., 28
classification problems, 54-57, 116
class struggle, 239, 285-287
clay tablets, 222, 225-228
cleaner fish, 158-5
clock, Universe likened to, 132
"closed" Origins mindset, xvii
clothes—worn only after the Fall, 63, 76
cloud systems formation, 83
coal, 65, 66, 101, 105, 142
coalified wood, 149-150
coelacanth, 143
Cohen, I. L., 33
Coliseum, the Roman, 222-8
colophon, 226
"collective unconscious," 327-332
comets, 151
Communism, 200, 282, 283, 286, 287, 316, 361
complementarity of sexes, 59, 60
complementary and or concurrent Creation, 47, 48, 53, 59, 60, 128, 137, 278
complexity, irreducible, 15-17, 120, 123, 129, 204
computer technology, 6, 44, 58

concordism, 49, 217, 259, 298
confusion, xxi, 23, 25, 92, 174, 184, 188, 218, 309-317, 351, 360
conscience, 167, 294, 316
consciousness, human, 41, 43, 85, 168, 174, 181, 286, 289-294, 304, 305, 313, 317, 324-327, 334, 342, 352, 361
conservative/liberal definitions, xxx
continents formation, 67-72
continental shelves, 72
contraception, 187, 205, 235, 311, 358
convergence, 55, 99, 113, 127, 205, 303
Copernican model—included circular orbits, not elliptical orbits, 234
Copernican revolution, 174, 228-240
Cosmic Microwave Radiation, 156
cosmology, 239, 265
Council of Trent, 45, 46, 88, 184, 188, 191, 196, 197, 219, 222, 249, 266
covenant, Flood, 251, 252, 275
Cram, Len, 139
Crane, Paul, 314
Creation
— clues to Creation events, 46
— comparison model, 10
— complementary, 53, 128, 137, 278
— Day, 7th day divine motive, 256
— Days 1 to 6 creation scenarios, 80, 156, 161-164, 254-256, 262-265
— deduction of existence of Creator is fully scientific, xxvii, 343
— events, not interventions, 47
— from nothing (*ex nihilo*), 45, 59, 296, 303
— groans in travail, 88, 187, 237, 273, 274
— legends in all continents, 85
— "mature" creation, 264
— partial revelation, 74, 264
— Progressive Creation, 270-278
— repeatedly deemed "good," 274
— two accounts (*Genesis* 1, 2), 220, 253
Creator, transcendent, 12, 39, 41, 44, 132, 174, 187, 289
Crick, Francis, 119
Cro-Magnon Man, 108
cross fertilization, 158-5
crossopterygians, 111, 112
cubit size, 249
cultural history of man, 145, 146
culture shock—brought on by great scientific innovators, 239
Cummins, Harold, 33-35
"cursed is the ground . . . ," 83, 88, 179

D

D'Abrera, Bernard, 57, 58
damnation, eternal, 338
"dark matter," 157
Darwin, Charles
— attempted to do without Creator, 39
— Darwin, xii, 3, 13, 15-17, 19, 21-22, 26, 32, 36, 39, 40, 99, 103-105, 128, 144, 233, 239, 285, 286, 290, 298, 330, 331
— Darwinian/Darwinism, xxiii, 26, 27, 36, 40-42, 61, 109, 110, 116, 124, 290, 298, 315, 331, 358, 359
— did not explain origins, 40
— failed his own test of credibility, 36, 40
— his own "gravest objection," 104
— impact on Freud's scientific generation, 331
— "innumerable transitional forms," 104
— origins—"pure philosophy," 40
— Social Darwinism, 42, 282-287
dating methods, 136-165, 243, 273
Davies, Paul, 14, 41, 47, 157, 169
Davis, Kenyon, Thaxton *et al.*, 22, 26, 27, 31, 53, 56, 57, 104
Dawkins, Richard, 14, 15
Dawson, Charles, 306
days of creation, 46, 47, 180, 241-244, 252-265, 268, 269, 272, 273, 297
death
— brought on only by sin of Adam, 62, 85, 88, 177, 178, 180, 187, 276
— death repugnant to God, xviii, 62, 85, 177, 179, 181, 186
— of human beings and of non-human life, xxv, xxxi, 59, 62-64, 85-91, 176-184, 187, 245, 275, 276, 292, 293, 325, 340, 344
— physical and spiritual, 186, 275
— vast pits on Earth, 70, 102, 141, 252, 274
deception, xxx, 58, 198, 264, 273, 330
de Chardin, Teilhard, xiv, 200, 290, 302-309, 327
de Lubac, Henri, 289

democracies, liberal, xxii, 43, 282
demythologize, 187, 209, 242, 313, 341, 348-351, 356, 358
Denton, Michael, 36, 37, 110, 113-116, 120, 121, 134, 358, 359
Deposit of faith, 185
dermatoglyphics, 33-35
Descartes, Rene, xxii, 360
descent, xviii, 7, 17, 18, 24, 30, 38, 40, 55, 128, 137, 201, 243, 271
Design
— argument from, xxix, 10-18, 29, 30, 34, 38, 39, 44, 49-61, 89-91, 92, 120, 121, 126-128, 135, 166-168, 180, 204
— similarities not proof of evolutionary descent, 18, 107
— unlike biological evolution, to thwart evolutionary explanation, 17, 49-53, 120, 158-2
— unlike planetary "evolution," 47-53
Designer: See intelligent Designer
dialectical materialism, 200, 285
dictatorship of the proletariat, 283
Dillow, Joseph C., 66, 73-76, 80, 81
dinosaurs
— general notes, ix, 68, 69, 73, 76, 76, 89, 105, 141, 179, 268, 269
— Hadrosaur (duckbill), bones not yet mineralized into rock, 142
— lung size, 69, 76
— maiasaur type—10,000 entombed in Montana, 141
diseases resistant to herbicides, antibiotics, xi, 127
"disruptive" young-agers, 137
dissenting theology, 309-317
diversity, 18, 24, 37
diversity thwarts evolutionary lineages, 17
divine intervention: See intervention, divine
Divino Afflante Spiritu, 210, 219, 224
"divorce" of science from religious faith, 228-240
DNA, xi, xii, 5-8, 15, 26, 31, 59, 119-121, 124-127, 133, 137, 158-4, 166, 175, 190, 194, 200, 205, 212, 217, 243, 267, 268, 296
DNA "fingerprinting," 33
DNA/RNA/protein interdependence, 6, 30, 31, 120, 121, 133, 137, 166
Dobzhansky, Theodosius, 42

Documentary Theory of the Pentateuch, 216, 225-228, 298
dog varieties, 24, 25
donkey/horse hybrid, 125
Doppler effect, 155
doubt cast on Christ's words, 250, 256
Down's Syndrome, 124
dragonflies, 57
dragons, 69, 219
dreams, 328-332
Dubois, Eugene, 108
ductus arteriosus, 29
Duggan, G. H., 306
"dust" account, 59-60, 190-194, 201

E

Eagle Nebula, 222-6
Earth
— age of, 158-15
— created to be inhabited and not to lie idle, 245, 246, 262, 263
— cultural center of Universe, 236
earthquakes, Flood legacy, 72
ecosystem, change of, xxv
Eden, Garden of, 61-64, 84, 90, 178, 302
Eddy, John A., 152
Einstein, Albert, 162, 170, 238, 239
Eldredge, Niles, 105, 109
elementary particles, 59, 190, 201, 265
embryology, 37, 116, 293, 300
empirical science, object of, xiv, 3
empiricism, 40
Encyclopaedia Brittanica, 39, 128
Engels, Friedrich, 285
Enlightenment, xxii, 238, 327, 360
entombed creatures, 64, 66, 69, 72-75, 139, 141
entropy, 91, 130-135, 180, 273, 274
environmental determinism, 325
eons of time, 47, 49, 101, 103, 177, 241-243, 245, 264, 269
epicycles, 229, 236
epistemology, 265
era, 252, 261, 273, 287
errors in Scripture: See inerrancy
errors—conceptual problem, 221-224
essence, 179, 191, 260
Eucharist, celebration of the, xix
eugenics, 41, 42, 295
euphemisms, to hide death, 294
euthanasia, xxiv, 205, 292, 295

Eve, origin of, xv-xvii, 59-60, 184, 190-199, 205
Eve was not product of evolution, x, 59-60, 191, 193, 206
"event horizon" of black holes/white holes, 161-163
evil, 39, 88, 90, 189, 276, 282, 295, 306, 309, 313, 326, 333-340
evil—moral/physical aspects, 335, 336
Evolution
— arguments against Designer/transcendent Cause, 10
— beliefs, double standards, 10
— birds—evolution?, 27, 114, 115
— cannot explain origin of life, 13
— complementary, concurrent evolution, need for, 32, 60
— conflicting versions of, 12, 37, 38
— confusing/misleading terminology, 23, 25, 37, 38
— contradictory mechanisms, 37
— credibility, 7, 35-38, 200, 201, 204
— definitions of, xxv, 13
— doctrinal error parading in scientific guise, xviii
— elastic, accommodates all data, 37
— evolutionism, 39-44, 168, 184, 187, 191, 200, 206, 281, 283, 287, 312, 359
— "fact" of, ix, xxvi, xxix, 17, 41, 42, 176, 222, 301, 305, 309
— human evolution rejected by the ordinary Magisterium of the Church, 193, 194
— illusion, xx, 7, 36-38, 40, 106, 129, 308
— is not science, 36
— life forms designed to look unlike evolution, 17
— missing mechanism, xxvi, xxix, 5, 13, 16, 21,23, 40, 92, 110, 175, 185, 200, 205, 253
— modern synthesis, 124
— myth of, xxvi, 36, 40, 317, 332, 358
— "newspeak" terminology, 299
— not official Catholic doctrine, 202
— not open question for Catholics, 195
— papal permission only to hypothesize about human evolution, 195, 199-202
— philosophy of origins, 40
— problem from von Baer's laws of embryology, 36

— reptile-bird evolution?, 114-115
— science fiction, 37
— solar system looks unlike planetary "evolution," 47-53, 154
— Theistic Evolution, xiii-xx, xxv, 38, 90, 175-184, 186, 204, 206, 211, 278, 316, 360
— "upward" change, 24, 124, 132, 177
— why God would not use evolution to create, xxvii
Ewing, J. Franklin, 197, 215
ex cathedra, 199, 204, 206, 230
existence, authentic/inauthentic, 342-344
existence/essence, xvi, xx, 41, 267, 328, 342-353
existentialism, 187, 206, 216, 341-353
experiences, subjective, 344-349
exploitation, 286
extinction, 25, 69, 104, 109, 142, 167, 275
eyes—need for repeated evolution in unrelated species, 14, 32
eyes, sight properties, 14, 15, 127-129
Eye-witness to Creation, divine, xxx, 9, 65, 178, 217, 254, 262, 264, 277

F

"facing the East," 84
faith, xxix, 5-10, 39, 42, 43, 92, 123, 167, 184, 184, 195, 203, 230, 233, 267, 281, 290, 311, 343, 350, 355, 356, 361
Fall, effects of the, 61-63, 87-92, 176-181, 273, 337
Fall of Man, xviii, xxvii, 61-63, 85, 88-92, 177-178, 196, 273, 325, 337, 345
falsifiable, 5, 10, 34
Fathers, Church, xvi, xvii, 91, 185, 191, 192, 217, 233, 256, 269, 299
Fatima, 220, 238
feathers, properties of, 26-28, 114, 115
feelings, 240, 312, 313, 329, 352
Fehlner, Fr. Peter D., xiii-xx, 191, 192, 232, 233, 259, 260
feminist theology, 216, 311
feral children raised by wolves, 183
fetus, 28, 163, 169, 173, 293, 300, 346
fidelity in marriage, xvii
Fides et Ratio encyclical (Pope John Paul II), 205

INDEX

figurative sense (figures of speech, using the language of appearance), 60, 80, 186, 214-220, 237, 258, 260
fingerprinting science—precedent problem facing evolution, 32-35, 158-11
firmament, 75-83, 221, 262-264
First Cause/Unmoved Mover, 45, 58, 132, 133, 169, 170, 267, 276
fishes
— evolution of?, 111-113
— fish-swallowing-fish fossils, 106, 158-14
— fossils, billions of, 70
— fossils on high mountains, 70, 106
— salinity survival, 68
— "walking catfish," 113
flight, need for repeated evolution of, 32, 114, 115, 129
flight, origins riddles, 57, 113-115, 129
Flood
— accounts around world, 65, 66, 84
— date of beginning of, 274
— Flood of Noah, xxiii, xxx, 48, 49, 64-83, 100, 101, 103, 110, 138-142, 149, 150, 153, 179, 184, 187, 205, 221, 242, 243, 248-255, 261, 263, 274, 275, 297
— global extent, 48, 64-88, 142, 150, 242, 249-252, 274
— timing and placement in strata, 49
footprints, human and dinosaur, 142
footprints, Laetoli fossil (human), 106
foramen ovale, 29
fossils
— formation process, 18-20, 64, 70, 103, 158-13
— fossil record, ix, xxvi, 13, 18-20, 29, 30, 37, 40, 49, 65, 66, 70, 99-117, 137, 143, 158-3, 252, 272, 274, 296
— fossil record incompleteness, 103-106
— hominids, 106-109
— "living," 143
— quarter million species, 104, 105
— trees, 100, 101, 143, 144, 158-14, 252
— vast museum of death, 102, 104, 178, 252
founder effect, 23
"fountains of the deep," 67, 74, 82
Fouts, David M., 253
Frankl, Viktor, 346, 347, 354
Frazer, Sir James G., 65

Free Will, 61, 174, 179, 276, 339, 342
Freud, Sigmund, 116, 330-332
frozen Arctic creatures, 69, 73, 75
fruit flies, 24, 88, 126
fundamentalism, xxix, 60, 210, 216, 219, 264, 357

G

Galapagos Islands, 23, 298
galaxies
— created by God, 47
— rotation problems, 155, 156, 264
— spiraling "arms," 152, 263
— "walls," clusters, 152, 156-160, 164
Galileo, xxx, 174, 228-240, 298
Gaudium et Spes, 337
genealogies, 242, 246-248, 261, 268
gene load, 125
gene pool, x, 24-26, 85, 125, 298
gene selection (Richard Dawkins), 13
Genesis
— foundational importance/integrity of, xxix, 48, 78, 87, 174, 186, 242, 273, 297, 356, 357
— historicity: See historicity
— not detailed science textbook, xxx, 243
— partial revelation of Creation, 74, 243, 264, 340
genetic
— drift, x, xii, 24
— drive, xii
— engineering, 42
— genetic recombination, xi
— information, increase in, x, 182, 261, 271
— limits to change, 126
— manipulation, 295
— mixing ("sex"), 60, 61
— plant population genetics, ix-xii
— science of genetics, 22, 23, 49, 118-129, 136, 185, 201, 204
— "switches," 87
— variation, 22-26, 125, 136, 158-10, 271
Gentry, Robert V., 147-150
geocentrism, 228-240
geological column, 18-20, 70, 101, 141-144, 158-12
geological column, importance to age question, 141

geology—evidence affirms Creation, 158-3
giant life forms before Flood, 55, 76
Giertych, Maciej, ix-xii
giraffes, 27, 28, 68, 158-7
Gish, Duane T., 54, 70, 112-116, 140
gnosticism, 327
God
— Creator of all, 46, 47
— existence of God can be known by reason, 92, 186, 343
— incompatible with error, xxix, 9, 221
— Principal Author of Scripture, xxx, 9, 48, 65, 85, 184, 207, 210, 216, 222, 245, 258, 269, 273
— "stretched out the heavens," 262, 263
— synonymous with humility and unselfishness, 340, 354
— "Evolutor"?, 357
— Creator back to center stage, xxx, 361
— timeless nature of, 45, 48, 158, 253, 255, 276, 339
— took on human form so that He could suffer torture in reparation for fallen mankind, 340
— trustworthy Eye-witness, incapable of deception, xxx, 9, 65, 85, 178, 186, 216, 240, 254, 264, 277, 358
— will of God—permissive/compulsive, 339
Goldschmidt, Richard B., 109
Gould, Stephen J., 109, 110, 306, 307
grace, xx, 63, 89, 179, 180, 188, 315, 335
Grand Canyon, 66, 67, 101, 158-13
gravity, 6, 49, 51, 154-157, 161-165
Grebe, John, 31
Green, William, 228
greenhouse effect, 76-83
Grigg, Russell, 116
Gross, Otto, 327, 329

H

Haeckel, Ernst, 116, 328-332
Hahn, Scott, 86, 87, 255
Haigh, Paula P., 266, 267
hailstones, 75, 81-83
Haldane, J., B.S., 35, 122
Haldane's Dilemma, 35

Hammes, John, 294, 295, 314, 325, 334, 355
happiness, 346, 352
Hardon, John A., 194, 195, 244
Harrison, Brian W., 192, 193, 220, 237
Hawking, Stephen W., 8, 202, 203
heart development, 116
heaven/heavens, 81-83, 180, 261, 262, 315
Hegel, Georg W., 290
Hegelian, xviii, xix, 290
Heidegger, Martin, 343, 349, 352
Heidelberg Man, 108
heliocentrism, xxii, 174, 228-240
Hell, 46, 184, 291, 297, 299-302, 306, 314, 315, 335-338
herbicides, xi
herbivores, 178-180
heresy, 309, 310, 313, 315
hermeneutics, 210
hibernation, 68
Higher Criticism, xxiii, xxix, 210, 211, 242, 309
Himmler, Heinrich, 285
Hinduism, 326, 334
historical theology, science of, 7
historical reductionism, 332
historicism, 206
historicity of *Genesis,* xvii, xxii, xxvii-xxx, 48, 85, 87, 186, 192-195, 206, 211-215, 222-7, 241, 242, 243, 248, 254-256, 269, 296, 310, 356, 357
history, mankind's cultural, 145-147
historical reductionism (the use of the past as a key to the present), 331
Hitching, Francis, 18, 19, 109
Hitler, Adolf, 283-285, 294, 308
Holt, Roy D., 49
hominids, 106-109, 176, 181, 201, 275, 284
Homo erectus, 107-109
Homo habilis, 108
Homo sapiens, 41, 107-109, 144, 275, 293
homologous organs, 57, 127
homosexuality, xxiv, 281
hope, 290, 344
horses—problem of how they could have evolved, 105, 125
Hoyle, Sir Fred, 31, 32, 235
Hubble Space Telescope, 159
Hulsbosch, A., 244, 301, 302
human beings

INDEX

— animated spiritually, xviii
— dominion over animals, 179
— not products of evolution, xvi, 109
— temple of the Holy Ghost, 337
— unique dignity of, xiii, xvi
— would not have died, 187, 188
— "wounded in natural powers" by Original Sin, 180, 187
Humani Generis encyclical, xxvi, 190, 193-202, 205, 210, 219
Humanism
— Christian, xxii, 288-290, 316
— Humanist manifestos, 291
— Secular, 43, 282, 290-295, 310, 358, 361
humility, 355
humming birds, 57
Humphreys, Russell, 78-79, 153, 160-164, 264
Husserl, Edmund, 352
Hutton, James, 137
Huxley, Aldous, 327
Huxley, Sir Julian, 123, 290, 291, 305
Huxley, Thomas, 233
Hydroplate Theory, 71, 72
hyperbaric pressure, 76

I

"I am who am," 48
Ice Age, 67-69
ice—huge amount fell in Flood, 74-76, 82
ichthyostegid, 111-113
idealism, xxii
illusion of evolution, xx, 7, 35-38, 40, 106, 129, 308
Immaculate Conception, 258, 357
immanence, 41, 266, 289, 309, 326, 351, 352
immunological adaptation, xi, 167
imperfection of fossil record, 103-106
Incarnate Saviour, xix, xx
incest, 85, 281
individuation, 327
inerrancy of Scripture, xxix, 9, 92, 173, 186, 207-240, 249, 270, 297, 350
infallibility, 193, 194, 199, 232
infanticide, 42
information, coded, x, 23, 25, 48, 58, 93, 119, 120, 123, 125, 126, 127, 135, 244, 260

Information Technology, 44
insectivorous plants, 62
inspiration of Holy Spirit, 207
instincts, animal, 181
integration—selfist, 346
intelligent Designer—evidence of, xxix, 5-7, 10, 12, 13, 16, 18, 28-36, 44, 49-61, 89, 92, 120-123, 126-128, 135, 166-170
interdependence in creation, 32, 53, 58, 60, 121, 129, 137, 160, 165, 166, 277, 278
intermediate stages (includes transitional forms), xxvi, 19, 30, 38, 53, 99, 103-107, 109-116, 120, 128, 129, 141, 245
interpretation, long ages, 250
intervention, divine, xix, 38, 47, 48, 67, 80, 136, 163, 175, 182, 183, 186, 190, 194, 203-205, 268, 270, 272, 276, 277, 316
invertebrates, 105, 134
in vitro fertilization (IVF), 187, 205, 235, 292, 293
irreducible complexity, 15-17, 120, 128, 204
Islam, 334
isolated populations, 23, 24
isostatic balance, 71
isotopic composition, 140, 149
ivory tusks in permafrost, 74

J

Jaki, Stanley, 217, 298
Janus and Epimetheus, 52, 222-4, 263
Jastrow, Robert, 157, 169-170
Java Man, 108
jawbone structure, 113
JEDP: See Documentary Theory
jellyfish fossils, 70, 103, 158-13, 252
Johanson, Donald, 109
Johnson, Paul, 238, 239
Johnson, Phillip, 40, 41, 44
Johnson, Wallace, 154
Jones, Alexander, 244
Joshua 10:13—the day that Earth stood still?, 237
Judaism, 334
judgment by God, 338
Julien, Pierre Y., 143
Jung, Carl, 116, 325-331

Jupiter, 51, 83, 229
Jupiter's diverse satellites, 52

K

Kant, Immanuel, xxii
karyotype, 124
Kepler, Johannes, 229, 230
"kill or be killed," 62, 167, 180, 292, 357
kinds, 23, 30, 51-58, 64, 67, 100, 105, 136, 137, 146, 182, 201, 204, 241, 243, 266-271, 277
Kitchen, K. A., 145, 146
Koestler, Arthur, 229-234
Kramer, William, 176, 203
Kreeft, Peter, 209, 210
Kuhse, Helga, 293
Kuiper belt, 152

L

Laetoli (human) footprints, 107
Lamarck, Jean Baptiste, 27, 330-332
land bridges before and/or after Flood, 69
language, 24, 43, 83, 84, 146, 147, 308
"language of appearance," 48
lava beds, 70
laws of nature, xxx, 5, 6, 8, 14, 17, 41, 48, 49, 91, 93, 130, 215, 268, 271, 341
Law, profession of, 35
Lazarus, 48, 90, 180
Leakey, Mary, 109
Leakey, Richard, 106, 108, 144
Le Grand, Leon, 301
Lejeune, Jerome, 123-125
Lescoe, Francis J., 348
levels of Being, 168, 169
Levitical priesthood, 87
Lewin, Roger, 29
Lewis, C. S., 43, 212, 213
Libby, Willard, 138
liberal theology, xxiv, 39, 282, 311, 351, 358
liberation theology, 216, 312
life beyond Earth?—conceptual problem, 236
life forms—designed to look unlike evolution, 17
life—what is it?, 48, 135

light before Day 4, 161-164, 256, 261, 262
light, nature/speed of, 155, 160, 255, 256, 261-264, 272
"like begets like," 205
limits to genetic change, 24-26
lineages, evolutionary, 17, 38, 307
links, still missing, 103-111, 146
Linnaeus, Carolus, xii
liquefaction, 71, 72
literal sense, xvi, xxix, 85, 186, 192, 209-220, 233, 234, 237, 252-263, 296
literal, two meanings of, 218
literary criticism, 212, 213
literary genres, 218
"living fossils," 143
localised Flood?, 250, 261
loess, 75
long ages, conceptual problems, 245, 254, 255
longevity, 76, 85, 86, 88
Long, Valentine, 87, 88, 91
Lubenow, Marvin, 106-109, 144, 275
"Lucy" (an australopithecine), 108
Lumen Gentium, 193
lungfish, 116
Lyell, Charles, 19, 137
Lysenko, Trofim, 287
Lyttleton, R. A., 1

M

Macquarrie, John, 349
macroevolution, x, xviii, 25, 29, 37, 40, 182
Magisterium, xxi, xxvi, xxxi, 48, 185-194, 197, 202, 206, 208, 209, 215, 223, 299, 302-304, 310, 359
magnetic field of the Earth, 79, 80, 152, 153, 167
magnetic field reversals, 80, 153
magnetosphere, 79, 80
Malcolm, David, 71
Malthus, Thomas, 21, 41
mammals, marsupial/placental, 56
mammoths
— dating, 137
— five million entombed, 70
— snap-freezing of, 70-77, 158-14, 251
mankind, cultural history of, 144-147, 245

INDEX

manna, 68
Marra, William A., 352
marriage, xvii, 21, 59, 85, 217, 248, 256, 292, 357
Mars, no life on, 50, 134
Mars-rock, found in Antarctica 50
Mass, Latin rite Catholic, xxiii, 84, 86, 189
Marxism, 239, 283, 310, 333
Mary, the Mother of God, xiv, xvii-xx, 222, 222-8, 223, 236, 258, 276
matter
— eternal?, 8, 10, 41, 261, 286
— mechanistic?, 48, 328
— nature of, 48, 131, 135, 166, 265, 286
— vitalistic?, 48, 328
materialistic philosophies, x, xii
mature creation, 155, 264
maturity, growth to, xxv, 25
McCarthy, John F., 3-10, 217
McKee, John, 223, 224
meaning in life, 174, 294, 295, 314, 321-332, 342, 346, 347, 355
mechanism of evolution
— contradictory mechanisms, 36, 37
— problems in, 5, 13, 16, 21, 36, 37, 40, 110, 125
— still missing, xxvi, 5, 16, 21, 36, 37, 40, 92, 110, 125, 177, 182, 185, 200, 205, 253
Medawar, Sir Peter, 307
meiosis, 61
melanin, 158-10
Melchisedech, 86, 87, 255
Mendel, Gregor, 22, 26, 194
Mercury, 51, 154
Mesopotamia, 65, 145, 146, 226, 261
Message Theory (Walter ReMine)
— intended by Designer to thwart evolutionary explanation, 17, 18
— is testable science, 17
metamorphosis, 57, 58
metaphysics, xiv, xv, xix, 98, 264-268, 286, 301, 306, 349, 359
metaphysics definition, 265
meteorites, 68, 79, 82, 83
meteoritic bombardment, 81, 82
microevolution, x, 25, 29, 36, 40
micrometeoroids, 152
Mid-Atlantic Ridge, 72
Midlo, Charles, 33-35
migratory habits of creatures, riddle of, 57, 158-9, 167
mimicry, 58, 59, 158-7
mind, nature of the, 13, 43, 167, 286
miracles, xiv, xvii, xxix, 76, 184, 190, 193, 210, 215, 220, 238, 270, 277, 297, 299, 316, 316
Mirandola, Giovanni Pico della, 289
missing links, 103-111, 183, 253, 268
"missing mass" of Universe, 156-158
missing mechanism of evolution, xxvi, xxix, 5, 16, 20, 22, 40, 92, 110, 175, 183, 185, 200, 205, 253
missing supernova remnants (SNR's), 164
Modernism, xxii, xxiii, xxx, 200, 210, 211, 297, 309-317, 352, 355, 361 (See also Appendix B)
modesty, innate sense of, 63
molecular biology, xxvi, 14, 120, 200, 204
Molnar, Thomas, 289-290, 295
monism, 200
monophyletism, 194
momotheism, 224
monotremes, 115
Montgomery, John W., 65
Moore, Ruth, 13
moral standards, 292, 313-316, 344-350, 360
moral sense, 209
Morris, Henry M., 24, 100, 102, 126, 127, 134, 142, 143, 156
Moses, 46, 81, 85, 214, 221, 224-228, 256, 269, 270, 298
Moses—wrote about Jesus Christ, 228
mosquitos—amazing complexity points to design, 158-4
Most, William G., 208, 224, 258
moths, peppered, 23
Mount Saint Helens volcano, 67, 68, 158-15
mountains, folded, 72
mountains formation, 65-67, 70-72, 141
Muggeridge, Anne Roche, xxiv
Mukerjee, Madhusree, 8
Muller, H. J., 119
museum of death, vast, 103, 104, 141, 178
mutations, x-xii, xxiii, 13, 22-36, 61, 97, 110, 123-129, 167, 176, 183, 245, 271
mutations, beneficial, 22-35, 61, 97, 110, 123-129, 167, 176, 182, 245, 271

mutations, point, 26, 123
mutton birds of Australia, 57
mysticism, Eastern, 326
myth of evolution, xxvi, 36, 40, 176, 275, 317, 358
mythology, xviii, xxvi-xxix, 36, 49, 85, 176, 184, 215, 242, 244, 250, 255, 269, 275, 297, 315, 328, 332, 349, 350, 356-358

N

natural selection, x-xii, xxiii, xxix, 13, 15, 17, 21-25, 27, 37, 42, 43, 97, 99, 103, 104, 112, 113, 126, 167, 175, 204, 271, 284, 325
naturalism/naturalistic philosophy, 3, 10, 12, 14, 17, 39-44, 50, 58, 59, 92, 97, 164, 175, 186, 195, 356, 358
Nazism, 239, 282-287, 361
Neanderthal Man, 108-109
Nebraska Man, 109
Nelson Ethel R. *et al.,* 84
Neolithic Period, 144, 245
Neptune, 51, 52
New Age "Christ," 326
New Age movement, 289, 311, 326-330
Newton, Isaac, 229, 239, 290
Nietzsche, Frederick, 285
Noah (Noe), xxiii, xxv, 62, 68, 76, 83, 85, 86, 100, 205, 221, 225, 227, 243, 248-256, 261, 263, 272, 274, 297
Noll, Richard, 328, 329
noogenesis?, 305
Norman, Trevor, 160

O

Oakley, Kenneth, 307
obedience/disobedience, 62, 63, 85, 88-91, 176-180, 186, 197-199, 236, 237, 245, 264, 266, 274-276, 289, 299, 349, 355
objective truth: See truth
odds question, 27-35, 60, 120, 122, 126, 165, 182, 245, 271
Oedipus complex, 332
oil, 65, 101, 143
Old Testament credibility, 249-252
"ontogeny recapitulates phylogeny" hypothesis 116

ontology, 205, 265
onus of proof is on long ages proponents, xxviii, 254, 268, 269
opals "growing" in bottles, 139
order aboard the Ark, 68
order imposed upon matter, 93
order in the Universe, 160, 165, 264
O'Reilly, Sean, 131, 132
organs
— general notes, 14, 15, 28, 60, 75, 87, 127, 129, 167, 271
— nascent, 129
— system of complex interdependent systems (i.e., eyes), 13
— vestigial, 129
Original Sin
— deprivation of original holiness and justice, 189
— general notes, xxi, xxvi, 46, 61-63, 85, 87-92, 176, 177, 184-202, 217, 273, 295, 297, 302, 309, 314, 315, 323, 329, 337, 340, 345, 349, 354, 357, 361
— inherited from Adam and Eve by generation not imitation, 188, 189
Origins question is theological, not scientific, question, xvi
Origins' relevance, xiii-xxxi, 10, 184, 184, 211, 354-361
Origins' three basic beliefs, 3
oscillating theory of Universe, problems in, 156, 157
Ott, Ludwig, 46
oxidizing atmosphere, 122
oxygen-ozone discussion, 133, 134
ozone layer, 122

P

paganism, 288, 326-329
pain, 39, 293, 337, 338, 340
paintings, cave, 145
pandas, giant and lesser (red), 56
pantheism, 41, 200, 266, 326
Palaeolithic Period, 144
Paley, William, 14, 15
Paluxy River footprints, 141
parables, 219
Parker, Gary, 24, 100, 126, 127, 156
Pasteur, Louis, 12
Patriarchs, hundreds of years age, 85-87, 247, 248, 254

INDEX

patristic exegesis, 216
peace on Earth, 357
pedagogy, 255, 351
Peking Man, 13, 108
Pelagius, 184
Peleg, 70
penguins, 158-9
Pentateuch, 186, 224-228, 269, 298
Pentateuch, computer study of, 228
peppered moths, 23
permafrost, 72-75
personality types, 329
pharyngeal arches, 116
phenograms, 17, 115
phenomenology, 352-353
Philberth, Bernard and Karl, 299-301
photosynthesis, 16, 32, 122, 133, 134
phyla, sudden appearance of, 19, 105, 106, 110, 123
phylogeny, xii, 18, 30, 37, 55, 109, 110, 116, 116, 123, 253, 328, 332
phylogeny descent problems, 30, 55
physics, higher than theological reality?, 170, 238, 272, 301
pictographs, Chinese, 84
Piltdown Man, 109, 306, 307
Pithecanthropus, 108
Planets, A Smithsonian Guide, 50
plants, riddle of existence, 53, 166
platypus, 55, 116
plesiosaur, 143
pliosaur, opalized, 139
plumes, vertical, 72
pluralist societies, xxii, 43, 44, 321-324
Pluto, 51, 154, 222-4, 263
polonium radiohalos, 147-150, 158-16, 245
polygamy, 329
polygenism
— conceptual weakness, 197-199
— general notes, xix, 86, 181, 186, 189, 194-199, 206, 300
— irreconcilable/impermissible in doctrine of Original Sin, 86, 194-199
polyphyletism, 194
polystrate fossilized trees, 100, 101, 142, 252, 274
Pontifical Academy of Sciences, xxvi, 188, 194, 202-205, 211, 212
Pontifical Biblical Commission, xxvi, 48, 188, 197, 211, 215, 216, 227, 245 (See also Appendix A)
Pope Benedict XIV, 232

Pope Benedict XV, 210, 219, 247
Pope John Paul II, 177, 184, 186, 202-206, 211, 212, 217, 338, 352-353, 360
Pope Leo XIII, xxviii, 183, 186, 192-194, 210-212, 219, 224, 257, 269
Pope Paul VI, xviii, 177, 188, 189, 196, 360
Pope Pelagius I, 193
Pope St. Pius X, xxiii, 222-8, 227, 282, 310-312
Pope Pius XI, 190
Pope Pius XII, xxvi, xxx, 188, 190, 194-202, 205, 210, 219, 224, 236
Pope Urban VIII, 231
population control tactics, 295
positivism, 3
potassium-argon dating method, 140
Poynting-Robertson effect, 152
"pre-Adamites," 300
precedent in law, 32
predators, 23, 62, 89, 90, 111, 113, 167, 178-181, 275, 287
predators—behavior wounded in natural powers?, 90-92, 180
"prehistory," 211, 299, 329
pride, 63, 314, 336, 355
priesthood, Catholic, 87
"primitive" tribes, 84
principles, absolute, 44
probability: See odds question
process theology, 47, 311
procreation, xvii
progenitor, common—ruled out 127
progress, upward?, xviii, 306, 345
Progressive Creation, 90, 177, 270-278
pro-life beliefs—hostility toward, 295
promiscuity, 292
proteins comparison, 125
protostars, 263
Providence, 215, 266, 338-340
Providentissimus Deus encyclical, 210-212, 219, 257
psychoanalysis, Freudian, 330-332
psychology, 330-332, 345
Ptolemaic theory, 229, 234-240
puffins, 57
punctuated equilibrium, 30, 109, 110, 182, 194

Q R

quicksand, 72
racism, 284, 285
Radday, Yehuda T. *et al.,* 228
radiation in Universe, 156
radioactive dating methods, 140, 158-15, 160
rafts, floating Flood debris, 68, 69
rain before the Flood?, 67
rainbow covenant, 82, 83, 251, 252, 274
Ramapithecus, 108
raqia, 78, 80
rationes seminales, 257, 260
Ratzinger, Joseph Cardinal, xxiv, xxvi, 216, 361
Raup, David M., 105, 109
reality, xxvii, 3, 4, 36, 53, 132, 204, 256, 266, 286, 295, 298, 303, 306-309, 313, 316, 321, 344, 351, 352, 353, 359
Real Presence, 208
reason, use of, 7, 9, 186, 343
recapitulation theory, 116, 328-332
recombination, genetic, 26
Redi, Francesco, 12
red in tooth and claw, 90
Red Sea crossing, 214, 215
red shift, 155, 160-164
re-education (concentration) camps, 286
relativism, moral, 43, 44, 184, 239, 240, 282, 289, 291, 294, 298, 313, 314, 350
relativity, theory of, 155, 160, 161-164, 235-240
religion—product of evolving consciousness?, 290, 291, 306, 310
ReMine, Walter, 6, 7, 17, 18, 37, 38, 49, 61, 112, 113, 122
Renaissance, xxii, 289
replication, DNA, 120, 121
reproduction, asexual, 61
reproduction rate of animals, 178
reptiles-birds evolution, 113-115, 134
reshuffling of genes, 26
resistance to antibiotics, herbicides, xi, 127
revelation, divine, 185, 207, 242, 264, 295, 312, 337, 338, 340
reverence, 211
revisionist theology, xxix, 190, 223, 255, 269, 277, 282, 297, 359-351
rib account of Eve's creation, 59-60

rights, human, 173, 282, 292-294, 316, 357
Romanes, George J., 330
Roman Liturgy, xxiii, 87, 190
Roman Martyrology, 247
Romans 5:12, 38, 187, 205, 276
Ross, Hugh, 270-278
rotations problems in Universe, 154, 156
Ruffini, Ernesto Cardinal, 185, 186, 189-191, 195, 199, 257-259

S

Sagan, Carl, 134
salinity—fishes' survival 68
salt deposits, 70
salvation history, xxix, 186, 187, 215, 220, 248, 250
salvation, mystery of, 63, 192
Sarfati, Jonathan, 50, 153, 164
Satan/devil, 63, 87, 88, 177, 184, 219, 281, 297, 302, 314, 334-340, 360
Satre, Jean-Paul, 343
Saturn, 51, 52, 222-4, 263
Saturn's co-orbital satellites, 52, 222-4
Scarrabelotti, Gary, 350
Schumacher, E. F., 42, 43, 168, 169, 281
Schwarzschild radius, 161
science
— equivalent to truth?, 41
— of historical theology, 7
— overlap with theology, 187, 205, 234, 235
— perceived divorce of science and religious faith, 234, 235
— scientism, 3, 39-41, 164, 205
— the study of reality, 3-10
scientific naturalism, 40, 41
Scopes "Monkey" trial, xxix
secondary causes, xxviii, 48, 80, 187, 204, 262, 266-268, 276-278
secularizing effect within pluralist society, 324
sediments formation during Flood, 66, 69-72, 100-103
sediment erosion (from granite and basalt) caused by Flood, 71
sedimentology, 20, 21, 101, 143
self-awareness, 43, 167-163, 168, 181, 286
selfism, 174, 291, 313-315, 341-353, 356

INDEX

Sennott, Thomas Mary, 231
senses of Scripture, 209-211, 215, 218-224
sentience, 293
Setterfield, Barry, 160
"sex"—genetic mixing from two parents, 60, 61
sexual dimorphism and polymorphism in butterflies, 158-7
Shakespeare, 212
Shang Di, 84
Sheehan, Michael, 201, 213, 214
Shem (Sem), 83-87, 247, 251, 255
"shepherd" satellites, 51, 222-5, 263
sight, repeated evolution of, 129
"signature," man's, 145, 146
similarities, 16, 17, 55-57, 107, 128, 129
similarity in function (analogous) or in structure (homologous)?, 57
Simpson, George G., 105, 146
sin, nature of, 302, 306, 334-340
sin, Original: See Original Sin
Singer, Peter, 293
Siri, Joseph Cardinal, 311
situation ethics, 313
Skinner, B. F., 325
skull "KNM-ER 1470," 144
Slusher, Harold S., 151, 157, 158
Smith, Richard F., 220
Smith, Wolfgang, 235, 239, 240, 265, 266, 303, 308, 329, 330
snails, best evidence for evolution found in?, 106
Snelling, Andrew, 71, 139
snow clouds height, 76
social justice, xxiv, 316, 357
solar system data, 50-53, 150-151, 222-2, 277
solar system designed to look unlike planetary "evolution," 154
Solzhenitsyn, Alexander, 333, 334
soul, rational, xiii, xvi, 47, 59, 63, 91, 173, 176-184, 190, 199, 236, 267, 281, 291, 299, 300, 315, 339
souls, salvation of, xxxi, 291, 338, 345
space = Heaven?, 83
Special Creation concept, xxviii, 45-92, 97, 100, 174, 184, 204, 322, 360
speciation, 124
speciesism, 293
species—quarter million fossil species, 104, 105
speech origins puzzle, 146, 147

Spencer, Herbert, 22, 330
Spergel, David, 159
spiders—behavior, 173
Spieker, Edmund M., 19, 20
Spiral Galaxy M100, 222-6
spiritual sense, 209
Spiritus Paraclitus, 210, 219
spontaneous generation, 12, 16, 305
squirrels, 23, 125
Stalachtis butterflies, 58, 59
stalagmites/stalactites, 138, 139
Stalin, Joseph, 283
starlight, rapid creation of—"stretched out," 155, 262, 263
stasis, 29, 30, 271
"Steady-State" theory, 132, 169
Steichen, Donna, 311, 326, 327
Steinheim fossil (human), 107
Steinmueller, John E., 227
St. Peter's Sandstone, 100
strata/sediments, 18-21, 49, 64-67, 69-72, 82, 100-103, 107, 137, 141-143, 252, 274
strata in folded mountains, 72, 141, 158-13
"stretched" out the heavens, 262, 263
Strauss, David Friedrich, 285, 327
Strickling, James A., 66
subjective reality: See truth
suffering, human: See evil
suffocation of mammoths, 75
Sulloway, Frank, 330-332
Sun shrinkage possibility, 152
supernova remnants (SNRs) missing, 164
superposition, principle of, 20
supraspecific groups, 38
survival of the fittest, 22, 31, 91, 109, 284, 357
survival of the luckiest, 22, 109
Swanscombe skull (human), 107
symbiotic relationships, 167, 158-5
symbolic, xvii, xix, 189, 199, 210, 215, 239, 243, 248-252, 296, 350

T

tautology, 37, 256
taxa, 38
taxonomy, xii, 55, 108, 116
Taylor, Charles V., 226, 227
Taylor, F. Sherwood, 234

Taylor, Paul S., 54
teaching of Origins—teach case for and against Evolution, 196
teleology, xv, xix
temperate early Earth climate, 73-79
theism, 44, 224
theology—can overlap with empirical science, 235
theology—"queen of the sciences," 7
the past as a key to the present, 331
thermosphere, high temperature, 76, 82
Theory of Everything (TOE), 8
"the present is the key to the past," 13, 19, 137
"thorns and thistles," etc., 88, 179, 180
Thornton, Elizabeth M., 332
tidal waves, 65, 67, 71, 72, 251
time
— arrow of time, 132
— atomic/dynamic concepts, 160
— different concepts of, 9, 161-164, 166
— had a beginning, 261
— irrelevant to evolution, 27
— linked to vast distances?, 155, 160, 254, 270
— never enough for evolution, 35, 126, 127, 137, 243
— time, and no time, at same time, 159, 166, 254
— timeless nature of God, 45, 48, 158, 253, 255, 276
— time scale, 20, 144, 254, 255, 272
— time warp in Universe?, 155, 161-164
Time-Life, 13, 111, 114, 119
totalitarian, 40, 42, 282, 294, 316, 348, 354
Tradition, Catholic, xxi, xxviii, xxx, 38, 176, 185-194, 198, 199, 206, 209, 215-219, 222-3, 227, 237, 253-258, 269, 273, 276, 299
traits, 22-26, 32, 35, 55, 57, 115, 126, 287
transcendent Creator/God, xxvii, 10, 12, 39, 41, 44, 132, 174, 187, 240, 290
tranquil Earth, 66, 68, 76-80, 87, 88, 177-181, 245
tranquility of order, 90, 181, 273, 357
transformism, 10, 194
transitional forms: See intermediate
transubstantiation, xiv
"tree," "bush" concepts of phylogeny, 18, 30, 55, 105, 123
"tree of life" and "tree of knowledge of good and evil," 62
trees, age of, 143
tree trunks, fossilized, 100, 101, 142, 158-14, 252, 274
trustworthy, reliable Creator who cannot deceive, xxx, 9, 65, 85, 179, 186, 217, 240, 256, 264, 277, 358
truth—objective/subjective, xv, xxvi, xxx, xxxi, 41, 44, 49, 97, 137, 173, 174, 184, 202-204, 210, 222-8, 240, 243, 253, 282, 286, 289, 305-308, 312-317, 325, 326, 332, 341, 342, 347, 351-353, 356, 359, 361
typology, xvii, 10, 40, 83, 115, 192, 222, 249-251

U

unconscious instincts, 41, 327-332
unconformity, 102
underground waters during Flood, 64, 65, 67, 71, 75
unfalsifiable, 5, 34
unification, 303
uniformitarianism, xxiii, 13, 19-21, 50, 70, 99-106, 137, 139, 149, 153-156, 243, 252, 254, 257, 269, 274, 296, 309, 314, 327
Universe
— boundary?, 78, 159-164, 203
— had no beginning?, 132, 203, 291
— exists in time-warp?, 155, 161-164
— groans in travail, 88, 187, 237, 273, 274
— like clock running down, 132
— must have had beginning, 132
— evidence from refutes "Evolution," 222-2
— rotation problems in, 154, 156
— still expanding?, 277
— vast distances, 78, 79, 136, 155, 158-160, 222-5
unification, Teilhard de Chardin's, 304, 307
uniqueness of life forms, 34, 53-59
Uranus, 51, 52
Ussher, Bishop, 152

V

"value-free" morality, 346
variety within kind, xxv, 22, 25-26, 38, 54, 85, 137, 177, 204, 268, 271, 277
Vatican Council I, 46, 191, 207, 224, 303, 343
Vatican Council II, 177, 208, 223, 224, 282, 311, 337, 343
Vawter, Bruce, 244, 298, 299
vegetarian diet originally intended, 62, 89, 90, 178
Venus, 51, 76, 154
vertebrates, 51, 56, 61, 70, 105, 112, 114, 115, 128
Vertesszollos fossil (human), 107
Viking spacecraft, 50, 134
Virchow, Rudolf, 108
Virgin birth, xvii
vision-days theory, 270
Vitz, Paul C., 344-346, 355, 356
volcanic activity, 64, 141, 252
von Baer's laws of embryology—problem for Evolution, 37
Von Hildebrand, Dietrich, 308, 309
Voyager spacecraft, 50

W

Wallace, Alfred Russel, 330
Walsh, John Evangelist, 306
wasps—behavior, 167
watchmaker argument, 14, 15
water boundary around Universe?, 161
water vapor canopy, 73-83
Watson, David C., 261
Watson, James, 119
Watters, Thomas R., 50
weaver birds, 57
Wellhausen, Julius, 225
whale, baleen fossil, 142
whales—mammals returned to sea?, 112
when and how of creation, 9
Whipple, Fred, 151
Whitcomb, John C., 66
Whitehead, Alfred North, 47, 327
white holes, 161-164
who and why of creation, 9
Wilders, Peter, xxiv
Wilder-Smith, A. E., 123
Williams, Emmett L. *et al.,* 130
Wilson, Clifford, 226, 261
winds, gentle before the Flood, 76
"windows of the heavens," 74, 81, 82
Wiseman, P. J., 226, 270
witchcraft, 326
Woodmorappe, John, 68
"wombats," giant, six feet high (now extinct) in Australia, 55
worms, up to 16 feet long, living in Australia, 55
Wysong, R. L., 142, 151-153

XYZ

yedoma ("rock ice"), 75
yom, xxviii, 215, 253-264, 269

If you have enjoyed this book, consider making your next selection from among the following . . .

The Death of Evolution. *Wallace Johnson*	12.50
Science of Today and/Problems of Genesis. *Fr. Patrick O'Connell*	18.50
Our Saviour and His Love for Us. *Garrigou-Lagrange, O.P.*	18.50
St. Jude Thaddeus. *Anonymous*	2.00
Holy Eucharist—Our All. *Fr. Etlin*	2.00
Sermons of Curé of Ars. *Vianney*	12.50
Novena of Holy Communions. *Lovasik*	2.00
Moments Divine—Before Bl. Sacr. *Reuter*	8.50
True Devotion to Mary. *St. Louis De Montfort*	8.00
Eucharistic Miracles. *Joan Carroll Cruz.*	15.00
The Four Last Things—Death, Judgment, Hell, Heaven	7.00
The Facts About Luther. *Msgr. Patrick O'Hare*	16.50
Little Catechism of the Cure of Ars. *St. John Vianney.*	6.00
The Cure of Ars—Patron St. of Parish Priests. *Fr. B. O'Brien*	5.50
St. Teresa of Avila. *William Thomas Walsh*	21.50
The Rosary and the Crisis of Faith. *Msgr. Cirrincione & Nelson*	2.00
The Secret of the Rosary. *St. Louis De Montfort*	5.00
Modern Saints—Their Lives & Faces. Book 1. *Ann Ball*	18.00
The 12 Steps to Holiness and Salvation. *St. Alphonsus*	7.50
The Incorruptibles. *Joan Carroll Cruz*	13.50
Raised from the Dead—400 Resurrection Miracles. *Fr. Hebert.*	16.50
Saint Michael and the Angels. *Approved Sources*	7.00
Dolorous Passion of Our Lord. *Anne C. Emmerich*	16.50
Our Lady of Fatima's Peace Plan from Heaven. *Booklet*	.75
Divine Favors Granted to St. Joseph. *Pere Binet.*	5.00
St. Joseph Cafasso—Priest of the Gallows. *St. J. Bosco*	5.00
Catechism of the Council of Trent. *McHugh/Callan*	24.00
The Sinner's Guide. *Ven. Louis of Granada*	12.00
Padre Pio—The Stigmatist. *Fr. Charles Carty*	15.00
Why Squander Illness? *Frs. Rumble & Carty*	2.50
The Sacred Heart and the Priesthood. *de la Touche*	9.00
Fatima—The Great Sign. *Francis Johnston*	8.00
Heliotropium—Conformity of Human Will to Divine. *Drexelius.*	13.00
Purgatory Explained. (pocket, unabr.). *Fr. Schouppe*	9.00
Who Is Padre Pio? *Radio Replies Press.*	2.00
The Stigmata and Modern Science. *Fr. Charles Carty*	1.50
The Life of Christ. 4 Vols. H.B. *Anne C. Emmerich*	60.00
The Glories of Mary. (Lrg. ed.). *St. Alphonsus Liguori*	16.50
The Precious Blood. *Fr. Faber*	13.50
The Holy Shroud & Four Visions. *Fr. O'Connell*	2.00
Clean Love in Courtship. *Fr. Lawrence Lovasik*	2.50
The History of Antichrist. *Rev. P. Huchede*	4.00
The Douay-Rheims Bible. *Paperbound*	35.00
St. Catherine of Siena. *Alice Curtayne.*	13.50
Where We Got the Bible. *Fr. Henry Graham*	6.00
Imitation of the Sacred Heart of Jesus. *Fr. Arnoudt*	15.00
An Explanation of the Baltimore Catechism. *Kinkead*	16.50
The Way of Divine Love. *Sr. Menendez*	18.50
The Curé D'Ars. *Abbé Francis Trochu*	21.50
Love, Peace and Joy. (St. Gertrude). *Prévot.*	7.00

At your Bookdealer or direct from the Publisher.
Call Toll-Free 1-800-437-5876.

Prices subject to change.